Improvements in reservoir construction, operation and maintenance

Proceedings of the 14th Conference of the British Dam Society at the University of Durham from 6 to 9 September 2006.

Edited by Henry Hewlett

Conference organised by the British Dam Society. www.britishdams.org

Organising committee: Jon Green (Chairman), Henry Hewlett, Andy Hughes, Mark Morris, Andrew Pepper and Jim Prentice.

In addition to members of the organising committee, the following personnel kindly assisted with the review of papers: Alan Brown, Ian Hope, Jim Millmore, Mark Noble and Paul Tedd.

Cover photograph: Kielder Dam at sunset, courtesy of Northumbrian Water

Published by Thomas Telford Publishing, Thomas Telford Ltd, 1 Heron Quay, London E14 4JD. http://www.thomastelford.com

Distributors for Thomas Telford books are
USA: ASCE Press, 1801 Alexander Bell Drive, Reston, VA 20191-4400, USA
Japan: Maruzen Co. Ltd, Book Department, 3–10 Nihonbashi 2-chome, Chuo-ku, Tokyo 103
Australia: DA Books and Journals, 648 Whitehorse Road, Mitcham 3132, Victoria

First published 2006

Also available from Thomas Telford Books
Long-term benefits and performance dams. British Dam Society. ISBN 07277 3268 4
A guide to the Reservoirs Act 1975. DETR and ICE. ISBN 07277 2851 2
Reservoir Engineering. Guidelines for practice. E. Gosschalk. ISBN 07277 3099 1
Risk and uncertainty in dam safety. D N Hartford and G B Baecher. ISBN 07277 3270 6

A catalogue record for this book is available from the British Library

ISBN: 0 7277 3470 9

Printed and bound in Great Britain by MPG Books, Bodmin, Cornwal

Preface

This book contains the proceedings of the 14th Conference of the British Dam Society, *Improvements in reservoir construction, operation and maintenance,* held at the University of Durham in September 2006.

There are 36 papers covering a wide variety of issues. Recent changes to reservoir legislation in England and Wales relating to enforcement of the Reservoirs Act 1975 and the requirement to prepare flood plans, and the introduction of the Controlled Activities Regulations in Scotland, are discussed. Following the upgrading of many spillways in the last 30 years, internal erosion is increasingly seen as the greatest threat to UK reservoirs, and a number of methods for the early detection of internal erosion are described. *The Interim Guide to Quantitative Risk Assessment of UK Reservoirs* was introduced at the Society's 2004 Conference, and there are reports on feedback in its use. Hydrological and hydraulic issues covered include: a study into the effect of changes in weir crest coefficient with head; computer modelling of the operational systems of reservoirs, and developments in dam break modelling. Various schemes and studies in Portugal, India, Kazakhstan, Georgia and Egypt are also covered. Grouting works at two reservoirs are described and there is a paper on the desiccation assessment of the puddle clay cores at several reservoirs. Various other works to refurbish and rehabilitate dams are described and illustrated.

The conference included the presentation of the biennial Geoffrey Binnie Lecture by Chris Binnie. The 2006 Lecture, entitled *'Dams, responding to society's needs'* is published in the Society's journal *Dams and Reservoirs.*

Contents

The influence of inspection and monitoring on the phased construction of the
Barragem de Cerro do Lobo
M CAMBRIDGE and M OLIVEIRA TOSCANO   419

1. Implementation and operation of UK reservoir Legislation

Reservoir Safety – Are things improving ?

DR ANDY HUGHES, Director of Dams & Water Resources, Atkins

SYNOPSIS
This paper seeks to look at reservoir safety over the years and how some
things have changed. The paper suggests that there have been some
significant improvements in reservoir safety such as the creation of the
Supervising Engineer role, the creation of the single Enforcement Agency,
but it also suggests that there are a number of areas where improvements
have not been made and in fact the situation might be getting worse. The
paper suggests some areas for improvements but also invites readers to think
whether the issues raised apply to their organizations and situation in a hope
that they will then bring about change.

HISTORICAL BACKGROUND
Reservoir safety on the whole in the UK has been driven by the legislative
framework which has been developed with time. The Reservoirs (Safety
Provisions) Act, 1930 bought in the regular and frequent inspection of dams
by Panel Engineers and was prompted by a couple of failures, which caused
loss of life – notably the failures in the Conway Valley at Eigiau and Coedty
Dams, and Skelmorlie on the Forth of Clyde. This set the definition of a
'large raised reservoir' as a reservoir containing 5 million gallons above the
level of the natural ground, adjoining the reservoir.

As a result of a number of further incidents and failures, particularly in
Europe, an ad-hoc committee of the Institution of Civil Engineers was set up
to review the legislation, and this committee made a number of
recommendations which eventually resulted in the Reservoirs Act 1975.
This Act brought in a number of new features, the main ones being the
creation of the role of the Supervising Engineer, the creation of the
Enforcement Authority and thus enforcement of recommendations, and
registration of reservoirs.

The UK has had no dam failures resulting in loss of life since 1925 although
there have been a number of failures, mainly of small dams and there have
been a large number of incidents/accidents – some more serious than others

– some would say that the UK has been fortunate in not having failures resulting in loss of life since 1925.

In 2006 the question I wish to pose is 'have we made progress in the field of reservoir safety since 1925 – what have we achieved in 81 years?'

In looking for the answer to this question there are a number of issues which I have considered and present in this paper;

THE SUPERVISING ENGINEER
There is no doubt in my mind that creation of the role of the Supervising Engineer has been a useful addition to improve reservoir safety. The Supervising Engineers have become the 'eyes and ears' of the Inspecting Engineers and certainly in recent years there seem to have been more inspections called for under Section 10 (2)(d) – by the Supervising Engineer.

It has been recently suggested in some quarters that the Reservoirs Committee has raised its standards, resulting in a number of failures of candidates seeking appointment and/or re-appointment to the Supervising Engineers Panel. Many of those failures have been cited as people who were nearing the end of their career, who had a low level of activity and/or could not display the commitment to continued professional development (CPD) or do not have confined space training.

In my personal opinion a number of people who have not been re-appointed are Supervising Engineers especially if one remembers that the original idea of the group formulating the legislation was that the Supervising Engineer would be the Reservoir Keeper at the site. However, the role of the Supervising Engineer is changing and I believe the role of the Supervising Engineer will become more onerous with time. Already, the Supervising Engineers are 'responsible at all times' when there is not a Construction Engineer and as time progresses are likely to get more involved with monitoring progress with recommendations in the interests of safety and perhaps calling for inspections as conditions deteriorate within the 10 year period set by the Inspecting Engineer. They are also likely to get more involved with the checking, exercising and even rehearsal of Flood Plans.

THE ENFORCEMENT AUTHORITY
The creation of an Enforcement Authority was undoubtedly an improvement and important addition to the legislation in the form of the Reservoirs Act 1975. Unfortunately, in the 1980's and 90's, the Enforcement Authority role was vested in 168 organisations, and still remains the responsibility of 32 different organizations in Scotland. However, the Scottish Executive has

seen the advantages of the new system in England and Wales and is consulting on change. This led to an enormous range of different standards throughout England, Wales and Scotland.

In the Water Act 2003, the Enforcement Authority role in England and Wales transferred to the Environment Agency. The Agency has set up an office in Exeter, which monitors compliance with the Act. Undoubtedly some would be critical of the system developed and the way the EA has been making decisions within the enforcement framework. However, I believe all would agree that there have been many advantages that have occurred as a result of the adoption of a single Enforcement Authority, even though some will have been subject to a significant amount of paperwork! Certainly since formation of the 'new' Enforcement Authority some reservoirs have been registered (195 are 'new' and currently a review of in excess of 400 potential reservoirs is under way). Inspecting Engineers have been appointed to over 40 reservoirs that had not been inspected and Supervising Engineers appointed to reservoirs reducing the number of reservoirs with no known SE from 379 to currently 19 – they would not have been appointed had the enforcement system not changed.

One of the roles that the Enforcement Authority undertakes is to ensure that recommendations in the interests of safety have been completed. This has been achieved by asking for copies of Certificates under Section 10(6) of the Act, often within 6 months of the registration. In recent times, an exercise has been undertaken to pursue owners with outstanding recommendations in the interests of safety which are more than 5 years old. In adopting a risk based approach to the backlog of non compliance, the approach of the Enforcement Authority has been to interpret that 'as soon as is practicable' means that a delay of 5 years or more is unacceptable. I understand that they will be turning their attention to those that are 3 and 4 years overdue shortly! In other cases where measures are outstanding for over 5 years, Enforcement Notices have been issued with time periods stated which have been agreed with a 'qualified civil engineer'. This has led to some differences of opinions associated with the phrase 'as soon as practicable' and 'problems' with tools such as 'portfolio risk assessment'. These issues will have to be resolved in order to make progress and the timely use of resources.

MAINTENANCE
The provision of maintenance and getting owners to undertake works has always been something difficult to achieve – depending on who the owner is. It is the case that some issues which would be classed as maintenance today, if not repaired, will become recommendations in the interests of safety. What I have noticed is that, as the major water undertakings have

changed and in particular where they have reduced manpower and outsourced works there is often a significant reduction in the level of maintenance and capability to carry out minor works. Whilst it can probably be accepted that the levels of staffing in the 1970's/80's were too high I would argue that perhaps they have gone too far the other way. No longer do we have the dedicated reservoir keeper with a pride in his site; very rarely do we have a maintenance team with a maintenance at reservoirs; usually we have to wait for the resources of an external contractor provided through a framework contract – usually selected on a lowest price basis. How often have we heard the words – I no longer have any staff to do the maintenance work?

I have witnessed in recent years; saplings growing on land adjacent to dams; grass up to a metre high; drains overgrown and malfunctioning; valves inoperable; turf ripped off the faces of embankments; stones missing from wave walls and upstream protection systems; broken windows to valve towers. In general there appears to have been a reduction in the frequency of providing maintenance and in the quality of that maintenance.

One area that seems to cause moderate amounts of problems is the cutting of grass on embankments. In most cases water companies have outsourced this and it is often the case that the grass is not cut at the correct time, is cut too frequently or not enough times, grass cuttings are not removed, inappropriate machinery is used causing damage to and rutting on the embankment, wet patches are just driven through. Long gone are our beautifully manicured embankments and perhaps they should but some do not reach an acceptable standard that allows inspection and examination! Contractors are also less likely to observe and understand the relevance of new damp patches, areas of settlement etc.

Valve operating – how many times are valves being fully exercised? Have we seen a situation where valves are being exercised less frequently or not over their full range resulting in valves which become stiff or inoperable? Some companies have had to increase the frequency of operation back to what they used to, to ensure the valves remain operable.

Historically we have often experienced a reluctance to operate scour valves because 'we might not get them shut, and also the EA will object to the discharge of dirty water'. Following the Rivington Incident where the scours were vital to the satisfactory resolution of the problem I now recommend a full scale scour test, with the water left running until the flow runs clear to try to ensure the scour facility is operable and does not silt up. The Environment Agency is being very helpful in developing protocols which will allow these tests to be carried out.

Are we seeing what would in the past have been called day to day maintenance not being carried out? Are there times when Supervising Engineers have said 'I've been trying to get this done for months' – I certainly have heard this on a number of occasions. Will we see a need for Supervising Engineers to recommend an inspection if maintenance is not provided?

PROCUREMENT OF ENGINEERING SERVICES

In many cases, and particularly in the large water undertakings, companies have embarked on procurement strategies which have led to 'outsourcing' and a reduction of 'in house' staffing and also a number of frameworks, alliances, and strategic partnering initiatives have been set up. These initiatives I have always understood to have been set up to make things easier, cheaper and quicker. Unfortunately I have experienced situations which it has been difficult, certainly more expensive, very protracted and in some cases the services provided have been inappropriate.

It seems to be impossible to procure the services of an Engineer or get an Engineer to procure services, as one used to some years ago, when the Engineers were trusted to carry out this role, devise a contract, get prices from three contractors and then arrange and supervise the works. Nowadays the formal procedures can take many months to procure a team to do the works and secure the finance. I have also seen the procurement of consultants and contractors who have little or no experience of doing work related to dam safety, but who have won frameworks on the basis of other skills offered. Then we get badly designed schemes which cost too much, take too long and in some cases after several millions of pounds of expenditure, and don't work – or we get protracted arguments over the skills and qualifications of staff and the rates at which they should be charged out!

I have also had personal experience of carrying out the role of Inspecting Engineer, making recommendations in the Interests of Safety and the spending time and clients' money in briefing yet another consultant to carry out the recommendations – double counting and double expense!

I personally have seen situations of poor communication and support within alliances where the consultants never seem to talk to, or work with, the contractor, and contractors working on reservoirs who have never worked on reservoirs before – just because they have insufficient work of other types from the alliance.

What about site supervision? How many times are we told that site supervision is not required because we trust the contractors or we don't want to pay for full time supervision? How many times are young inexperienced staff asked to supervise works? How many times have you seen mistakes made, inappropriate materials used and poor workmanship not only accepted but also paid for? Are we getting the level of site supervision correct?

In cases where I have requested surveys, site investigations, and leakage investigations, many organisations can no longer procure these quickly and it can take weeks to get this information which is essential if one is to make decisions regarding safety or any design works.

OPERATIONAL RISK

I detect that there is also a distinct lack of willingness to take operational risks, in some of the water undertakings. For example in reducing water levels to carry out works or taking a service reservoir out of service in the summer. There have been many problems, which have affected a company's ability to programme and carry out works "in the interests of safety" which has, in some instances, resulted in Enforcement Notices being served.

Is this because there is a lack of communication within the organisation? Is it because reservoir safety is seen to be the poor relation of the organisation? Is it because the managers either do not have the knowledge and confidence in their own abilities and in that of their systems? Is it because they are not engineers, or is it because a blame culture exists within the organisation? Is it fuelled by media reaction to problems? Whatever it is, I perceive there is a lack of understanding within organisations about the need for planning and execution of works, required to meet the recommendations made in the interests of safety in a timely manner and often a reluctance to give sufficient regard to reservoir safety.

RISK ASSESSMENT

Risk assessment is now a major part of ensuring the safety of our dams and one which seems to impact on all aspects of our lives these days. In fact the inspection process itself is, and has always been, an observation based risk assessment. I am sure all would agree that the Reservoirs Act 1975 needs to be modified to embrace a risk based approach, if nothing else in the definition of what a reservoir is – i.e., not one based on retained capacity alone.

However, are some of our risk assessments techniques too complicated and too costly to make them universally acceptable? Are owners, and consultants for that matter, using risk assessment for the right reasons and then using it in the right way? Are the opportunities and improvements that risk assessment brings being utilised in terms of improving reservoir safety, bringing about organisational change, directing research etc? To the last question I should say – not often!

RESEARCH
As problems are experienced and conditions change there is a need for research in a number of areas of reservoir safety. There are a number of research organisations, universities, and companies undertaking in house research, specialists offering enhanced services based on research, funded by agencies including Defra, the EA, NERC and others. Yet we do not seem to have a coordinated approach to research. Indeed, there are areas of research that many of us will not even know are being carried out.

In addition, there are cases where there are research needs and there are insufficient funds from one body to meet the needs of that research. I believe that there is a clear need to firstly understand, and communicate to all of the profession, what research is being undertaken, that there is a requirement to develop a list of prioritised research needs, there is a need to attract funding sufficient to carry out that research and then there is a need to communicate the results of that research to the profession. Unfortunately I believe we are falling short of the mark in all areas.

Our judgements and decisions are often supported by Guidance Documents. We must continue to review and update those guides as new information becomes available and ensure there is transfer of information from other sectors, e.g. coastal engineering technology applied to waves and wave impact forces on wave walls etc.

INCIDENTS
Whilst we have had no recent failures, we have experienced a number of incidents some of which have been quite serious. Years ago there were many professional papers written on these incidents, for example one can remember papers on Balderhead, and very more recently Upper Rivington and Ogston and Carsington, where the consultants and owners were prepared to present information/'air their dirty washing' for the benefit of the profession. I congratulate those prepared to do this but there are others who are more concerned about company reputation, share price etc. who are more secretive about the technical information associated with the incidents and the way they have managed the incident. In these cases we 'cannot learn from our mistakes'.

The proposal to have a system of 'Incident Reporting' is an initiative that I consider must be supported. It might be, as I have suggested in the past that the report will have to be done by an Inspecting Engineer who commands the confidence and respect of the owner, and it may result in a report which 'sanitises' the incident without mentioning the dam but we <u>must</u> get reports of incidents reported to the owners and Panel Engineers so that we can learn.

I fully support a voluntary incident reporting system, managed by the Enforcement Authority, in order to learn from our mistakes – if owners will not co-operate then it must be made a mandatory requirement.

TRAINING AND SUCCESSION PLANNING

It becomes clear that we must ensure that the engineers associated with reservoir safety are properly trained. We must ensure Supervising Engineers can not only exercise the judgement necessary to call for a statutory inspection but also review and rehearse flood plans, monitor progress with respect to recommendations in the interests of safety etc.

The average age of our Supervising Engineers is 56 years – are we doing enough to train the prospective Supervising Engineers of the future? In 2000 we had 94 SE's under the age of 50, now we have 54. Are we doing enough to train the Inspecting Engineers of the future? The average age of our All Reservoirs Panel is 60 years. In 2000 we had 63 All Reservoirs Engineers, in 2005 only 53. How are we going to provide for the future?

Recommendations in the Interests of Safety must be enforceable – in other words they must be certifiable – and hence well defined and not open ended. Yet, examples of actual recent recommendations in the interests of safety have included:

- 'Regularly clear weed growth and vegetation from around the main circular overflow and the secondary concrete weir overflow to maintain a clear water area to and around the cill. Keep both spillways clear of debris.'

- '...... stoplogs may be installed between 1 April and 30 September each year ....to a level not more than 180mm above the sill of the main spillway. The stoplogs must be removed not later than 30 September each year and must not be reinstalled before 1 April.'

- 'I recommend in the interest of safety that the above points of maintenance should be continued......'

- 'No residential caravans should be sited in the area where the natural ground is below the water level in a 10,000 year flood (assuming no breach).'

- 'No homes to be built on plateaux immediately downstream of the dam.'

Currently the Inspecting Engineer inspects a reservoir and has no idea of the condition of the reservoir next in the cascade or in the next valley. Consequently he has no idea whether resources, often limited, should be directed towards the reservoir being inspected or another in the owner's stock. Portfolio Risk Assessment (PRA) seeks to address this problem enabling an owner to reduce the total risk he faces as quickly as possible. The risks can be measured in a number of ways; - probability of failure, consequence of failure, in terms of life, or economic loss, security of supply to customers (single source supply reservoirs) etc. Portfolio Risk Assessment can be used to direct limited resources in a way that reduces the risk posed by an owner's reservoirs. However, for the system to be of use it will need both Inspecting Engineers, who are making recommendations, and those who are enforcing to understand the concepts and take account of the assessments – in other words more education is needed.

CONCLUSIONS

In my opinion, we have undoubtedly made progress in some areas of reservoir safety. Our enforcement system is undoubtedly better; our Supervising Engineers are carrying out a very useful role – but there are many areas where we are not doing well and I suggest we are not improving– we can do more. I have not made an attempt to answer some of the questions I have posed – they have been posed to generate thought and debate. However, I do believe we need to:

- Ensure the adequate training and assessment of Supervising Engineers.
- Ensure the adequate training and assessment of Inspecting Engineers.
- Improve the quality of maintenance of our reservoirs.
- Review the methods of procuring the services of all work associated with reservoirs.
- Educate those associated with the operation and maintenance of reservoirs.
- Raise the profile of reservoir safety in owner organisations.
- Educate the profession in general about risk assessment – its advantages and disadvantages.
- Achieve an integrated, well funded programme of research.
- Establish an incident reporting system.
- Engage in a programme of succession planning.

In conclusion, yes, we have improved in some areas, but in some areas things are worse and certainly we can do better.

NOTE
The views expressed in this paper are the personal views of the author and not necessarily the views of Atkins Ltd, Defra, ICOLD or the British Dam Society.

REFERENCES

HMSO, Reservoirs Act 1975

HMSO, Reservoirs (Safety Provisions) Act 1930

Development of the requirements for Flood Plans under the Reservoirs Act 1975 (as amended)

A J BROWN & J D GOSDEN, Jacobs Babtie

SYNOPSIS   The Water Act 2003 amended the Reservoirs Act 1975 and gives the Secretary of State power to direct that the owner of a reservoir regulated in England and Wales under the 1975 Act shall prepare a Flood Plan (emergency plan).   This paper describes the value of such plans followed by the various factors taken into consideration in the development of both the proposed specifications for Flood Plans, and the accompanying Engineering Guide. It also discusses how these would be expected to contribute to ensuring the continuing safety of UK reservoirs

INTRODUCTION
It has been recognised for many years that effective emergency planning can prevent or reduce the impacts of dam failure, with owners of major dams including such plans as part of their dam safety management system. Additionally several countries have passed legislation which requires dam owners to produce such plans.

Elements of emergency planning have been applied to reservoirs in the United Kingdom for some time and can include
   i)   the prescribed Form of Record for a large reservoir, established by statutory instrument under the Reservoirs Act 1975 includes details of access to the dam and the maximum rate of discharge of water from outlets.
   ii)   The Department of Environment (now Defra) funded development of DAMBRKUK (Binnie & Partners, 1986, 1991), which several major dam owners used to produce inundation maps for their dams.
   iii)   Owners of major dams also maintain on-site plans
   iv)   periodic Inspections under Section 10 of the Reservoirs Act 1975 generally consider the ability to lower the reservoir in an emergency.

Section 77 of the Water Act 2003 amended the Reservoirs Act 1975, by addition of new Sections 12A and 12B. This gives the Secretary of State

Improvements in reservoir construction, operation and maintenance, Thomas Telford, London, 2006, 13–25

power to direct that the owner of a reservoir regulated in England and Wales under the Reservoirs Act 1975 shall prepare a flood plan (emergency plan). This direction will specify "matters to be included" and require preparation to be in accordance with specified "methods of technical analysis".

The authors have been working with Defra and others, under a research contract between 2002 and 2006 (novated from KBR to Jacobs Babtie in December 2005) to identify the structure and content of such plans and, using a risk based approach, which reservoirs should be required to have part or all of the elements of such a plan (KBR, 2004). This was followed by drafting of the two proposed specifications associated with the proposed direction and an Engineering Guide to Emergency Planning for UK Reservoirs. The latter includes examples of the various elements of a flood plan as appendices to the Guide.

This paper summarises the key factors determining the structure and content of the specifications and the accompanying Engineering Guide. These documents developed over several years starting in September 2003 and details are being further refined at the time of writing this paper. The development process included meetings with Defra, other government departments, the Environment Agency, reservoir owners and panel engineers, as well as attending the Cabinet Office Civil Contingencies Secretariat course on Management of Flooding and other severe weather incidents.

THE VALUE OF FLOOD (EMERGENCY) PLANS FOR RESERVOIRS
In the United Kingdom since 1975 although there have been a relatively high number of emergency drawdowns (three a year, Gosden & Brown, 2004), to date there have been no failures with loss of life. This demonstrates the usefulness of, and need for, effective planning of emergency action to avert failure, and that this should become routine for all reservoirs which could cause loss of life, rather than being limited to a few of the major owners.

Continuing research in the United States (BOR, 1999) has shown that effective warning can reduce the fatality rate in a medium severity flood from 15% for no warning to 1% with a precise warning more than 60 minutes in advance. For high severity floods the fatality rate with no warning is suggested as 75%. This confirms the value of having impact assessment already available in the event of a serious structural problem, to facilitate effective warning and evacuation of those at risk in the event of a dam failure. Other benefits of impact assessment include for the dam owner in terms of quantifying the consequences, and thus the risk posed by his dam, and by the emergency services for scenario planning.

PRECEDENT FOR CONTENT OF EMERGENCY PLANS
The Water Act does not specify the format or content of a flood plan. Thus the first task was to review how these were approached in other industries and other countries.

In relation to overseas practice for emergency planning for dams no single overall summary of requirements was identified. Although ICOLD published Bulletin No 111 on dam break flood analysis in 1998, there have not yet been any Bulletins on other aspects of emergency planning. In France legislation requires that for major dams the dam owner installs sirens within the 15 minute zone (Royet P, & Chauvet R, 2000); however in France dams are generally larger with a greater predominance of concrete dams (which generally fail faster than embankment dams) than in the UK. In Norway dam break warning systems were installed in the Second World War, abandoned but then resurrected in the 1970's (Svendsen, 1997, ICOLD Q75, R20; Konow, 2004).

In Australia Emergency Management Australia published Guide 7 on "Planning for floods affected by dams" (2004), whilst ANCOLD have published "Guidelines on dam safety management" (2003). In the United States following new legislation in 1996 the Federal Emergency Management Association (FEMA) has published (1998) "Federal Guidelines for Dam Safety: Emergency Action Planning (EAP) for dam owners".

Existing legislation or guidance in other high hazard industries in UK is summarized in chronological order in Table 1. A key document in relation to management of the safety of high hazard industries in UK is the HSC Policy statement on "permissioning regimes" (HSC, 2003). This notes that the responsibility for managing the risk lies firmly with the owner of the hazardous installation and the duty of care they owe to everyone who is put at risk by the existence of that hazard. In particular the legislation is not prescriptive, but requires owners to think through their operations, and describe, demonstrate and document how they manage risks. This was discussed in McQuiad (2002) and Brown and Gosden (2002). This principle has been adopted in drafting the requirements for flood plans, as described below.

Table 1 - Summary of emergency planning in high hazard UK industries

1997	Further guidance on emergency plans for major accident hazard pipelines. ISBN 0717613933 HSE 1997
1999	A Guide to the Control of Major Accident Hazards Regulations. 125pp
2001	A Guide to Radiation (Emergency preparedness and public information). Regulations 148pp
2004	Civil Contingencies Act
2004	Fire And Rescue Services Act

MATTERS TO BE INCLUDED IN A FLOOD PLAN

Strategy

Defining what has to be included in a flood plan has been tested throughout against both the objectives of a flood plan and the experience of dam owners who already have emergency plans in place. The objectives of a flood plan are to:

- minimise the probability of failure in the event of a structural problem at a dam,
- contribute to minimizing the loss of life and injury to those in the potential inundated zone, both through the direct results of the dambreak and its consequential effects

These should both provide real benefits to the dam owner and the community in reducing the risk to life and property posed by a reservoir.

Roles and responsibilities

Under the Reservoirs Act (as amended) the undertaker, where so directed, is responsible for preparing the flood plan in accordance with the direction. Although non-compliance is an offence, there is no power for the enforcement authority to prepare the Flood Plan themselves, in the event of a default by the undertaker. This contrasts with other aspects of the Reservoirs Act, where the Enforcement Authority has the power to take actions themselves to assure dam safety, for example in relation to periodic inspections and the implementation of matters in the interests of safety. As the Water Act amendment to the 1975 Act does not explicitly refer to a qualified civil engineer, it has been agreed that it will be recommended that a flood plan is examined and signed off by an independent qualified civil engineer (Inspecting Engineer) as defined in the Reservoirs Act 1975.

<u>Elements comprising a Flood Plan</u>
Examination of precedent for emergency plans, as identified above, shows that there are generally three sections, an assessment of the consequences if the hazard escaped from the owner's land, the on-site plan and an off-site plan. The latter two are separated partly because the lead is generally taken by the hazard owner and emergency services respectively, and also because legal powers to take actions vary depending on the owner of the land where the actions are being taken.

For Flood Plans under the Water Act there is no power to require emergency services to prepare off-site plans, or to otherwise cooperate. Additionally the Civil Contingencies Act 2004 was being developed in parallel with the Water Act Flood Plan powers, the former setting out new responsibilities for emergency services in relation to planning for all forms of emergency. It was therefore decided that off-site planning under the Water Act 2003 would be limited to a plan relating to the interfaces of the reservoir owner with the emergency services.

<u>Content of each element of a plan</u>
Following the principle of permissioning regimes the contents of a Flood Plan have been specified as a series of mandatory headings and issues which should be covered under each heading, illustrated in Table 2 with the headings for the On-site Plan. It is then up to the owner to document how he would manage an emergency. The experience of Hydro-Tasmania (Barker, 2003) was noted, who found that producing a plan for each one of their 54 referable dams involved disproportionate cost/ resources, and instead have developed a generic dam safety emergency plan. This includes trigger levels for automated warning of floods (> 20 year return period) and a commercial arrangement where the Seismology Research Centre determines seismic intensities with Modified Mercalli Intensity > 4 at Hydro Tasmania dams. Thus the draft specification allows a generic main text, with information on individual dams given in appendices.

Table 2: Schedule of headings required in On-Site Plan

1	*Objectives, scope and administration of the On-site plan*
2	***Management of emergency by Undertaker***
2.1	*Undertaker's procedures and authorised personnel*
2.2	*External communication*
2.3	*Checklist for those attending the emergency*
3	***Description of the reservoir and retaining dam(s)***
3.1	*Situation*
3.2	*Detailed records*
3.3	*Physical dimensions and features*
3.4	*Other facilities relevant to on-site operations*
3.5	*Access to dams*
3.6	*Communications*
3.7	*Welfare facilities*
3.8	*Normal operation*
4	***Actions by undertaker on site***
4.1	*Situation assessment*
4.2	*Undertaker's Resources relevant to on-site activities*
4.3	*Reservoir drawdown*
4.4	*Other measures*
4.5	*Off-site impacts of site activities*
4.6	*Assistance from external organisations with on-site measures*
5	***Measures at other installations***
5.1	*Interaction with other reservoirs in cascade (where present)*
5.2	*Measures at other installations*
6	***Maintenance of the On-site plan***
6.1	*Training of staff*
6.2	*Periodic testing of existing outlets (and any other measures for emergency lowering of reservoir)*
6.3	*The level and frequency at which the on-site plan shall be exercised*
6.4	*Review and updating of the plan*

METHOD OF TECHNICAL ANALYSIS

Consideration was given to specifying which software should be used for dambreak analysis. However, it was recognised that there is a wide range of scenarios which would need to be analysed, ranging from narrow steep valleys to wide flat floodplains and areas around non-impounding and service reservoirs, for which the appropriate software was likely to vary. In view of both this and the relatively rapid development of software it was decided that it would be more appropriate to adopt an end product specification, where the analysis was specified in terms of the output required, and that the Flood Plan must state the assumptions made in the analysis. The key elements of the output required are summarised in Table 3.

The other key issue is how to manage the wide range of possible dam failure scenarios and assumptions for each of those scenarios. It was decided that the minimum requirement would be to estimate the inundation and consequences for a single "standard analysis scenario" to achieve consistency in the methodology, this being defined in the Engineering Guide. Two scenarios were defined, a rainy day and sunny day scenarios, with the minimum requirement being to model the rainy day scenario, as a conservative estimate of the likely extent of inundation in the event of a dam failure

Table 3: Summary of output required from Impact Assessment

1	Table of peak breach outflows for different cascade failure scenarios, to identify which combination of dam failures would give highest peak discharge into each of watercourses into which the reservoir could escape
2	Tables for points at intervals down each valley with • maximum discharge, velocity and depth of flooding • time of onset and peak flooding • total population at risk and likely loss of life in length represented by that interval
3	Figures showing • flood hydrographs at points in '2' • how peak flow varies down valley for dambreak flood, and 1% and 0.1% annual probability floods with no dam failure • longitudinal section showing peak inundation water level, ground level and position of significant infrastructure embankments
4	Tables with total population at risk and likely loss of life
5	Maps showing (not required for Rapid Analysis as Interim Guide to QRA) • locations of hydraulic model cross sections and structures • extent of inundation, damage category and properties flooded • plans for use in an emergency, suitable for photocopying at black and white and at a map scale no smaller than 1: 10,000

MAINTENANCE OF FLOOD PLANS

To be effective emergency plans need to be regularly reviewed, updated and exercised, so they remain valid and effective at all times. The level of exercising varies from checking that telephone numbers and other contact details are correct through to full scale site exercises. The level and frequency of exercising is likely to be a major component of the cost of Flood Plans. Following a proportionate cost approach the level and frequency of exercising and other maintenance tasks recommended in the Guide was related to the consequence class of the dam, as defined in the

Interim Guide to Quantitative Risk Assessment (QRA) (Brown & Gosden, 2004). The more major elements of maintenance, such as seminars or site attendance could also cover a group of reservoirs, defined as reservoirs in reasonable geographical proximity (maximum of one hour's drive apart)

PUBLICATION OF FLOOD PLANS

The requirement for publication of a flood plan to "persons likely to be interested" is still under development, noting the national security powers in Section 12B of the Reservoirs Act. It will include appropriate information to local authority emergency planners and the emergency services, may include local authority planners (development control) and may include access to view by members of the public likely to be affected in the event of a dam failure.

STRATEGY FOR PRODUCTION OF ENGINEERING GUIDE

A key part of testing the robustness of the two specifications was to draft accompanying guidance and to test this by preparing example plans for real dams, and comparing these proposed plans with existing plans prepared by the owners. Individual owners were therefore approached and suitable existing reservoirs identified for which a plan would be produced which conformed to the specifications. These reservoirs already had a form of emergency plan, so that the new format could both build on these, and any changes envisaged could be tested for the value added. For the impact assessment this allowed comparison of three methods:

- a DAMBRKUK analysis carried out in 1997
- ISIS within Infoworks carried out as part of this study
- the Rapid Method given in the Interim Guide to QRA (Brown & Gosden, 2004).

For the on-site plan an assessment of possible emergency scenarios was carried out on site, followed by discussions with the water company Reservoir Safety Manager and a Control Room Duty Manager.

WHICH RESERVOIRS SHOULD BE REQUIRED TO HAVE A PLAN

The approach used to determine which reservoirs should be required to have flood plans was a combination of reasonableness and an "As Low As Reasonably Practicable" (ALARP) analysis. The latter compares the estimated costs of a plan with the anticipated benefits of the plan, to see if the cost is proportionate to the benefits obtained.

The wide ranges in both probability of failure and consequences of failure of reservoirs which come under the regime of the Reservoirs Act 1975 should be noted, both varying by several orders of magnitude. Consideration was given to basing the specification of which reservoirs should have plans, on estimates of risk (annual probability of failure x consequences), but this was rejected partly because techniques for estimating probability are still developing, and because of the practical difficulties of enforcement. The requirement was therefore based on the consequences of failure using the Consequence Classes defined by the quantitative estimates given in the Interim Guide to Quantitative Risk Assessment (Brown & Gosden, 2004), with the proposed application summarised in Table 4.

Consideration was given to having varying levels of complexity of Flood Plans. However a specification defined by a list of headings offers sufficient flexibility and it was impractical to define different levels of headings. The exception was the impact analysis, where the technical specification differentiated two levels of analysis, a standard analysis including hydraulic modelling and production of GIS maps and a rapid analysis limited to Excel spreadsheet calculations with no maps.

Table 4: Normal minimum level of Flood Plan required for UK dams

Highest Consequence Category of dam[1] retaining a given reservoir	Element of Flood Plan		
	I	II	III
	Impact assessment[2]	On site	External Interfaces in an emergency
A1	Standard	Required[4]	Required
A2	Standard	Required[4]	Required
B	Rapid method	Required[4]	Required
C	Rapid method[3]	Not required	Not required
D	Rapid method[3]	Not required	Not required

Notes
1. As given on Sheet 11.2 of the Interim Engineering Guide to Quantitative Risk Assessment for UK Reservoirs (2004)
2. Rapid method of inundation analysis means a simplified rapid method designated in the method of preparation of a Flood Plan (e.g. the method in the Interim Guide to QRA for UK Reservoirs, 2004)
3. Required as part of every periodic Inspection under Section 10 of the Reservoirs Act 1975, to confirm the Consequence Category of the dam
4. The recommended level of exercising will vary with the Consequence Category.

The estimates of cost are not repeated here, as they are to be presented on the Defra website with the draft Guide. The ALARP analysis assumed that the existence of a well maintained on-site flood plan would reduce the probability of failure by a factor of 5. In regard to the effectiveness of off-site activities, it was assumed that the impact assessment and external interface plan would reduce the fatality rate in the event of a failure by a factor of 2.5. Clearly these values will vary for individual reservoirs, but these were considered to be reasonable median values.

There was some discussion over whether flood plans should be limited to impounding reservoirs. It was concluded that having adopted a risk based approach it would be logical to also apply the requirement to non-impounding reservoirs.

DISCUSSION – THE CONTRIBUTION OF EMERGENCY PLANNING TO THE CONTINUING SAFETY OF UK DAMS

There is no reason to be complacent about the good public safety record of dams in the UK, and this is one of the reasons behind the new requirements for reservoir owners, stipulated in new legislation. The new requirements will extend what many responsible owners are already doing to be a requirement for all reservoirs in England and Wales which could cause loss of life. Flood Plans should significantly reduce the probability of a failure through an effective on-site plan, and if a failure does occur reduce the fatalities through increased warning time and better targeted evacuation. The Scottish Executive are monitoring developments and may well promote similar requirements in Scotland.

For the benefits of emergency planning to be fully realised it is essential that the plans are maintained, including training, exercising and regular review and updating. As well as the direct demands on reservoir owners, it will increase the scope and demands on panel engineers, on Inspecting Engineers in including emergency planning as one of the tools for dam safety management and on Supervising Engineers in checking ongoing maintenance of the Flood Plan.

Preparation of on-site plans will, in addition to the direct benefits of facilitating actions in the event of an emergency, also provide indirect benefits in encouraging consideration of the credible failure modes of a dam as part of the preparation of the plan. This should in turn provide feedback to other tools of dam safety management, including

- more effective surveillance, both in terms of the issues which are monitored and the frequency of monitoring
- any physical rehabilitation or safety improvement works being focused on the items most relevant to the safety of the dam

There were extended discussions regarding the need for reservoir specific off-site plans. In drafting the flood plan requirements it was anticipated that each Local Resilience Forum, as defined in the Civil Contingences Act (HM Government, 2005) would assess the risk posed by the reservoirs in each area, entering these on the Community Risk Register. It would then allocate resources appropriately across all the risks to the community, resulting in the production of either generic or reservoir specific off-site emergency plans for dam failure.

One of the considerations dictating the effectiveness of off-site generic plans is the amount of warning time that the emergency services would get of a potential dam failure. Where this was significant, because of notification at an early stage of a potential dam failure, or because a breach took several hours to develop from the initial instability, then generic plans would provide significant risk reduction. Where no warning was given, for example overtopping failures, or failures of concrete dams, then generic plans may be of reduced benefit. At the time of writing it is anticipated that the need for site specific off-site plans for very high consequence reservoirs would be reviewed a few years after the Flood Plans power has been fully implemented, and if appropriate additional legislative powers sought. In the meantime the need for the early notification of a potential problem at a reservoir is emphasised in the Guide.

An indirect, but equally important aspect of off-site activities, is how to increase the awareness of the general public of the risk from dams without unnecessarily raising alarm, noting that although the consequences of failure could be very high, the corresponding probability is generally extremely low. Discussions are ongoing as to the extent to which simplified inundation maps should be made available in Local Authority or Environment Agency offices for inspection by the public living downstream of dams.

CONCLUSIONS

This paper has described the key issues determining the content of the possible direction and proposed specification under the Flood Plan power under Section12A of the Reservoirs Act 1975 (as amended), and associated Engineering Guide to Emergency planning for UK Reservoirs. These have been structured to follow the key principles of a permissioning regime where the reservoir owner is responsible for the management of the safety of his dam. He is assisted by an independent qualified civil engineer, who provides advice to the reservoir owner, and who certifies that in his or her professional judgement an aspect meets minimum standards.

Flood plans should provide real benefits to reservoir owners and the community by reducing the risk from reservoirs. However, to remain effective Flood Plans will require ongoing maintenance, and should be viewed as one of the tools in the toolbox available to a reservoir owner in managing the safety of his reservoir.

ACKNOWLEDGEMENTS
The work described in this paper was carried out as a research contract for Defra, who has given permission to publish this paper. However, the opinions expressed are solely those of the authors and do not necessarily reflect those of Defra.

The advice and assistance of the Steering Group, whose members consist of Alex MacDonald (Chair), Andrew Robertshaw, Jonathan Hinks, Ian Hope (Environment Agency, Reservoir Safety Manager) and Tanya Oldmeda Hodge (CLA, Derek Holiday stood in for Meeting No 3), who reviewed the various stages of the Engineering Guide on behalf of Defra is gratefully acknowledged.

REFERENCES
ANCOLD, 2003, Guidelines on Dam Safety Management.
Barker G, 2003, An innovative approach to dam safety emergency planning by an owner of a large portfolio of dams. ANCOLD Conf
Binnie & Partners, 1986, Modes of dam failure and flooding and flood damage following dam failure. Final Contract Report to DoE Contract No PECD/7/7/184.
Binnie & Partners, 1991, Estimation of flood damage following potential dam failure: guidelines. 1989 Report for DOE Contract no 7/7/259. Published by foundation for Water Research FR/D 0003 March
Brown AJ & Gosden JD, 2002, A review of systems used to assess dam safety. Proc BDS Conf. pp 602-619
Brown AJ & Gosden JD, 2004, The Interim Guide to Quantitative Risk Assessment for UK Reservoirs. Thomas Telford. 161pp
Bureau of Reclamation, 1999, A Procedure for Estimating Loss of Life Caused by Dam Failure, *DSO-99-06*. Author Wayne Graham. Sept. 1999, p.43.
Emergency Management Australia, 2004, The Australian Emergency Manual Series. Vol 3 Guidelines. Part III Emergency management Practice. Guide 7 - Planning for floods affected by dams. On Internet at www.ema.gov.au/emalnetenet.nsf/AllDocs
FEMA, 1998, Federal Guidelines for Dam Safety: Emergency Action Planning (EAP) for dam owners. Prepared by the Interagency Committee on dam safety. Available at www.fema.gov/fima/damsafe/eap_pref.shtm.
Gosden, J.D, and Brown, A.J., 2004. An incident reporting and investigation system for UK dams. Journal of the British Dams Society, Dams and Reservoirs, Vol.14, No.1.
HM Government, 2005, Emergency preparedness. Guidance on part 1 of the Civil Contingencies Act 2004, its associated Regulations and non-statutory arrangements. 230pp. www.ukresilience.info/home.htm

HMSO, 2004, Civil Contingencies Act

HMSO, 2004b, Fire And Rescue Services Act

HSE, 1997, Further guidance on emergency plans for major accident hazard pipelines. ISBN 0717613933

HSE, 1999, A Guide to the Control of Major Accident Hazards Regulations. 125pp

HSE, 2001, A Guide to Radiation (Emergency preparedness and public information). Regulations 148pp

HSE, 2003, Policy statement on permissioning regimes. March. 11pp

ICE, 1996, Floods and Reservoir Safety. 2nd Edition

ICOLD, 1998, Dam-Break Flood Analysis. Bulletin 111. 301 pages

KBR, 2004, Note on Flood Plans under the Water Act. Rev 04 July. Internal report for Defra

Konow T, 2004, Monitoring of dams in operation - a tool for emergencies and for evaluation of long-term safety. Proc BDS Conf. Thomas Telford

McQuaid J, 2002, Risk assessment – its development and relevant considerations for dam safety. Proc BDS Conf. pp 520-533

Royet P, & Chauvet R, 2000, Preparation of a specific emergency plan for Bimont dam and information to the public. ICOLD Q76 R37. Pp547-568

Svendsen VN, Molkersrod K, Torblaa, 1997, Emergency action planning for major accidents within river basins in Norway. ICOLD Florence Q75, R20. pp261-270

A New Incident Reporting System for UK Dams

A L WARREN, Halcrow Group Ltd
I M HOPE, Environment Agency

SYNOPSIS.   This paper describes the development of a system of incident reporting for UK dams, to be administered by the Environment Agency. Drawing from recent research for Defra by KBR Consultants, and building on an existing database developed by the BRE, the paper describes the development of the system specification for the particular requirements of the Environment Agency and the reservoir industry. The main aim of the new specification is to provide an effective system for the reporting, storage and analysis of information on incidents at dams and related structures, and related remedial measures. This information will then be used to inform the industry on vulnerabilities and trends in incidents and to improve reservoir safety through the lessons learnt. It may also inform future research priorities. Some of the key issues addressed in the paper are: what constitutes an incident; who reports an incident; confidentiality and liability. The paper also discusses proposals for the investigation of major incidents by suitably-qualified engineers.

INTRODUCTION

The concern of society in responding to incidents or accidents is often influenced by the character of the incidents. The general public appears to take great interest in serious incidents at establishments capable of causing significant number of deaths and destruction and where victims have no influence or control over the accident or its outcome, especially where the number of people at risk is large, even if no one was hurt. In this respect, dams can be likened to the nuclear industry, which also has a very good safety record but one that still captures the public's imagination. The UK dam safety record since the introduction of legislation in 1930 can be considered as good. However, major incidents continue to occur and the industry should learn, and be seen to learn, from such events to minimize the likelihood of dam failures.

Improvements in reservoir construction, operation and maintenance, Thomas Telford, London, 2006, 26–36

Incident reporting and investigation in several UK industries has recently been studied by the Royal Academy of Engineering (RAE, 2005). Of the conclusions reached by this study, three are of particular relevance to the general aims of the incident reporting and investigation specification development:

- Incidents that by chance fall short of developing into major accidents should attract an equal intensity of investigation if they are to serve as sources of insight into causes and allow future accidents to be prevented that may not benefit from the same fortuitous chance.
- The primary aim of any post accident investigation into cause should be to allow accidents having similar causes to be prevented for the future.
- A powerful contributor, possibly the most important one, to preventing accidents is by companies and individuals learning from those that do happen, digesting their causes and consistently applying them throughout their own organisation wherever it is relevant to do so.

The potential benefits of incident reporting to the UK reservoir industry have recently been explored by Charles (2005).

The development of a database for recording and analyzing incidents at UK dams can draw from the experience gained from the development and administration of the Building Research Establishment (BRE) database (Tedd, 1992). This BRE database was primarily developed with the following objectives:

- To provide a register of dams that come within the ambit of the Reservoirs Act 1975;
- To identify research needs and provide background information for the government's reservoir safety research programme;
- To assemble data on dam failures and incidents and remedial works to allow some form of risk assessment to be carried out.

Prior to the development of the new system, there has been no formal request to supply information to a central agency and the information has been acquired from a number of sources including responses to questionnaires, published information and private communications. It is likely that the majority of all serious incidents in recent times are contained in the BRE database as they will be in the public domain. Analysis of incident data has been reported by Tedd *et al* (2000), Charles *et al* (2000), Skinner (2000) and Brown *et al* (2003). The BRE database has also been used in the preparation of a number of UK engineering guides.

Defra funds the great majority of the Environment Agency's expenditure on flood risk management and gives financial support to improvement projects undertaken by local authorities and internal drainage boards. Defra has a number of Service Delivery Agreement (SDA) targets relating to flood defence. One of these targets (SDA 26) aims, by the encouragement of sustainable defence measures (including timely and effective flood warning systems), to have no loss of life through flooding. Clearly, investment in reservoir safety is relevant to this aim.

In 2002, Defra commissioned a research contract for KBR Consultants to develop a specification for incident reporting and investigation, under the direction of a steering group. The need for such a system is underlined by the recent RAE report. This included a questionnaire to the dam industry and progress on this work was reported by Gosden and Brown (2004) and, in the context of other developments in reservoir safety research contracts, by Brown and Gosden (2004[1]). Access to the completed Defra specifications is available through the Defra website.

The incident reporting and investigation system will be administered by the Environment Agency. The database will be developed through 2006 and shall be available to the industry from early 2007. A current contract between the Environment Agency and Halcrow Group Ltd aims to develop the Defra specifications through:

- Consideration of the Environment Agency's particular requirements for interfacing the database with it's Reservoir Enforcement and Surveillance System (RESS) (Hope and Hughes, 2004);
- Further consultations with the reservoir industry through 2006;
- Development of the database through trials using data on hundreds of incidents from the BRE database and major incidents as they arise through 2006;
- Consideration of how the system will be administered and managed.

The formalization of incident reporting for dams in the UK conforms with international best practice. ICOLD Bulletin 59 states:

The operator should be obliged to immediately inform the government agency of any occurrence, distress or deficiency that affects or may affect the safety of the dam or reservoir.

It can be argued that the underlying need for this provision is not only to render the reservoir safe, as provided for by the Reservoirs Act 1975, but to allow others to learn from the experiences gained.

INCIDENT REPORTING
Incident Definition
Various categories of incident were considered in developing the Defra specifications and similar definitions will be used by the Environment Agency for incident management and reporting purposes. These include dam failures and 'near-miss' incidents that would have had a high probability of leading to dam failure had not prompt corrective action been taken. Reportable incidents will generally be instigated by:

- external threats (e.g. a flood);
- internal threats (e.g. progressive internal erosion); or
- human error (adverse changes to operating, maintenance or surveillance provisions or procedures).

Routine reservoir safety measures, as may be carried out following statutory reservoir inspections, would not normally be considered as reportable incidents.

Use of panel engineers and Undertakers in reporting incidents
There is currently no legislation to enforce incident reporting. It is highly likely that future reservoir safety legislation will provide for this if the reservoir industry does not support the new system by providing information voluntarily. Defra has made it clear in a recent communication to panel engineers and reservoir undertakers that panel engineers and Undertakers should support the incident reporting system. Supervising Engineers are usually ideally placed to report on incidents through their technical training, familiarity with the dam and its history, and (in many cases) good knowledge of how an incident arose and the measures taken. In some cases, the responsibility could reasonably be passed to an Inspecting Engineer involved in dealing with the incident.

In many cases, especially where incidents arise at dams owned by major Undertakers, it is anticipated that the dam owner will wish to take a lead role in the reporting process.

It is proposed to provide an internet site dedicated to reservoir incident reporting to provide appropriate contact details of the Environment Agency team who will guide and support them through the reporting process.

<u>Scope for prosecution</u>
The Environment Agency intend to operate the database using staff drawn from its Reservoir Safety team. Relevant information held on RESS in relation to panel engineers and dam characteristics will be transferred onto the incident database. However, there is no intention of using the incident database for the purposes of prosecution as non-compliances will already have been detected by RESS. However, where significant damage and/or loss of life arises due to gross negligence, there may be reasonable grounds for prosecution by, for example, the Health and Safety Executive. In such cases, contributing to the incident database is highly unlikely to increase the risk of prosecution from third parties. Furthermore, in any prosecution it may be in the operator's interests to demonstrate compliance with industry initiatives on safety management.

<u>Confidentiality</u>
The system will make provision, as far as practicable, to ensure that data is only released for the purposes of analysis. In most cases, analysis will not require the identity of the reservoirs to be released. The Environment Agency is aware that some dam owners are sensitive about information being released which may, for example, affect their share price. It is proposed that information that identifies reservoirs should only be released to third parties with the consent of the dam owner.

<u>Coverage</u>
The database shall be used for incidents arising at reservoirs within and outside the ambit of Reservoirs Act 1975 (the Act). It is expected that only a small number of serious incidents arising at reservoirs outside of the Act is likely to be captured, but the database should not be developed to exclude the possibility of learning from incidents at small reservoirs on these grounds.

The intention is that the database should capture information from the whole of the UK. Northern Ireland (NI) is not covered by the Act, but the Act is often applied in spirit, with Supervising Engineers appointed accordingly. We therefore propose that incidents at NI reservoirs should be captured by the system.

Information on Scottish reservoirs is not held on RESS. The Scottish Executive is currently considering proposals to follow England and Wales in forming a single enforcement authority for Scottish reservoirs under the Act. In this eventuality, it should become possible to populate the database with the same basic information on dam characteristics currently available from RESS for English and Welsh reservoirs.

THE DATABASE

Incident Details

The database development draws on some concepts developed for the *Interim guide to quantitative risk assessment for UK reservoirs* (Brown and Gosden, 2004[2]). It is appropriate that the database clearly distinguishes between:

- the external and internal threats acting on the dam structure; and
- the mechanism(s) of deterioration.

For some incidents, there may be more than one threat that contributes to an incident being declared. The database will need to clearly identify the primary and secondary threats (external or internal) leading to the incident, as well as documenting:

- the indicators that led to the incident being declared;
- the immediate measures taken to deal with the situation;
- the lessons learned, including any implications for surveillance frequency, reservoir operation, instrumentation requirements, etc;
- details of any longer-term remedial measures carried out in response to the incident.

Dam Characteristics

To make full use of the incident data, it is desirable to record dam characteristics to a greater level of detail than that held on statutory records or those held on RESS or the BRE database. The Defra Specification includes fields for dam characteristics which have been reviewed and developed as part of the current work. The intention is to complete these fields for reservoirs at which incidents arise with the assistance of the contributor and/or the dam owner or designate. The aim is to ensure that the details of the incident are recorded within the context of the dam details which will assist when evaluating trends or apparent vulnerabilities.

The long term aim is to complete the dam characteristics fields for all 2,600 reservoirs under the Act in the UK, and for any reservoirs not under the Act at which incidents are recorded. This represents a significant challenge for the industry. Dam owners and Supervising Engineers represent the two groups most capable of providing this information. The benefits of achieving this goal are:

- an improvement in the level of incident analysis and an associated improvement in the effectiveness of the reporting.

- a step-change in the industry's ability to access detailed information on UK dams for research and development activities (beyond the capabilities of the existing BRE database);
- the ability in the long term to provide estimates of the annual probability of various mechanisms of deterioration arising at certain dam/structure types. This should prove useful for the purposes of quantitative risk assessments.
- to enable well-informed debate on the safety of the UK dams industry thereby instilling confidence in it.

INCIDENT INVESTIGATION

In the event of a dam failure, it is likely that an independent investigation will be instigated either by the owner or an independent panel of engineers. In some cases there may be a public inquiry. Incident investigation, as detailed in the specification under development, is concerned with the full capture of information regarding the incident for the sole purpose of ensuring complete capture of relevant information. It will therefore aim to ensure that the root causes of accidents are correctly identified in sufficient detail for the purposes of the database. It will have no remit to apportion blame for the incident or to criticize reservoir operation, surveillance and monitoring regimes.

When serious incidents arise, the Environment Agency (in the case of England and Wales) will appoint a suitably qualified engineer (normally a Panel AR engineer) to investigate such incidents. It is important that the investigation is, and is seen to be, as thorough, objective and impartial as possible.

Incident investigations will usually need to consider:
- any previous history of incidents at the dam;
- any recent works or change in regime leading to, or contributing to, the incident;
- the root cause(s) of the incident and any contributing factors;
- the main lessons to be learned;
- the nature of any remedial works or changes in regime arising from the incident;
- the possible implications for other dams/structures of the same type or configuration.

Investigations should not only contribute to our understanding of how incidents arise and how they are managed, but should also raise the profile of the dams industry through demonstration of a professional response from the industry as a whole as well as from those immediately involved with the incident.

The RAE concluded that incidents that by chance fall short of causing death and destruction should, depending on the circumstances, be regarded equal to those that do.

It is axiomatic that incident investigations should be as independent as feasibly possible. Under any new legislation this is likely to be made mandatory.

In some cases the reporting will need to capture the human side to the incident. Individuals involved in incidents may experience an array of emotions ranging from guilt to denial. Incidents may arise where the proper behaviour is either known to the individual but not practiced, or is not known by those concerned. The reporting should seek to capture and address such weaknesses.

Investigators may find it difficult to correctly pitch the level of investigation, especially where it becomes evident that there were several different types of deficiency each contributing to the incident. In such cases it may be difficult to be specific about the lessons to be learned and the benefits to be gained from investigation may be closely linked to the time taken in unravelling the full causal chain of events.

The investigation should aim to commence as soon as possible following an incident to prevent deterioration in the quality of the information made available to the investigator.

The investigation of serious incidents should aim to separate the factors that could have prevented the accident (design, operation, training, etc) from the measures that would have reduced the severity of the accident (evacuation procedures, etc).

REPORTING

By taking lessons learned from one incident and checking whether these lessons apply in other situations, similar incidents can be avoided by making suitable changes in dam design and management.

There is no intention to report on incidents in a manner that would disadvantage any potential contributor or associated business interests. It is widely understood that the number of incidents associated with a particular region or owner is likely to reflect the intrinsic condition or age of dams more than the procedures and policies in place for dam monitoring, surveillance and remedial work.

It is currently proposed to make annual reports available to major dam owners and panel engineers. These would not provide detailed information on any particular incidents recorded or provide information that would allow the identification of the dam, or persons or organizations linked to them. The aim would be to provide information on the lessons learnt and the number and nature of incidents recorded.

The database will be subjected to a detailed statistical analysis at intervals of a few years. The value of such work will partly depend on the implementation of suitable measures to populate the dam characteristics database fields.

CONCLUDING REMARKS

It is important to acknowledge that our past and future performance in dam safety is only the product of learning through incidents and a determination to improve methods and procedures. The development of the new incident reporting and investigation procedures aims to provide a robust, formalized framework which can contribute to our further understanding of how incidents arise at UK reservoirs.

Voluntary reporting of incidents in the UK through learned society meetings and publications has probably captured a significant proportion of the most serious dam incidents to date. The British reservoir industry has long shared its experiences openly. However, it can be argued that a formalized system of incident reporting and investigation for the UK dams industry forms a natural partner to the provisions of the Act. The performance of the UK dams industry, especially over the last twenty years or so, can be regarded as good in terms of the number and severity of incidents. The industry should not make the mistake of viewing this situation as a stable one. The average age of the UK stock of dams exceeds 110 years. Incidents still arise regularly, and almost invariably from causes that have been experienced in the past at other dams. Dams, especially embankment dams, are very complex structures and it is possible that incidents may arise due to new, unexplained causes. Such incidents are likely to be very rare but very valuable as they can contribute much to our understanding and learning. Remedial and upgrading works following incidents or design reassessment have the potential to not only improve the structural integrity of the dam population but to also reduce the probability of incidents occurring.

One of the difficulties faced with voluntary reporting is an entrenched belief in modern society that when incidents arise, somebody must be to blame.

Even when incidents are reported, there may be suspicions that not all of the facts have been disclosed.

The value of incident reporting to any one individual can be difficult to grasp. The RAE report states that:

"It is not easy to inculcate a desire to learn from others' misfortunes. The human default position seems to be resistant to this. Many reasons are advanced for not looking at the experience of others – it is perceived to reflect poorly on what has already been done, it hints at lack of knowledge on the part of individuals, it takes time that often does not exist. Most managers would readily agree that if the extra work of learning from accidents would definitely allow an accident to be prevented it would certainly be worthwhile. But the implicit belief is often that taking on this extra work will not prevent anything because nothing was going to happen."

Great care is needed in how incident data is managed to protect the interests of the contributors and to ensure that the time and effort spent in reporting incidents, and the trust placed in the system managers, is repaid through careful and effective use of the incident information.

ACKNOWLEDGEMENTS

The authors wish to thank the Environment Agency for permission to publish this paper.

The authors also wish to acknowledge Paul Tedd of BRE for his support and advice in the development of the new specification for incident reporting, and his assistance in conducting trials of the system using historical incidents from the BRE database.

REFERENCES

Brown, A.J. and Gosden, J.D., 2004[1]. *Developments in management of reservoir safety in UK.* Risk Assessment and reservoir management. Proceedings of the 13th Conference of the BDS held at the University of Kent, Canterbury. Thomas Telford.

Brown, A.J. and Gosden, J.D. (KBR), 2004[2]. *Interim Guide to quantitative risk assessment for UK reservoirs.* Thomas Telford.

Brown, A.J. and Tedd P., 2003 *"The annual probability of a dam safety incident at an embankment dam, based on historical data"* Hydropower & Dams. Issue 10, Vol. 2, pp122-126.

Charles, J.A., 2005. *Use of incident reporting and data collection in enhancing reservoir safety.* Journal of the British Dams Society, Dams and Reservoirs, Vol.15, No.3.

Charles J.A., Tedd P. and Skinner H.D. (1998). *The role of risk analysis in the safety management of embankment dams.* The prospect for reservoirs in the 21st century. Proceedings of 10th British Dam Society Conference, Bangor. Thomas Telford, London.

Defra, SDA 26: www.defra.gov.uk/environ/fcd/policy/sda2627/default.htm

Gosden, J.D, and Brown, A.J., 2004. *An incident reporting and investigation system for UK dams.* Journal of the British Dams Society, Dams and Reservoirs, Vol.14, No.1.

Hope, I.M. and Hughes, A.K., *Reservoirs Act 1975 – Progress on the implementation of the Environment Agency as Enforcement Authority.* Proceedings of the 13th conference of the BDS held at the University of Kent, Canterbury. Thomas Telford, London.

The Royal Academy of Engineering, 2005. *"Accidents and Agenda".*

Skinner, H.D., 2000. *The use of historical data in assessing the risks posed by embankment dams.* Dams & Reservoirs, Vol. 10, No. 1.

Tedd, P., Holton, I.R. and Charles, J.A., 1992. *The BRE dams database. Water resources and reservoir engineering.* Proceedings of the 7th conference of the British Dams Society, Stirling. Thomas Telford, London.

Tedd P., Skinner H.D. and Charles J.A. (2000*). Developments in the British national dams database.* Dams 2000. Proceedings of 11th British Dam Society Conference. Thomas Telford, London.

Reservoir hazard analysis and flood mapping for contingency planning

J.C. ACKERS, Technical Director, Black & Veatch
R.V. PETHER, Senior Engineer, Black & Veatch
F.R. TARRANT, Chief Scientist, Black & Veatch

SYNOPSIS.   Under revisions to the Reservoirs Act 1975 introduced through the Water Act 2003, the Secretary of State has powers to direct reservoir undertakers to prepare flood plans. These are required to help the emergency services and others in providing an effective response – with regard to evacuation and other precautions – in the event of a threatened or actual dam failure, or some other uncontrolled escape of water.

This paper comments on current practice in producing flood plans to aid contingency planning in the UK, and how this may develop in the future. The paper also provides some interesting results from statistical analyses of reservoir hazard assessments, drawing from over 300 reservoir hazard analyses undertaken since 1990.

LEGISLATIVE BACKGROUND

Sections 74 to 80 of the Water Act 2003 are concerned with reservoirs and contain a number of changes to the Reservoirs Act 1975. Of relevance to this paper is Section 77, which inserts a new section into the 1975 Act that allows the Secretary of State to direct an undertaker to prepare a 'flood plan' for a large raised reservoir. The flood plan is intended to set out 'the action they would take to control or mitigate the effects of flooding likely to result from any escape of water from the reservoir'.

Although the wording in the Act is to do with controlling and mitigating the possible escape of water, the section goes on to say that the direction may (*inter alia*):
- 'specify the matters to be included in the flood plan';
- 'require the flood plan to be prepared in accordance with such methods of technical or other analysis as may be specified by the Environment Agency'; and
- require the flood plan to be provided to the Environment Agency.

Improvements in reservoir construction, operation and maintenance, Thomas Telford, London, 2006, 37–46

Part 1 of the Civil Contingencies Act 2004 sets out duties on 'Category 1 responders' to (*inter alia*):

- assess the risks of emergencies occurring;
- use the risks to inform contingency planning;
- put emergency plans in place;
- make information available; and
- disseminate warnings.

The Category 1 responders are those, such as the emergency services, local authorities and the Environment Agency, who are considered to be at the heart of planning for and responding to emergencies. Category 2 responders are those that would be involved in incidents that affect their sector and include such bodies as the water undertakers. It is the water companies who of course own the majority of large reservoirs in the UK, but they are defined as Category 2 responders, not because they own reservoirs that might fail and cause an emergency, but because of their infrastructure and other assets that could be affected by an emergency arising from a variety of causes.

Reservoir failure is identified as one of the risks covered by the emergency planning duties imposed by the Act. The Environment Agency's role under the Reservoirs Act 1975 (as modified by the Water Act 2003), both as the enforcement authority for England and Wales and as the recipient of the flood plans prepared by the undertakers, dovetails neatly into its duties as a Category 1 responder under the Civil Contingencies Act 2004.

PROPOSALS FOR FLOOD PLANS

Draft proposals for the format and content of flood plans, prepared on behalf of Defra by Kellogg Brown & Root in March 2005, anticipate that the flood plan will consist of up to three elements:
1 Inundation and consequence analysis
2 On-site emergency plan
3 Draft notification to local authority of imminent dam failure

The first element is where dambreak flood inundation and potential damage mapping first appears in the flood plan, as it is used to assess the consequences of a possible dam failure and determine the appropriate 'consequence class' – the successor to 'dam category' in the ICE guide 'Floods and reservoir safety' (ICE 1996). Although the determination of consequence class is intended to follow the methodology in the 'Interim guide to QRA for UK reservoirs' (Brown & Gosden, 2004) (and, in due course, its successor document), there is no reason to suppose that inspecting engineers will not continue to exercise their judgement over whether the methodology produces an appropriate answer.

The results of the inundation and consequence analyses are also referred to in the draft notification in Part 3 of the flood plan, which it is anticipated would be required only for category A and B reservoirs. Presumably, the draft notification to the local authority will be a standing document, so that the recipient of a 'real' notification will already be familiar – before the emergency arises – with the broad contents of the notification, and will already have determined what emergency measures they would need to implement.

These proposals will have been developed further by the time that this paper is presented and will also have been discussed at the seminar at ICE on 11 July 2006, which is planned to follow the issue of the draft 'Guide to emergency planning for UK reservoirs'. However, the implications for dam failure analyses and the associated flood mapping are likely to be broadly the same as anticipated in the March 2005 inception report by KBR. In that report, KBR include among the challenges faced in the formulation of the guidance:

- the need to keep the cost of preparing and maintaining emergency plans proportionate to the reduction in risk that might be realised; and
- the need for due recognition of plans that have already been prepared.

It may therefore be expected that previously completed dambreak studies would generally be accepted as the basis for determining the consequence class and for providing the inundation mapping for flood plans for at least an interim period. In many cases, the results from these studies will be amenable to suitable digitisation and processing to convert the results to the requisite format. There could, however, be a procedure for determining the period before the next review, taking account of such matters as how recently the study was undertaken, the category of the dam, when the next Section 10 inspection is due and the timetable for a range of contingency planning in the locality.

DAMBREAK METHODOLOGY

In common with other organisations carrying out dambreak and inundation mapping, we have moved forward in the approach that we use, in response to advances in digital mapping and flood routing software, although we still normally use what is substantially the original Dambreak UK model (but not the breach module) to estimate the outflow hydrograph from the failing reservoir. For the routing of the floodwave down the valley, however, we now use either ISIS or HECRAS, as these provide facilities for interfacing with digital mapping information, using software such as 12d.

In the last decade there have also been advances in knowledge regarding the breach mechanism itself – including studies into case histories – that have led to the development of improved empirical formulae. There have also been programmes of field trials, with the purpose of calibrating breach formation models that seek to integrate the hydraulic and soil mechanics processes involved in the initiation and development of the breach (see, for example, www.impact-project.net).

In due course, the numerical modelling approach will hopefully provide the way forward for breach formation and the corresponding floodwave hydrograph. But it must be recognised that this approach is necessarily limited for dams where there is little on no information about the internal structure and material properties. In the meantime, we have, for the last few years, based the dam breach geometry and development time used in the Dambreak UK module on the empirical formulae by Froehlich (1995). These were found by Wahl (1998, 2001 and 2004) to provide the best fit when applied to a dataset of 108 dam failures.

There are also a large number of empirical methods that predict the peak outflow from simple parameters for the dam and reservoir, typically the volume (V_w) and maximum depth of water impounded (H_w) at the time that the breach is initiated. In order to act as a broad check on the dambreak results and to demonstrate the degree of uncertainty that may apply, we also, as a matter of routine, present a table that compares the peak breach outflows by a total of about 15 methods. Most of these are as quoted by Wahl (1998–2004) and two of them are of particular interest:

$$Q_p = 1.154 \left(V_w H_w \right)^{0.412} \quad \text{after MacDonald \& Langridge-Monopolis (1984)}$$

$$Q_p = 0.607 V_w^{0.295} H_w^{1.24} \quad \text{after Froehlich (1995)}$$

The first of these is used as part of the rapid impact assessment procedure recommended in CIRIA C542 (Hughes *et al*, 2000), and the second in the corresponding procedure in the 'Interim guide to QRA for UK reservoirs' (Brown & Gosden, 2004). Both apply to earthfill dams only. (It should be noted that the MacDonald & Langridge-Monopolis relationship is presented in graphical form in their paper, so some differences arise in its conversion to equation form by different authors.)

For the majority of UK reservoirs, where there is a fairly narrowly defined flood route downstream, a one-dimensional model is sufficient and appropriate for dambreak modelling. In some cases, there are 2D effects that have to be taken into account in the interpretation of the results, that is in the process of transposing the 1D modelling results into the requisite inundation and damage plans. But there are cases where this is unlikely to be adequate, such as reservoirs whose failure would results in very wide areas of inundation, or where there is the likelihood of separate flood paths being formed following different routes. In these situations, two-dimensional (depth-integrated) modelling would normally be the appropriate choice. As the floodwave routing and mapping software develop, it is inevitable that increasing use will be made of 2D analyses, and these can be expected to become more commonplace over the next decade.

We would suggest that flood plans and the associated inundation maps be reviewed at the time of the Section 10 inspections and perhaps at more frequent intervals if significant changes are suspected. These reviews would examine whether any changes have occurred downstream of the dam in terms of flow routes and land use, but would not be expected to involve repeating the analyses, unless significant changes are found or there has been a recognised advance in the accuracy of the analysis method since the flood plan was prepared.

STATISTICAL ANALYSES

In their paper at the Bath conference, Tarrant and Rowland (2000) presented a number of anonymous case studies, together with a series of graphs which attempted to establish if a simple correlation could be devised between the basic reservoir characteristics (dam height and storage capacity), the characteristics of the downstream valley (gradient and shape) and the extent of total property destruction and partial structural damage. Such a relationship could be useful in the planning and competitive bidding for these studies, particularly in cases for which there is no obvious downstream boundary, such as the sea, but might also provide a useful screening tool for deciding whether a detailed dambreak study is required.

The best of the simple correlations, for reservoirs with a capacity of at least $1 \times 10^6 \, m^3$, was between dam height and the extent of partial structural damage, but this treatment still left a wide range of results and the authors concluded that the relationship was 'tenuous'. They went on to conclude that the extent of damage and inundation can only be determined by a full dambreak assessment, but that the simple relationship could be useful for planning the extent of the dambreak model required.

For this paper we have updated the statistical analyses carried out in 2000, and have also widened the compass to include reservoirs in our dataset with a capacity greater than $0.5 \times 10^6 \, m^3$, so that a total dataset of over 100 embankment dams was used. For the same dataset we have also tabulated the relationships between the peak dambreak outflows determined in our detailed dambreak studies and the peak outflows determined by the two empirical methods referred to above.

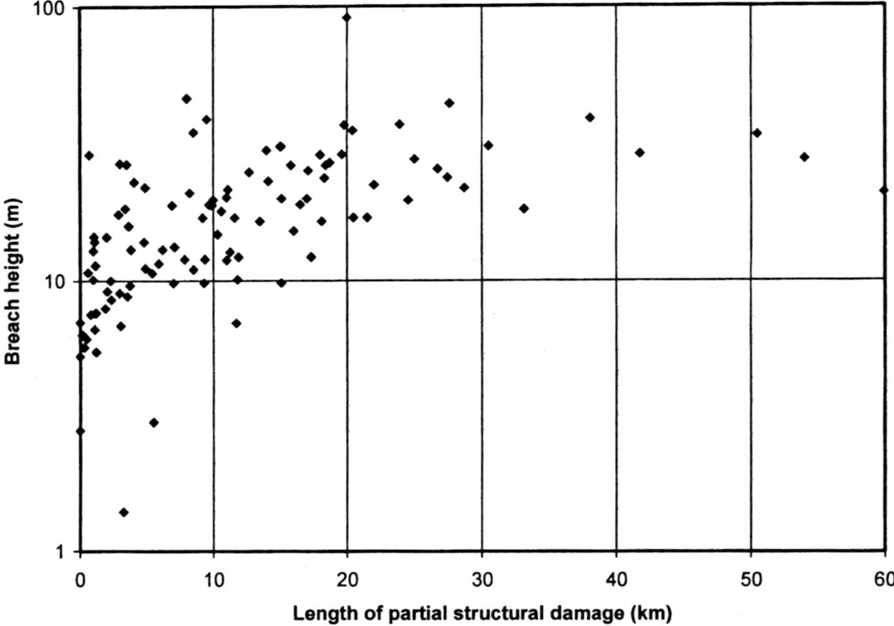

Figure 1.   Damage length as a function of breach height for over 100 dams

Figure 1 shows the best of the previously tested relationships, that is between the breach height and the length of partial structural damage, updated to include the current dataset. We experimented with a number of alternative formulations, involving various combinations of dam height and reservoir capacity and found no noticeable improvement over this relationship. As may be noted, there is a great deal of scatter in these results, even with the breach height plotted to a logarithmic scale.

Figure 2 shows the relationship between the peak reservoir outflow determined in over 100 detailed dambreak studies with those estimated using the empirical peak flow relationships suggested by MacDonald & Langridge-Monopolis (1984) and by Froehlich (1995). The empirical relationships are expressed as the ratios to the outflows determined in the detailed studies and it should be noted that they are plotted in log-log form.

What is immediately apparent in Figure 2 is the wide variation in the values obtained from the empirical formulae, encompassing a band of between about one tenth and seven times the value determined in the detailed study. Of course, some of the variation must be due to inaccuracies inherent in the breach geometry and timing parameters that were employed in the detailed analyses, and to the fact that a number of incremental changes have been made to the approach over a period of more than 15 years. It is also clear that there is a distinct trend for both of these empirical methods to over-estimate the discharge for the cases with smaller breach outflows and to under-estimate the larger discharges. This is perhaps fortuitous, as the rapid approach in the 'Interim guide' (Brown & Gosden, 2004) is likely to be applied to the smaller reservoirs, with a full dambreak study being most common for the larger reservoirs. Although both empirical flow estimation methods, on average, produce results that are similar to those from detailed studies, the Froehlich (1995) equation exhibits a lesser spread than the MacDonald & Langridge-Monopolis (1984) equation. This tends to support the adoption of the Froehlich equation for the 'Interim guide'.

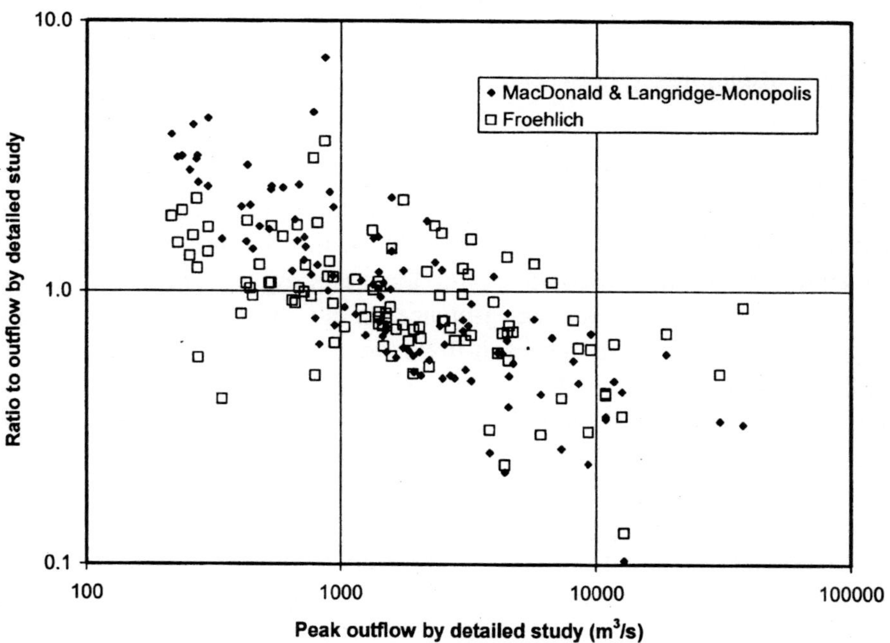

Figure 2.   Comparison of empirical peak flows with those from detailed dambreak study

We have also attempted to find a correlation between the average bed slope of the downstream flowpath and the 'attenuation length scale' (L_a), as used in CIRIA C542 (Hughes et al, 2000) and the 'Interim guide'. The attenuation length scale is the distance over which the peak discharge of the floodwave falls to 37% of the dam breach peak outflow. Again, such a relationship would be of use when planning the extent to which the downstream flowpath must be modelled. Figure 3 shows the relationship between bed slope and L_a for 20 dams with breach outflows between about 100 and 30 000 m³/s.

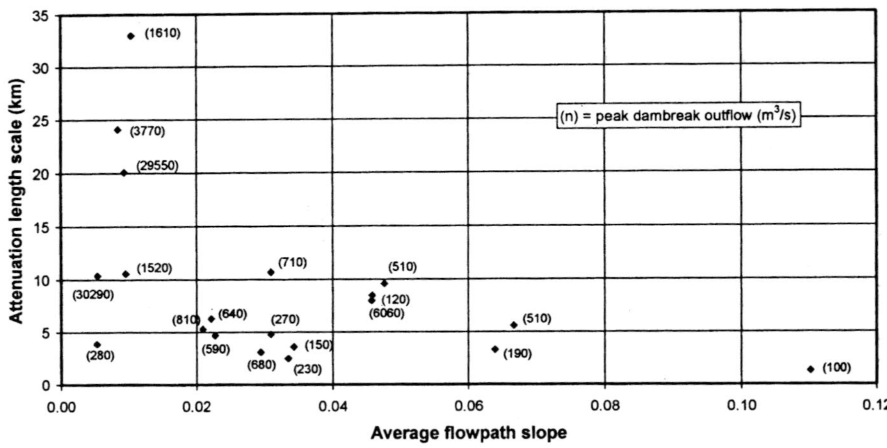

Figure 3   Relationship between L_a and average downstream slope

Again, there is a large scatter of results, with dam breaches of a similar size and average flowpath gradient having very different attenuation rates. This is because other factors – such as the width of the floodplain, the presence of obstructions to the floodwave and the resulting volume of available floodplain storage – also have an influence on the attenuation rate. There is a tendency for the larger dam breach floods to attenuate over a longer length than the smaller breach floods, which is no surprise, because the larger the flood, the more quickly the floodplain storage is exhausted, so that a larger proportion of the flow must continue downstream.

Although the results in Figure 3 do not appear to offer a useable correlation for general application, they can be referred to for additional guidance. Further investigations, involving the analysis of a larger number of case studies, together with consideration of different combinations of parameters, might lead to the development of a improved empirical approaches to assist in the planning of reservoir hazard analyses.

CONCLUSIONS
The new provisions for 'flood plans' in the Reservoirs Act 1975 (added by the Water Act 2003), together with the requirements of the Civil Contingencies Act 2004, create an impetus for improving and standardising approaches to dambreak modelling, inundation and consequence mapping in the UK. In order to maintain and enhance standards, it must be recognised that, where new studies are required, they must be carried out over a period of some years, taking account of such factors as the category of the dam, when the next Section 10 inspection is due and the timetable for developing contingency planning for a range of risks in the locality.

A key issue is how the dam industry presents the uncertainties to the emergency planners. We would support the adoption of some standard assumptions to obtain consistency between the analyses by different reservoir owners and consultants, recognising of course that these may vary in the future as efforts are made to obtain a closer representation of reality. The Impact research (www.impact-project.net) has provided some more data towards improving understanding in the subject, and there is clearly scope for providing standardised guidance for older reservoirs where little is known about their construction. Nevertheless, the preparation of flood plans should not be divorced from the requirement for dams to be assessed individually by experienced dam engineers.

A pragmatic approach should be adopted with regard to the large number of reservoirs for which dambreak studies have already been undertaken – some of them over 15 years ago – accepting that differences in approach and presentation from the current 'ideal' do not necessarily require the studies to be repeated in their entirety.

This paper presents the results of limited statistic analyses of flood damage lengths and the rate of floodwave attenuation in relation to simple parameters concerned with the reservoir and its setting. Nevertheless, it remains the case that there is 'no simple solution' (Tarrant & Rowland, 2000) to the question of how to relate the distance over which the peak breach outflow attenuates, or the extent of damage, to simple parameters such as the reservoir volume, dam height and stream gradient. Analysis of a wider range of parameter combinations and a greater number of case studies could lead to the development of an empirical formula for use during the planning of future dam failure impact assessments.

REFERENCES

Brown, A.J. and Gosden, J.D. (2004). *Interim guide to quantitative risk analysis for UK reservoirs*. Thomas Telford, London.

Froehlich, D.C. (1995). Peak outflow from breached embankment dam. *Journal of Water Resources Planning and Management*, Vol 121 N° 1. ASCE

Kellogg Brown & Root (2005). *Engineering guide for emergency planning for UK reservoirs – Inception report*, Revision A02. Defra, London.

Institution of Civil Engineers (1996). *Floods and reservoir safety*, 3rd edition. Thomas Telford, London.

Great Britain, *Reservoirs Act 1975, Elizabeth II, Chapter 23*, The Stationery Office, London.

Great Britain, *Water Act 2003, Elizabeth II, Chapter 37*. The Stationery Office, London. (http://www.opsi.gov.uk/acts/acts2003/20030037.htm)

Great Britain, *Civil Contingencies Act 2004, Elizabeth II, Chapter 36.* The Stationery Office, London.
(http://www.opsi.gov.uk/acts/acts2004/20040036.htm)

Hughes, A., Hewlett, H., Samuels, P.G., Morris, M., Sayers, P., Moffat, I., Harding, A. and Tedd, P. (2000). *Risk management for UK reservoirs.* CIRIA, London.

MacDonald, T.C. and Langridge-Monopolis, J. (1984). Breaching characteristics of dam failures. *Journal of Hydraulic Engineering*, Vol 110 N° 5. ASCE.

Tarrant, F.R. and Rowland, A. (2000). Prediction of downstream destruction following dam failure: no quick solution. *Dams 2000*, Proceedings of the British Dam Society Conference, University of Bath, 14–17 June 2000. Thomas Telford, London

Wahl, T.L. (1998). *Prediction of embankment dam breach parameters – a literature review and needs assessment.* Dam Safety Research Report, DSO-98-004. U.S. Department of the Interior, Bureau of Reclamation, Dam Safety Office, Denver.

Wahl, T.L. (2001). *The uncertainty of embankment dam breach parameter predictions based on dam failure case studies.* USDA/FEMA Workshop on Issues, Resolutions and Research Needs Related to Dam Failure Analysis, 26–28 June 2001, Oklahoma City.

Wahl, T.L. (2004). Uncertainty of predictions of embankment dam breach parameters. *Journal of Hydraulic Engineering*, Vol 130 N° 5. ASCE.

Impact of the Controlled Activities Regulations on Dam Construction, Maintenance and Operation in Scotland

C. W. BERRY, GHD Pty, Australia (formerly of W A Fairhurst and Partners)
K. M. H. BARR, W. A. Fairhurst and Partners, UK

SYNOPSIS. Regulatory pressure is being increasingly focused on engineering activities under the cover of the Water Framework Directive. Consideration of the environment and social impacts of what reservoir engineers and owners do in the future will play an important part in the management of our reservoir stock. From April 2006 all activities in or near the water environment in Scotland will be controlled under the Controlled Activities Regulations with the exception of very minor work with low risk to the environment. This new regime will form an important consideration for reservoir owners, inspecting engineers, consultants and contractors and is likely to have a significant effect on the way engineering works are prescribed and implemented in the future. This paper provides a practitioner's view of the new regulations highlighting the potential effects of these new statutory obligations and the additional burden on reservoir owners. This paper outlines the new regulatory framework and explores the impact on existing and future engineering activities in relation to reservoir works. Two recent case studies illustrate the potential impact drawing on practical experience from the hydro power industry, reservoir owners and engineering works.

INTRODUCTION

The European Union Water Framework Directive (WFD) aims to improve the quality and status of our water environment and recent significant, wide ranging environmental legislation has been enacted with a view to achieving the goals and aspirations of the WFD. Historical human intervention to watercourses and development within our river catchments has placed significant pressure on natural ecosystems. The new Controlled Activity Regulations (CAR) aim to provide a mechanism to ensure the adequate mitigation of the negative impacts created by the construction, operation and decommissioning of impoundments in Scotland. The delivery of such regulation under the ambit of the WFD needs to retain a focus on sustainability, in terms of achieving, not just environmental improvements, but delivering current and future community needs and demands.

The fundamental principle of the WFD is to improve the ecological status of rivers and water bodies, taking into account the socio-economic context and the demands of different stakeholders. There is general agreement that action is required to target the key factors affecting the status of rivers and water bodies such as water quality (point and diffuse pollution) and morphological impacts. Abstraction and storage of water needs to reconcile other ecological and social demands upstream and downstream in the same catchment.

The WFD was enacted into Scottish law through the Water Environment and Water Services (Scotland) Act 2003. This provides regulatory controls over activities in order to protect and improve Scotland's water environment. In fulfilling their duties Scottish Environment Protection Agency (SEPA) regulate activities such as abstraction, impoundment and engineering activities, as well as discharges, under the Water Environment (Controlled Activities) Regulations 2005 (CAR). A recurring theme of the European Union has been differences in how directives are brought into national law in member states. This is now occurring within the UK, with responsibility for the WFD falling to the Scottish Parliament. There are significant differences between the CAR and the proposals for England & Wales.

BACKGROUND

The introduction of the CAR provides the regulator SEPA with additional and wide ranging powers to influence engineering activities such as impoundments and river works in Scotland. SEPA in the past have influenced such activities through a number of existing legislative frameworks as statutory consultees, such as:

- Town and Country Planning Act 1997.
- Environmental Impact Assessment (Scotland) Regulations 1999.
- Environmental Assessment of Plans and Programmes (Scotland) Regulations 2004.
- Water Orders under Water (Scotland) Act for abstractions (may cover impoundments).
- Flood Prevention Orders (may also cover impoundments).

The introduction of the CAR is in addition to any consent that may be required through the planning process and provides SEPA with much more direct control over future activities.

The new regulations aim to provide a consistent approach to works affecting rivers and water bodies. In the past these works have tended to take a rather narrow, non-holistic approach to environmental mitigation. It has been shown that engineers have historically taken some cognisance of the environmental impact of reservoir construction, including various forms of mitigation (Turpin, 2000). Such work has often been undertaken unconsciously and unrecognised as part of good design practice. Turpin, 2000 describes factors such as increasing environmental regulation and legislation as having a "design squeeze" on reservoir design construction and operation over the last century. The CAR places additional pressure on reservoir practitioners and operators to draw from their historical role and develop a more holistic approach to reservoir management.

REGULATORY FRAMEWORK
The CAR will affect owners, designers, planners and contractors for both existing and proposed activities. The level of authorisation and control will be based on the risk (to ecological quality) posed by various activities (Fig.1). Significantly, activities such as abstractions (over $10m^3$ per day), impoundment and engineering in rivers will require some form of registration or licensing. This would apply to both existing and proposed impoundments (over 1m high) regardless of the volume impounded. The Reservoirs Act's principal focus is on the protection of people, the CAR key driver is protection of the water environment.

Authorisation will be required for all future and existing:
- Discharges
- Abstractions
- Impoundments
- Engineering Activities

'Engineering Activities' will cover all work on or near rivers, lochs, wetlands, etc but not transitional waters or coastal waters (covered by an existing licensing regime). Works requiring licensing will include developments in the vicinity of a river or loch and set-back embankments.

It remains unclear clear what "in the vicinity" means here but it has been suggested that this could encompass any activities in the "functional floodplain". As such in some cases the CAR may influence works some distance form the controlled water body. Compliance with CAR is required even if the works are covered by other consents e.g. planning consent or road construction consent.

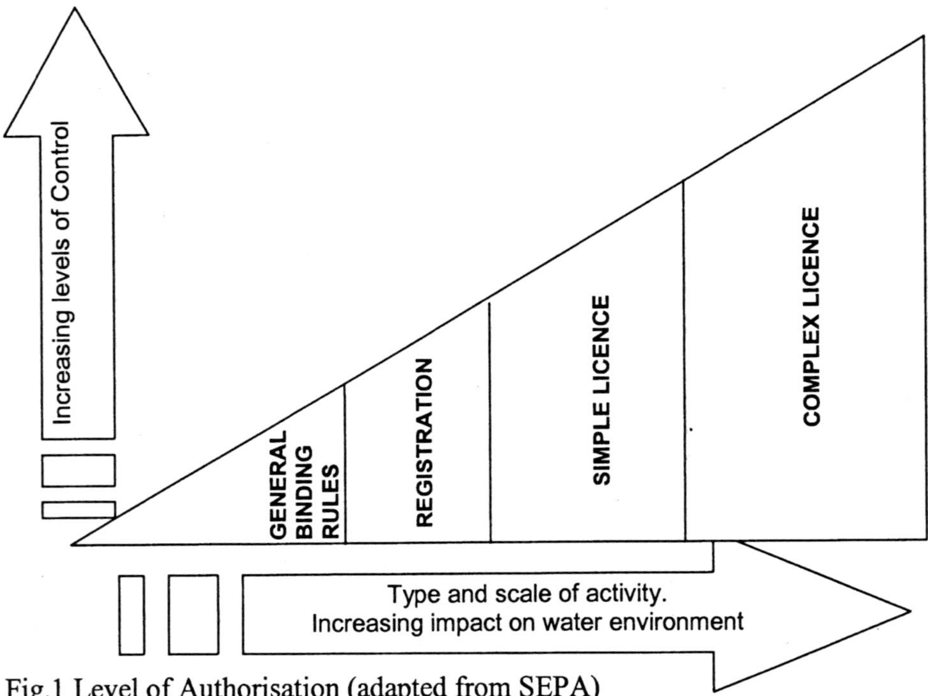

Fig.1 Level of Authorisation (adapted from SEPA)

SEPA will regulate activities in three principal ways depending on the type and size of work and risk to the environment (Fig 1):

i. Compliance with the **General Binding Rules** (GBR) – This is the simplest level of control covering minor activities with low environmental risk e.g. weirs less than 1m high if constructed before April 2006, abstractions of less than $10m^3$ per day, small scale bank protection works etc. SEPA do not need to be informed of such activities.

ii. **Registration** – SEPA will use registration for small-scale activities which individually pose a small environmental risk but which cumulatively can result in environmental harm, e.g. bank protection, sediment management, small river crossings, etc.

iii. **Licence** –A licence (simple or complex) will be 'person'-specific in that it will require the identification of a 'responsible person' who will be responsible for ensuring that the terms of the licence are complied with. A responsible person can be an individual, a company or a partnership. Licences will be required for both existing and proposed activities (including impoundments, discharges and abstractions) and where such activities are currently licensed (e.g. discharge consents) application should be made to transfer these over to the new system.

Organisations or people with existing discharges, impoundments or abstractions already consented were required to apply to SEPA during the transitional period between 1st October 2005 and 31st March 2006 to transfer their consents to the new regime.

From April 2006 works must comply with the GBRs or be registered or licensed by SEPA. Within the Regulations SEPA have 4 months from the date of application to make a decision regarding a licence and 30 days for a registration. It is understood that engineering works completed by 30 September 2006 will not require authorisation. However, works likely to continue after September 2006 will require authorisation to proceed. In such cases, where prior environmental impact assessments have not been undertaken through the planning process, significant delays could be incurred to allow SEPA to process the applications and possible additional constraints imposed. Discharges, abstractions and impoundments were enforced from April 2006. SEPA approval could present significant delays in starting or continuing work on site (separate from any planning application periods).

Consultation on proposed new abstraction and impounding licensing regulations to apply in England and Wales and proposed changes to the Water Resources (EIA) Regulations 2003 was undertaken in 2005 (DEFRA, 2005) and the results of the consultation published in March 2006. The level of regulation proposed is thought to be less prescriptive than in Scotland. Specifically, the new CAR will require regulation of all impoundments, maintenance, construction, decommissioning and operations (Table 1).

Table 1. Levels of Authorisation required for Impoundments

GBR	Simple Licence	Complex Licence
Existing passive weirs less than 1m high, not affecting fish passage	All other existing weirs and raised lochs more than 1m high	All new impoundments and operation of all other existing impoundments

Various charges will be applied by SEPA to process applications depending on the type of license and the number of activities involved. Charges will be incurred for each specific activity and many engineering works and reservoirs will have multiple activities, e.g. impoundment, abstractions, river training works and crossing. Annual subsistence charges will also be applied by SEPA to maintain the various licenses e.g. subsistence charges will be applied to impoundments which do not provide an "Environmental Service". Annual subsistence charges can be waived where the scheme is shown to provide an "Environmental Service" e.g. the primary purpose of the impoundment is to provide a fishing loch.

IMPACT OF CAR

The CAR will have a significant impact on existing reservoir owners, on any proposed remedial works as part of works in the interest of safety and will encompass impoundments currently not falling within the ambit of the Reservoirs Act 1975, including weirs and dams impounding less than $25,000m^3$.

SEPA define impoundments as "any dam, weir or other works by which surface water may be impounded but also include any works diverting surface waters in connection with the construction or alteration of any dam or weir" (SEPA 2005). This would include works to raise the level of a natural loch but not a pond created by excavation below existing ground levels. Non impounding reservoirs are not included in this definition.

The impact may:
* Introduce regulation of far more structures than are currently registered under the Reservoirs Act; owners will need to carefully consider their asset portfolio and likely mitigation measures that may be imposed.
* Affect maintenance of structures; previously un-regulated works will need to apply for approval if not covered by GBRs.
* Management of discharges and abstractions may have to be revised to reflect downstream ecology and morphological factors; available flow for hydro power generation may be affected e.g. compensation flows and control of hydro peaking flows.
* Hydro-power assets may require investment to provide mitigation; installation of fish passes, screens, 'fish friendly' turbines, etc.

Some of the proposed mitigation measures that may be applicable, particularly to impoundments and hydro schemes may involve non-structural solutions such as a change in operation and discharge regime. However, other measures such as changes to overflow characteristics may be significant and may required alterations that will need supervision of a qualified engineer under the Reservoirs Act 1975. It is clear that the benefit of such work in term of improvement of good ecological status or potential of the water environment should not be considered in the absence of the statutory requirements of the Reservoir Act in terms of safety. Similarly, the primary function of a facility (power generation, flood control) needs to be considered and where such benefits outweigh the environmental mitigation, derogation of the CAR should be sought.

The extent and scope of works required to satisfy SEPA in meeting their objectives is unclear. At the time of writing, SEPA were in the process of developing guidance on best practice for engineering works including impoundments. SNIFFER and the Environment Agency have however already produced preliminary guidance on management strategies and mitigation measures for impoundments which presents a step by step approach to identifying appropriate mitigating measures for existing and proposed impoundments (SNIFFER, 2004). The approach considers ecological, socio-economic and financial issues and constraints in the selection of measures. The costs and impact of any measures needs to be balanced against the primary benefit and function of a particular impoundment and the management and decommission of the impoundment. It is this comparison of potential disproportionate cost of mitigation and primary benefits of the structure that has the potential for derogation to be applied for a particular site and equally raises the potential for conflict between the regulator and owners.

The forthcoming guidance will have significant importance in the application and impact of the CAR. It is hoped that that guidance will provide clarity and consistency in the application of the rules both for the regulator and for owners/engineers in assessing what is acceptable. Failure to achieve clear guidance could result in protracted negotiations with SEPA and costly delays in implementation of engineering works. The environmental measures that may be imposed through the CAR need to recognize the primary function of many reservoirs and engineer's recommendations under the Reservoirs Act 1975.

Any activity to mitigate the environmental impact of an impoundment must reconcile the protection of the local aquatic ecology with the socio-economic benefits of impoundments and must fit within the overall River Basin Management Plan (RBMP) process (SNIFFER, 2004). It is the extent of such impacts and the level of mitigation as it relates to the overall RBMP that is currently being developed by the regulator.

EXAMPLES
The following examples of recently completed works are discussed to demonstrate the potential effect the CAR would have had on the outcome of the as built schemes. Two example have been presented, the first a new hydro power scheme, the second a new flood storage reservoir.

<u>Garrogie Hydro Scheme</u>
Garrogie hydro scheme consists of a 2.7MW run-of-river project constructed near Fort Augustus, Scotland, upstream of the Fechlin Intake on the River Fechlin. The project involved construction of a 9m high concrete gravity dam (Fig.2) serving as the intake together with a 2.6km penstock leading to the turbine house. The penstock followed the course of the river with two significant bridge crossings spanning the river.

Fig. 2 Garrogie Hydro Scheme Intake Dam

Consultation with SEPA and SNH as part of the planning process established various environmental condition and mitigation measures, including compensation discharge rates. Provision for fish passage was not considered necessary given the existing downstream barriers and this was agreed as part of the consultation process.

An environmental impact assessment was undertaken as part of the scheme however it is likely that this would require to be expanded as part of the CAR process with more focus on mitigation activities such as river enhancement and wildlife habitat provision. In addition, the license may in the future impose greater conditions on compensation discharges to the river during low flow with the potential to reduce the available flows to the turbines.

The following list indicates the various activities that would be applicable to license the scheme.

- Impoundment.
- River training works (<250m of bank).
- Construction of temporary river diversion works.
- Tailrace.
- Abstraction.
- Discharge (from tailrace).
- River crossing, 2 (road bridge and pipe bridge).

Where there are multiple activities as part of a single scheme charges can be reduced. Annual subsistence charges are also applicable to the scheme, as the primary function of the project does not offer any "Environmental Service". Whilst charges are small, around £8,000 compared to the capital outlay and annual operating costs of a scheme of this size, the costs to smaller private owners and operators could be significant.

<u>Corselet Road Flood Storage Reservoir</u>
Corselet Road Flood Storage Area is located in Darnley, in the south of Glasgow on the Brock Burn, a tributary of the White Cart Water. The storage area provides compensatory storage for floodplain storage lost by during the redevelopment of Pollok town centre. As part of the construction of the new Silverburn shopping centre at Pollok an area formerly used as a recreation ground was infilled. The recreation ground was retained as floodplain when the Brock Burn Flood Prevention Scheme was constructed. A low lying area upstream of Corselet Road, Darnley was identified as a suitable site for a compensatory flood storage area. The storage area was constructed in 2004-2005 and is registered as a large raised reservoir under the Reservoirs Act 1975.

The flood storage area was created by construction of a homogeneous clay embankment 4m high and about 240m long. It is designed to retain 96,000m^3. The Brock Burn passes through the embankment in a corrugated metal culvert. The storage area is normally dry with flood flows beyond the 1 in 5 year return period flow being retained by a reinforced concrete orifice control structure at the upstream end of the culvert (Fig 3).

The embankment is designed to overtop in events beyond a 1 in 500 year return period. A length of 180m of the embankment acts as a spillway and is designed to pass the Probable Maximum Flood. The downstream face is therefore reinforced with erosion control mat with hollow concrete blocks at the toe. A number of ecological enhancement features have been included including wildlife ledges in the culvert and the creation of a wetland area upstream of the embankment.

The areas of high flow velocities precluded the use of 'soft' engineering measures downsteam. The balance between achiveing environmental enhancement and ensure the safety of the embankment to cope with extreme events was discussed with SEPA.

Fig 3. Corselet Road Flood Storage Reservoir

The following activities would require a Complex Licence under the CAR:
- Impoundment.
- Culvert.
- River training works and bank stabilisation.
- Construction diversion works.
- Dredging to remove material from channel bed.

It is understood that an annual subsistence charge will be applicable although the structure provides an environmental service i.e. wildlife and wetland habitat and public amenity, because its primary function is flood storage.

FUTURE ROLE OF RESERVOIR ENGINEERS
A pragmatic approach is required from both the regulator and practitioners in delivering the goals of the CAR. Long term sustainability of solutions and engineering works need to be considered not just from an ecological view but also considering durability, safety of maintenance and operational costs.

There is a pressing need to strike a balance between the principles of flood management, maintenance and safety of reservoirs against social equity, economic efficiency and the need to conserve natural resources. It is believed that the CAR provides the opportunity to better manage this process.

Appreciation and correct application of ecological and hydraulic concepts can lead to decreased costs and environmental risks as well as added landscape, economic and community value. Whilst a proactive approach is desirable the current subjective nature of selection of river enhancement measures and application may lead to resentment amongst some stakeholder groups. Early consultation and collaboration between stakeholders is essential in agreeing the level of mitigation measures required.

It is suggested that engineers can play an important role in raising awareness of some of the issues involved with the CAR and in the development of clear and concise guidance on the best practice. Guidance should not be prescriptive but allow for some flexibility and creativity to suit the site and environmental constraints and practicalities. Systems and solutions can be placed under exhaustive scrutiny and analysis but the importance of careful detailing cannot be understated. An adaptive learning approach is suggested applying the 'best' system within the current level of knowledge and data available and allow for further improvements after its implementation and collection of actual performance data.

A holistic approach has been widely adopted by reservoir practitioners in the past and will continue to be practiced. Indeed many schemes currently promoted apply significant mitigating measures and would be unchanged by the CAR. It is therefore important that such experience and best practice is brought to the table during consultations and development of schemes

CONCLUSIONS
The introduction of the CAR will play an important role in ensuring future and existing activities and impoundments take due consideration of river and loch ecology and morphology. As a result, reservoir engineers and owners will need to take a more holistic approach to the operation and maintenance of reservoirs and impoundments. The successful implementation of the CAR will undoubtedly be a challenge for both SEPA and practitioners and places an additional burden on smaller private owners in addition to annual costs of monitoring of reservoirs under the Reservoirs Act 1975.

There is no doubt that previous engineering activities and neglect of our watercourses has removed ecological habitats and value and that to redress this problem more proactive regulation and implementation of mitigating measures is required. However, short term environmental impacts should be off set against future gains through careful design and innovation in engineering works. Use of 'soft' engineering techniques provides a useful additional tool but such measures need to be sustainable in relation to long-term stability and management. Greater understanding of the geomorphological processes will be essential and engineers will have an important role and opportunity in developing innovative functional solutions.

The aspiration of improving our environment and the sustainability of projects is shared by all. The key is to ensure that environmental improvement remains the key focus rather than the management of the regulatory process itself.

ACKNOWLEDGEMENTS

The authors wish to thank colleagues at W. A. Fairhurst and Partners for their help and advice in the preparation of this paper. The views expressed are those of the authors and not necessarily those of W. A. Fairhurst & Partners, SEPA or GHD Pty Ltd

REFERENCES

Defra, 2005, Consultation on new abstraction and impounding licensing regulations to apply in England and Wales and proposed changes to the Water Resources (Environmental Impact Assessment) (England and Wales) Regulations 2003, Defra, September 2005.

ICE, 1996, Floods and Reservoir Safety, 3rd Edition. Thomas Telford, London.

SEPA, 2005, Consultation on the Water Environment Charging Scheme 2006 (Water Environment (Controlled Activities) Fees and Charges (Scotland) Scheme 2006), SEPA.

SEPA, 2005, Levels of Authorisation under the Water Environment (Controlled Activities) (Scotland) regulations 2005, SEPA.

SNIFFER, Environment Agency, 2004, Management Strategies and Mitigation Measures Required to Deliver the Water Framework Directive for Impoundments, WFD29/Vol.1 - Preliminary Guidance Document, FWR.

Turpin, T.J., 2000 Environmental Impacts of Dams: the changing approach to mitigation, Dams 2000, Thomas Telford, London.

2. Hydraulics and hydrology

Improved reservoir level assessment through the mathematical modelling of weir crest coefficients

PETER MASON, Black & Veatch Ltd
KENNY DEMPSTER, Scottish & Southern Energy plc
JAMES POWELL, Black & Veatch Ltd

SYNOPSIS.   An accurate assessment of reservoir levels is a key part of any dam safety review.   Assessing the catchment hydrology will yield incoming floods but their accurate routing to assess associated reservoir levels will also require an accurate assessment of outlet work discharge characteristics.   In the case of simple overspill weirs the discharge coefficients are all too often guessed or estimated as constant values, whereas in fact they are more likely to vary with head.

Recent statutory inspections at Loyne and Cluanie dams revealed that weir discharge coefficients of 1.57, 1.63, 1.71 and 2.00 had all been used at different times in the past, by different engineers for essentially the same structures. For the inspection in 2005 the free flow surfaces over both weirs were simulated using computational fluid dynamics (CFD). This enabled the weir discharge coefficients to be assessed for a range of flows. The use of these for subsequent flood routing reduced reservoir levels at both dams.

The paper describes this work and gives recommendations for more simplified, assessments at other dams in the future.

INTRODUCTION

The Loyne and Cluanie dams were completed in 1960. Both are concrete gravity structures with similar design details and profiles. The upstream and downstream faces of both dams were formed using pre-cast units to retain an internal hearting concrete. The upstream faces of the dams are vertical and the downstream faces slope at 0.7 on 1.0. The central sections of both dams have long, un-gated spillways featuring profiled crests with flat upstream extensions. The spillway crest length at Loyne dam is 68.58m and the effective length at Cluanie, 129.39m, with bridge pier widths subtracted. Both dams are owned and operated by Scottish & Southern Energy (SSE).

Improvements in reservoir construction, operation and maintenance, Thomas Telford, London, 2006, 61–71

Stability analyses, carried out in recent years for both dams, as part of SSE's portfolio seismic and PMF stability reviews, indicated that safety factors are marginal and sensitive to the maximum water level reached during floods. In view of this a number of probable maximum flood (PMF) variants have been mathematically routed through the reservoirs in order to establish a probable maximum reservoir level. Surprisingly, however, there has been little refinement, or indeed consensus, on the associated discharge characteristics to be used for the overspill crests. The original designers suggested weir discharge coefficients of only 1.566 for Loyne and 1.626 for Cluanie. These are surprisingly low, in fact lower than a basic broad crested weir coefficient value of 1.71. In view of this, the 1995 statutory inspection adopted a value of 1.71 in checking freeboard adequacy. Even this, however, is lower than one might reasonably expect from a hydraulically profiled crest. Subsequent reviews of reservoir level for stability calculations adopted a discharge coefficient of 2.00 for both dams. The spillway crest of Loyne dam is shown on Figure 1.

Figure 1. The overspill crest of Loyne dam

Clearly the flood levels achieved will be directly dependent on the weir coefficient and it is surprising that such a wide range of values has been used for these dams over the years. In view of this, and with the agreement of Scottish & Southern Energy, the most recent statutory inspection also featured a more accurate derivation of the discharge characteristics of these crests using computational fluid dynamics, (CFD) modelling techniques. This was used to simulate the spill over the dams for a range of flow rates. The paper describes how this was done, the results obtained and how these affected derived reservoir levels in both cases.

MODELLING METHOD

The concept of CFD modelling

The CFD models were assembled using CFX5.7 which is a CFD commercial code widely used in the aerospace, nuclear energy, automotive and marine industries. CFD involves the numerical modelling of fluid motion based on the application of basic physical principles. The first step in building a CFD model is to set up a three-dimensional mesh which splits the fluid into a large number of small elements. The behaviour of these elements is then predicted using the Navier-Stokes set of simultaneous differential equations describing fluid flow. There are normally five sets of equations covering:

- Conservation of mass
- Conservation of momentum (three equations, one for each dimension)
- Conservation of energy

In order to simulate a free surface, there is the additional complication of solving for two different fluids (air and water) and tracking the interface between the air and water. CFX5.7 does this using a model based on the "Volume of Fluid" method which calculates the volume fraction of each fluid in each element for each time step. A limitation of free surface modelling with CFD is that it is computationally intensive and models typically take several times longer to run than an enclosed (single phase) model with a similar mesh size.

The model domain or geometry

The models were initially assembled in imperial units. This was done for convenience as the available drawings were in feet and inches. However all results were then converted to metric units for further use. The model domain (geometry) is shown in Figure 2. There are a number of notable features:

- The model is 2 dimensional (one element wide). This reduces the computational time.
- The actual reservoirs are several kilometres long and the dams 21m and 40.5m high in the cases of Loyne and Cluanie respectively. To prevent excessive computational run time, the effective reservoir reaches were modelled as 50 ft upstream of the dams and with depths of 20 ft below the crests. When truncating a model, it is important to ensure that such dimensions do not overly influence the solution.
- The downstream profiles of the dams were modelled in full to just beyond the point at which the profiles transferred into constant batters. A curve was then included to return the flow to the horizontal. This curve was not a real feature of the dam, but was included to simulate supercritical flow at the model outlet and hence simplify the set-up of boundary conditions.
- Inlet conditions were simulated as an expanding taper with a separate air inlet above. This was done to simplify the set-up of boundary conditions as keeping the inlet fully submerged also keeps the inflow as 100% water. The taper allows the water depth to gradually increase to the free surface depth as determined by the model.

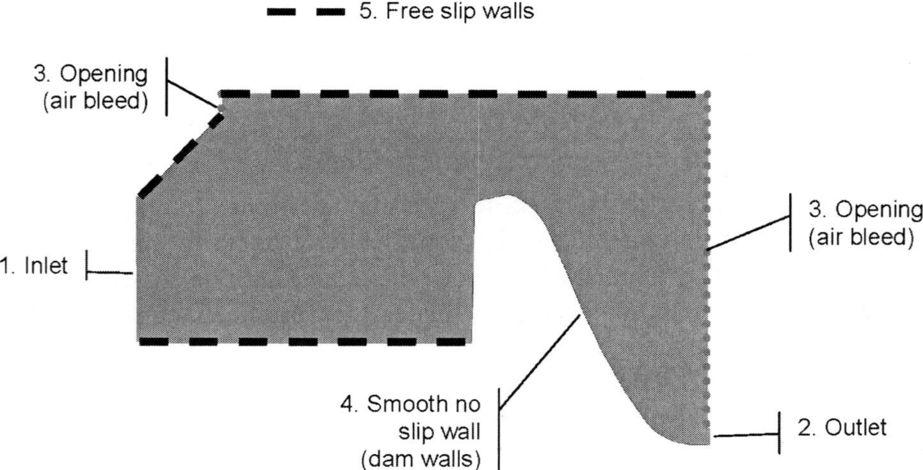

Figure 2. Model domain and boundary conditions

The mesh

The simulations were first run on a coarse mesh to obtain an initial approximate solution. A second set of simulations was then run using a finer mesh with the results from the initial runs used as the starting conditions. This two stage approach reduces the overall computational time and has the added benefit that preliminary results are quickly available. Key details of the meshes are summarised in table 1.

Table 1. Mesh Details

Feature	Coarse mesh	Fine mesh
Predominant mesh type	Unstructured prismatic wedge mesh	Unstructured prismatic wedge mesh
hexahedral inflation off the dam walls	5 layers to depth of 1 foot	5 layers to depth of 1 foot
Global mesh size	1 foot	0.5 foot
Refined mesh near free surface	0.3 foot	0.1 foot
Angular resolution	12	12
Approx. number of elements	20,000	90,000

An automatically generated unstructured mesh was used due to the speed with which it could be set up, although it would have been possible to build a more efficient structured mesh if minimising model run time had been a priority.

A partial mesh for one of the flow rates tested is shown in Figure 3. The meshes were set up with smaller mesh spacing in the vicinity of the anticipated water surface. This enables more accurate calculation of the free surface profile. For this reason a different mesh was set up for each flow rate simulated.

The Boundary Conditions

The following boundary conditions were specified (see Figure 2 for location):

1.   Inlet: Mass flow of water
2.   Outlet: Hydrostatic pressure profile – set at estimated height of downstream water level. The height used is not critical to the model performance provided the predicted flow remains supercritical.

3.   Openings: Inflow or outflow unspecified. Water and air are free to flow out of the domain. All flow into the domain is specified as 100% air. These openings are effectively air bleeds.

4.   Free slip walls (no restrictions to flow).

5.   No slip smooth wall (velocities forced to be zero at the walls but without any assigned, wall roughness value).

Coarse mesh

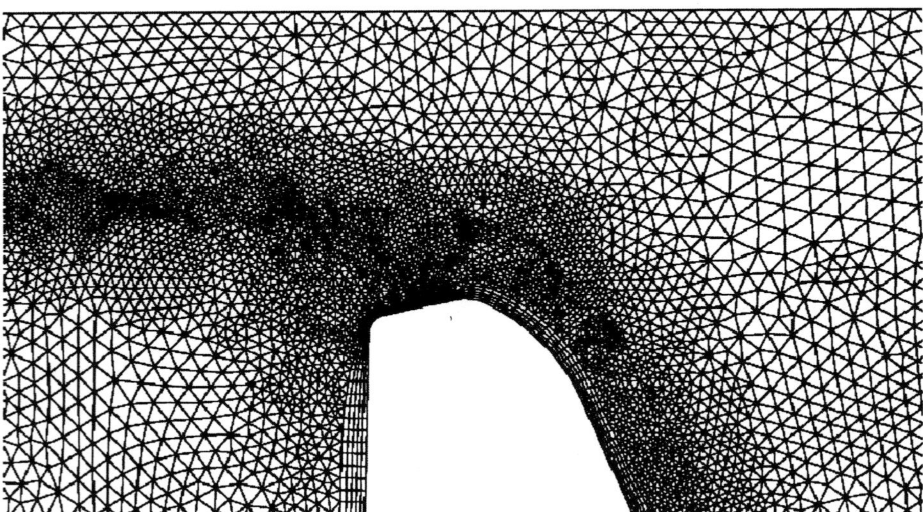

Fine mesh

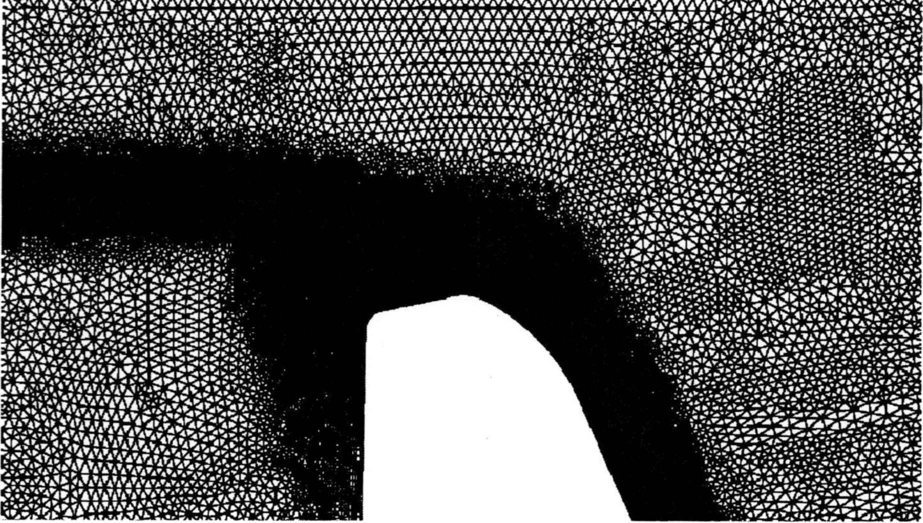

Figure 3. Meshes

Set-up and run time details

The simulation was set up in a fairly routine way for free surface modelling of water flows, the relevant factors are included here for reference but are not discussed in detail:

- Steady state
- k-epsilon turbulence model
- Homogeneous free surface model (based on the VOF method).
- Convergence criteria 1×10^{-4} RMS error

Free surface modelling is often constrained by computing power. To accelerate the convergence a high physical timescale was used at the start of the simulation. This value was set by trial and error, but could be up to 100 times higher than the default timescale. To help determine when the free surface had stabilised, the pressure at several locations upstream of the dam was monitored throughout the simulation.

The coarse mesh models typically took 2 to 4 hours to run, the fine mesh models took up to 24h to run on a twin 2.4Ghz processor computer. Interestingly it was found that the additional mesh refinement gave less than 1% difference in the predicted water depth above the weir crest.

Model output

Four different crest flows were simulated using CFD modelling. In addition it was assumed that at the very lowest heads over the crest the discharge coefficient would approximate to the broad crested weir value of 1.71. These five reference points in each case then enabled additional values to be derived by interpolation. The discharge coefficient curves, derived by curve fitting through the obtained points, are shown in Figure 4 for both Loyne and Cluanie. It can be seen that both crests exhibit discharge coefficient values greater than 2.00 as the heads on the crests increase.

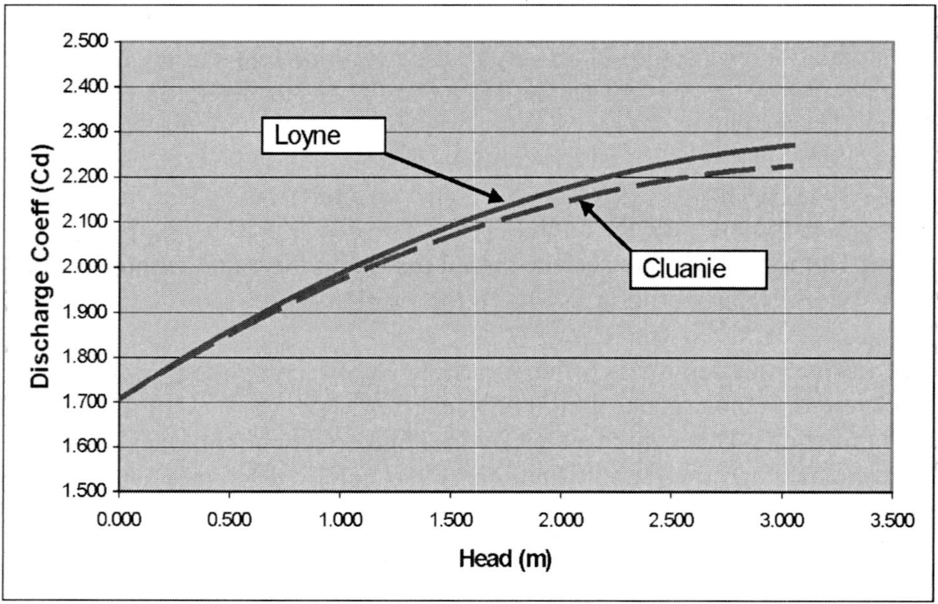

Figure 4. Derived discharge coefficients plotted against head

It should be noted that the suspected "design" heads for the Loyne and Cluanie crests were 4ft and 5 ft respectively and that for a standard ogee type crest profile and vertical upstream face the metric discharge coefficient at the design head will be in the order of 2.18. In fact the Loyne and Cluanie crests achieved 2.06 at these heads.

RESULTS OF FLOOD ROUTING USING THE REVISED DISCHARGE COEFFICIENTS

The previous PMF flood assessments had been carried out by others in 2004 using a non-variable crest discharge coefficient of 2.00 for Loyne and Cluanie dams and by routing PMF inflows in conjunction with snow-melt rates of both 1.75 mm/hr and 5.00 mm/hr. Both cases were re-run for Loyne and Cluanie dams using the variable discharge coefficients derived by CFD modelling. The results are shown in Table 2.

The general standard "base" cases are those featuring a snow-melt rate of 1.75 mm/hr. In those cases the use of a variable discharge coefficient produced a small, but useful, reduction in the maximum still-water reservoir level reached. In both cases it also, inevitably, produced a marginal increase in the maximum outflow, something which should be reflected in any downstream inundation studies.

Table 2. Derived Maximum Still-Water Reservoir Levels

Snow-melt (mm / hr)	Head over Crest (m)		Head reduction		Maximum outflow (m³/s)	
	2004 studies	with variable Cd	(m)	%	2004 Studies	with variable Cd
Loyne						
1.75	1.82	1.77	0.05	2.7%	336	349
5.00	2.29	1.98	0.31	13.5%	475	421
Cluanie						
1.75	1.59	1.44	0.15	9.4%	426	458
5.00	2.15	1.58	0.57	26.5%	534	533

The additional cases featuring snow-melt rates of 5.00 mm/hr produced particularly useful reductions in head over the crests, however, these were partly due to hydrological assessment errors in the 2004 studies.

MORE READILY ACCESSIBLE ALTERNATIVE METHODS

Clearly it is advantageous to accurately model the discharge characteristics of weirs as part of any routine flood safety assessment. However, not all will have ready access to CFD modelling facilities and in most cases it will, anyway, be sufficient to use a, simplified and approximate method.

In the case of a standard ogee crest, discharge coefficients at the design ahead are readily available from sources such as *"Design of Small Dams"* published by the USBR. These are based on experimental and prototype data and are repeated in a number of specialist textbooks. Furthermore the same sources will indicate how the discharge coefficient will vary according to the slope on the upstream face of the weir crest and also according to how upstream levels vary above and below the design head (*Hd*).

For a standard ogee crest with a vertical upstream face a discharge coefficient of 2.18 is generally assumed at the design head. This will increase by approx 7% at 1.6 times *Hd* and reduce by 10% at 0.40 times *Hd*. Equations can be developed to describe this relationship. An approximate one in metric units, and covering the range from +/- 50% of *Hd* gives a value for Coefficient *C* of:-

$$C = Cd\,[\,0.87 + 0.125\,(\,H\,/\,Hd)\,]$$

Where
H = the head on the weir
C = the effective discharge coefficient at head H
Hd = the design head
Cd = the discharge coefficient at the design head

However it is also necessary to assess the original "Design Head" in order to use this. This can be approximated, as well as possible discharge coefficients, using publications like USBR Monograph 9 on *"Discharge Coefficients for Irregular Overfall Spillways.* Alternatively the basic equation for an ogee crest is:-

$$Y = X^{1.85} / (2 . Hd^{0.85})$$

Where
X = the horizontal distance downstream of the crest apex
Y = the vertical distance down from the crest apex

Taking a number of X,Y distance measurements on a given crest, from the crest apex, it should be possible to back-analyse a value for Hd which can then be used in subsequent flow equations. In the case of Loyne and Clunie it was possible to compare the ogee profiles in this way to standardised profiles and derive values for Hd of 4ft (1.219m) and 5ft (1.524m) respectively. However the upstream crest extensions at both dams led to doubts about how accurate the direct use of ogee based values would be and hence led to the use of CFD modelling. As discussed earlier, in fact a potential ogee crest discharge coefficient of 2.18 was reduced to 2.06 at these "design" heads, by the upstream extensions.

Lastly it should be noted that the derivation of equations is not always necessary in order to carry out flood routing checks. Modelling programmes such as Micro-FSR can route a weir outflow using tabulated values of discharge against head and interpolate between them to obtain any necessary intermediate values.

CONCLUSIONS

It can be seen that the use of variable weir coefficients not only reflects engineering reality but can lower the maximum reservoir levels obtained from routine flood safety assessments. However, such a reduction may also be accompanied by an increase in maximum outflow which should be reflected in any downstream inundation studies.

In the case of overspill crests with standard shapes, published data can be used to derive appropriate discharge characteristics as well as to assess how those characteristics will vary with head over the weir. In the case of non-standard crests, CFD modelling techniques can be used to make an assessment of the weir discharge characteristics much more rapidly and inexpensively than would be the case using a physical hydraulic model.

Although the benefit in terms of water level reduction for the standard PMF scenario may seem small, it may prove of significance in ongoing stability reviews for both structures. Should more extreme PMF (5 mm/hr snowmelt scenarios) be adopted then the benefits will be greater.

Yuvacik Dam: Improvements to dam operation utilizing an integrated atmospheric-hydrological model

TOLGA GEZGIN – RWE Thames Water Turkey, Izmit, TURKEY
PROF. DR A. UNAL SORMAN - Middle East Technical University, Ankara TURKEY
ASST. PROF. DR. AYNUR SENSOY – Anadolu University, Eskisehir, TURKEY
ASST. PROF. DR. A. ARDA SORMAN – Anadolu University, Eskisehir, TURKEY

SYNOPSIS. The paper describes the methodology employed to control water levels within Yuvacık Dam near Izmit, Turkey. It also provides details of a current study, results of which are expected to improve the operational control of this water supply and flood retention dam. The study includes the application of a hydrological model coupled with atmospheric forecast data; the project is currently active and being carried out by Thames Water International in conjunction with Middle East Technical University and Anadolu University. The study proposes to use of semi-distributed hydrologic modelling coupled with numerical weather prediction and GIS technology. The predicted precipitation and temperature values obtained by the numerical weather prediction model will be used as input to the hydrological model for near-real time runoff estimation to the Yuvacık dam reservoir for better and efficient operation.

INTRODUCTION

Turkey Thames Water, working with local partners, was awarded the first build, operate and transfer water supply project in Turkey in 1995. The Thames led project to supply water to Kocaeli Province in Turkey is the largest privately-financed water project in the world. Opened on 18th January 1999, the Izmit scheme provides high quality water to over 1.2 million people along with the rapidly growing industrial sector in the province. The project involved construction of a 101m high earth fill dam (Yuvacık Reservoir), a water treatment plant and a trunk main system of 146 kilometers including branch lines in Izmit, to the northeast of the city of Istanbul. The reservoir has a capacity of 60.6 million m^3 and an estimated annual yield of over 142 million m^3. The treatment plant has a designed production capacity of 480 million litres per day and the pipeline feeds 38 service reservoirs and includes 6 pumping stations.

Thames Water will operate and maintain the new system for 8 years before handing it over to the Izmit Greater Municipality free of charge with all the

Improvements in reservoir construction, operation and maintenance, Thomas Telford, London, 2006, 72–83

assets. Until the handover on 18th January 2014, Thames Water is committed to transfer knowledge of the best skills and technology between the projects and to the local community. In August 1999 the area was hit by a massive earthquake, measuring 7.4 on the Richter scale and devastating the local community. The dam, pipeline and treatment works withstood the shock and remained undamaged.

Water control managers who regulate projects need access to timely and accurate information on which to base sound decisions. During a flood event, decisions sometimes have to be made within minutes or hours from the onset of rising river stages. During a drought, decisions can affect water availability for months into the future. Water control managers need continuous real-time data observations from field sites, as well as reliable watershed modelling tools, to provide appropriate responses to changing hydrologic conditions. Accurate real-time forecasts of natural inflows to reservoirs are of particular interest for operation and scheduling. A variety of methods have been proposed for this purpose including conceptual (physical) and empirical (statistical) models (WMO, 1994). As a result, TW identified a pressing need to develop a uniform water management system integrating communication networks, centrally developed and supported software, and state-of-the-art hydrological and hydraulic models.

The semi-distributed hydrological model is applied on near-real time basis with the integration of one day ahead forecast data from a regional atmospheric circulation model to provide an improved decision support system for the operation and management of Yuvacık reservoir. Using this methodology streamflow data contributing to the reservoir is estimated, the extreme flood events are managed and the storage is maximised for water supply during the uncertain periods in the spring months of each year. Data from the existing and newly established network of rain gauging stations could be used for this purpose in accordance with the outputs of atmospheric circulation model.

COMPONENTS OF THE SYSTEM
The solutions to manage a strategy plan for these kinds of problems are the followings: Rainfall-runoff modelling approach based on physical principals, near-real time model application using near-real time data obtained from the data management system network. The project software package can be regarded as a puzzle consisting of five interlocking pieces, as shown in Figure 1 (Fritz et al., 2002). The four pieces around the perimeter all deal with some aspect of data management. The key middle component, which interacts with all of the data management activities, is watershed modelling. It allows water control managers to use observed real-

time data and National Weather Service (NWS) forecast precipitation to simulate future conditions.

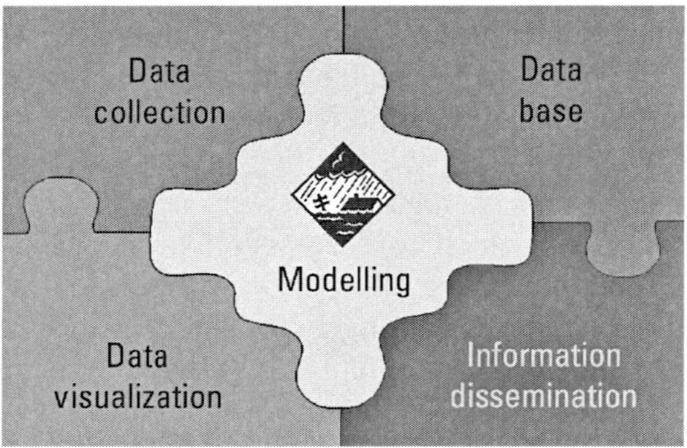

Figure 1: Components of the system (Fritz et al., 2002)

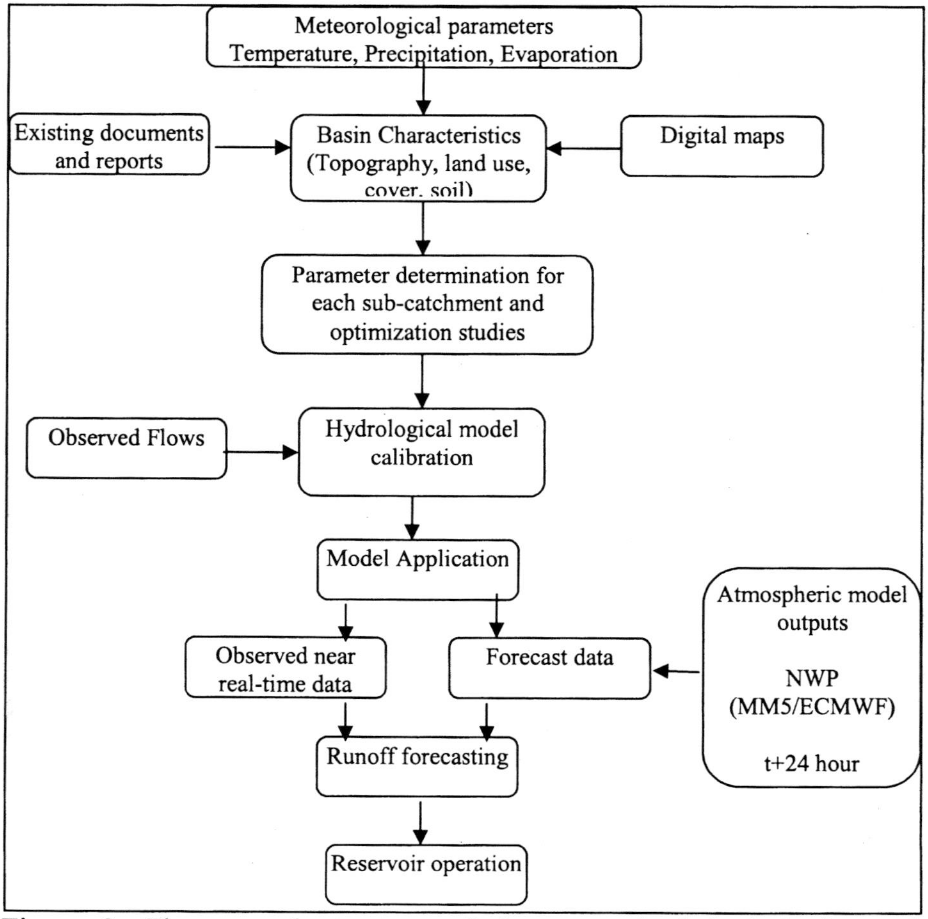

Figure 2: The general structure of the atmospheric-hydrologic model integration and hydro-meteorological components

Steps to be followed for the defined work include; Data retrieval from the existing gauging network archive and their preprocessing, selection of new instruments and establishment of the new network, installation and data transferring through fixed and mobile stations, snow data collection (snow depth data) and manual snow water equivalent data collection with ground observations at the selected sites, selection of hydrological models calibration of selected model, model verification, the development of interface programs and the integration of a regional atmospheric model with the selected hydrological model and model application (Figure 2).

EXISTING DATA AND PREPROCESSING

The digital and the analog maps are required for the physically based modelling studies. Geographical Information System (GIS) tools are also used to carry out analysis and enhance graphical displays. Aerial photography, digital maps and other GIS data allow the geo-referencing of critical features in the watershed. The maps basically include digital elevation model (DEM), land use and land cover, soil types and geological maps. DEM (Figure 3) of the basin is basically used to evaluate the performance of the existing rain gauge network which has been upgraded with new equipment. In addition, DEM is essentially used for the watershed delineation and formation of sub-basins. The slope and aspect maps are also derived to emphasize the basin characteristic of the region. Land use/land cover and soil/geological maps are used in the cross quarry analysis that will be required later in the study for the optimization of infiltration parameters of hydrologic models. All these maps form the geospatial background of the hydrological modelling.

Thames Water (TW) data can be classified as either real-time data or static information. Real-time data include time-series observations, such as river stages (3), precipitation (6) and reservoir elevations (2). Gauges are provided at key locations and they record observed conditions such as river stages and precipitation. Typically, gauges are equipped with telemetry units which record data in 5 minutes time-steps and transmit the information. The data are then downloaded to TW computer servers in the water control management center (DMS). DMS has a comprehensive system incorporating the acquisition, transformation, storage, display of information to support TW real-time water control mission. The incoming real-time data include hydrological information (river stage, reservoir elevation), meteorological information (observed and forecast precipitation), and other hydro-meteorological information (such as water quality data). Observed data, shown in Table 1 and Table 2, are used to view the current status of the watershed. Watershed modelling programs are used to forecast runoff, reservoir response and operations. A number of future precipitation and reservoir operation scenarios can be evaluated.

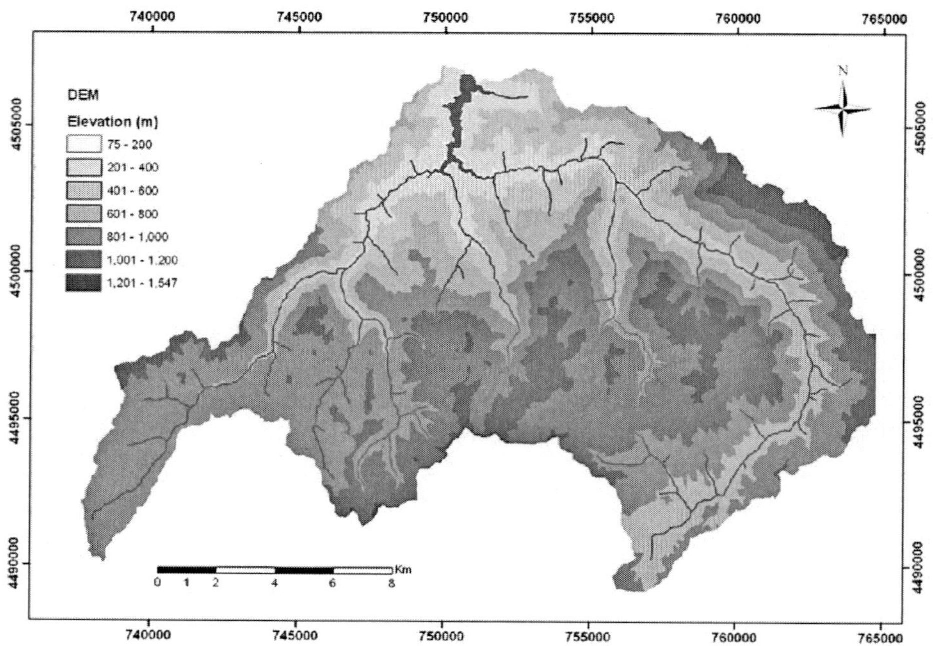

Figure 3: Digital Elevation Model of Yuvacık basin with dam reservoir and river network

Table 1: Available meteorological records for the project

Meteorological Records				
Organization	**Station**	**Data Type**	**Data Interval**	**Elevation**
State Meteorological Org. (DMI)	Kocaeli (2000-2005)	Precipitation Snow Depth Temperature	Daily Daily Daily	76 m
State Hydraulic Works (DSI)	Hacıosman (2000-2004)	Precipitation Snow Description	Daily Daily	900 m
State Hydraulic Works (DSI)	Kurtköy (1987, 1989, 1991)	Precipitation	Daily	40 m
Thames Water (TW)	DMS (2001-2005)	Precipitation	5 Minutes	320 m 170 m

Table 2: Available hydrological records for the project

Runoff Records				
Organization	**Station**	**No.**	**Data Interval**	**Elevation**
Thames Water (TW) (2001-2005)	Kirazdere	FP1	5 minutes	185 m
	Kazandere	FP2	5 minutes	180 m
	Serindere	FP3	5 minutes	200 m
	Reservoir	FP4	5 minutes	190 m
State Hydraulic Works (DSI) (1963-1992)	Kirazdere	02-06	Daily averages	37 m
State Hydraulic Works (DSI) (1987, 1989, 1991)	Kirazdere	02-06	Instant peak flow discharges	37 m

INSTRUMENTATION

Yuvacık basin has an area of 258 km^2 and an elevation ranging from 75-1547 m, as depicted in Figure 3, with a mid-altitude of around 800 m. The existing network of six rain gauges in Yuvacık basin was not capable of representing the spatial variation of precipitation amount in terms of the whole catchment and its sub-basins (Kirazdere, Kazandere, Serindere). Hence, it was decided to renew and upgrade the existing rain gauge network with new instrumentation. Numerous site trips have been carried out with a Global Positioning System in order select possible locations for the new stations based on certain criteria (location, communication, electricity, accessibility, elevation, aspect, land cover and security). According to the site conditions, four of these locations are chosen as fixed stations whereas the other three are marked as mobile stations. The reason for constructing mobile stations at certain locations instead of fixed ones arose mainly from electricity and communication incapability of the points although the position of the stations would represent valuable precipitation data. In the future, if some of these mobile stations are thought to be important representative points, then stronger measures could be taken to solve the current incapabilities and turn them into fixed stations. The locations of the new stations along with the old precipitation and flow plant (stream gauging) network can be seen in Figure 4. Views from one of the fixed and mobile stations are shown in Figure 5.

In the new stations, besides collecting precipitation data only, other important meteorological and snow measuring sensors (temperature, humidity, snow depth) are also installed in order to collect data about the region. Especially as the new stations are positioned at higher locations, the form of precipitation could be different at these locations (such as snow) as compared to the lower altitudes (rainfall) at certain time periods. This process may have a significant effect on the discharge reaching the dam reservoir when enough energy is present to melt the deposited snow on the higher grounds of the basin. To be able to follow the snowmelt process in the catchment, snow depths as well as air temperature are also monitored at the stations. Also at certain intervals of the winter season, manual snow depth and density measurements are conducted around the fixed stations to the stations to compare with the automatically collected data in the stations.

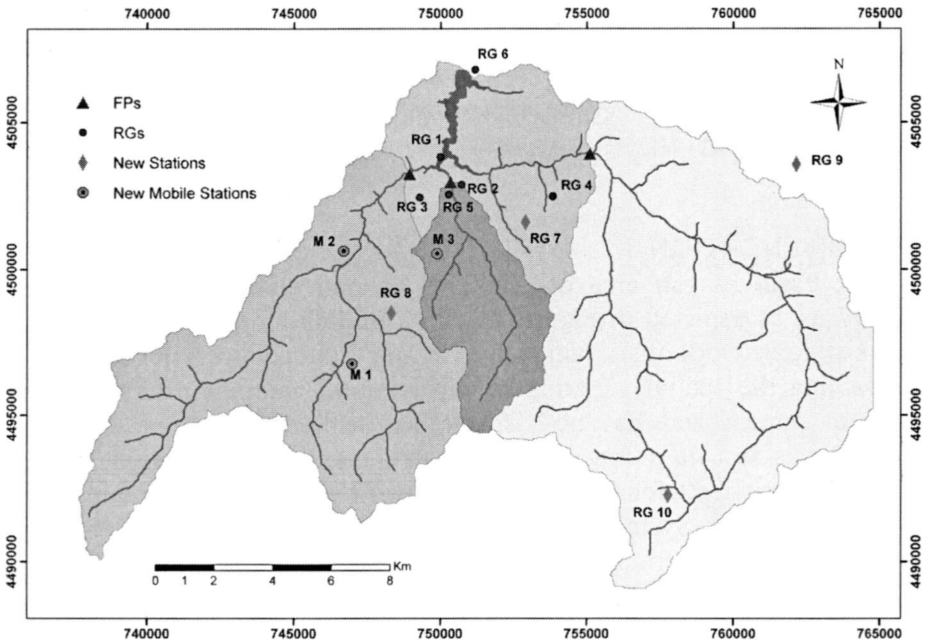

Figure 4: Location of the stations in the basin

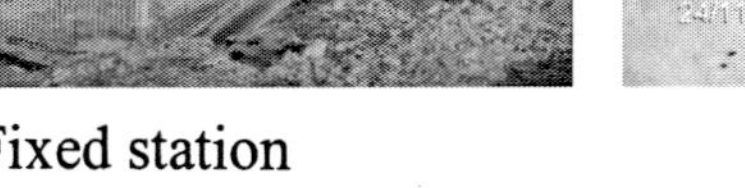
Fixed station   Mobile station

Figure 5: Views from a fixed and a mobile station in Yuvacık basin

ANALYSIS OF EVENTS

This analysis forms the main part of any hydrological study, since all types of modelling strategy need both qualitative and quantitative data. TW finds that an enormous amount of time is required in the retrieval and quality control associated with hydrological data collection. At the end of the analysis, there are problems especially for the runoff data set that should be controlled and verified. There are a few events where the three runoff stations collected accurate data simultaneously; therefore the calibration work mainly concentrates on sub-basin terms instead of the whole basin.

This part of the study is especially important since it constitutes the main input of the model calibration step. A query is carried out to derive the basic sub-groups of events from the existing data (September 2001 - June 2005) concerning the double criteria including both total daily precipitation amount (P_T) and the reservoir inflow volumes (R_I). For each event, precipitation (rainfall from all of the stations including DMS, KE and HO and snow depth collected at KE and HO stations) and runoff values (from FP stations) are analyzed in detail. Events are categorized in terms of rainfall/snowmelt – runoff relation.

In some events, there are discontinuities at water stages noticed in the collected data. These mistakes are thought to have occurred from the incorrect measurement of the level (stage) data in the stations. In addition, zero recordings are observed in the stations which cause frequent data jumps in the discharge values. Moreover, in some events although the runoff station seems to work throughout the event, there are missing parts especially in the flood events (peak values). Also sometimes, the radar and ultrasonic flow measurements differ from each other for the same event. Due to the large elevation range of the basin, snow becomes an important factor. In the past very minimal snow data has been collected which makes snowmelt modelling difficult besides rainfall. The HO station operated by DSI is the only snow data source.

HYDROLOGICAL MODEL STUDIES

Thames Water provides an integrated suite of modelling programs which represent the atmospheric conditions and hydrological aspects of the watershed. The installed modelling programs include HEC-HMS (Hydrologic Modelling System, USACE, 2005) interfaces for data inputs. A real-time simulation uses observed data up to the present time and forecast precipitation. The integrated models use those data to forecast the watershed's response to the anticipated precipitation.

HEC-HMS simulates watershed hydrology using observed and predicted precipitation to compute runoff. Runoff is combined with base flow to generate flow hydrographs at various points within the watershed.

The value of each parameter must be specified to use the model for estimating runoff hydrographs. Some of the models that are included in HEC-HMS have parameters that cannot be estimated by observation or measurement of channel or watershed characteristics. Calibration uses observed hydro-meteorological data in a systematic search for parameters that yield the best fit of the computed results to the observed runoff. To compare a computed hydrograph to an observed hydrograph, HEC-HMS computes an index of the goodness-of-fit. The key to automated parameter estimation is a searched method for adjusting parameters to minimize the objective function value and find optimal parameter values. In HEC-HMS, one of six objective functions can be used, depending upon the needs of the analysis. The goal of all calibration schemes is to find reasonable parameters that yield the minimum value of the objective function (Table 3).

The total catchment area is divided into four sub-basins and model calibration studies are applied one by one. Precipitation input is entered into HEC-HMS by meteorologic module, which is one of the major components of a project. The meteorologic module has three main components such as precipitation, evapotranspiration, and snowmelt to be used during simulations. Precipitation values are input with gauge weights which are determined by the inverse distance method. The first event set includes hourly hydrographs produced from rainfall events therefore snow data were not used in this part of the study.

After preparing the rainfall input for the hydrologic model, a transformation component for the model must be specified. The synthetic unit hydrographs are derived by State Hydraulic Works (DSI) from historical storms. These unit hydrographs are checked using historical storm events and used in the study.

Two different loss methods are selected in the study; the first one is initial and constant uniform loss method and the second one is exponential loss method. Although the former calibration results show consistent parameter ranges even for events of different seasons, constant loss causes most of the excess rainfall to be lost and hence, the simulated hydrographs do not have an accurate match to the observed measurements. The latter method, specific to HEC models, simulated better fitting results between the observed and calculated hydrographs in the model calibration step. Early summer event data set calibration is given below as an example:

Table 3:
Calibration results of rainfall events with different objective functions

	11-15 Jul 02	9-11 Jun 04	19-21 Jun 04	23-25 Jun 04	31 May-6 Jun 05	4-9 Jul 05
	Sum Squared Residuals	Percent Error Peak	Percent Error Peak	Sum Absolute Residuals	Percent Error Volume	Peak-Weighted RMS Error
Initial Range (mm)	14.183	16	23.905	21.16	19.589	16.397
Initial Coef. (mm/hr)	0.93891	0.98	0.96341	0.9604	0.93445	0.95593
Exponent	0.96953	0.99	0.98	0.97	0.95922	0.97496
Initial Discharge (m³/s)	0.37	1.5	1.4	1.37	1.15	0.92
Recession Constant	0.545	0.9	0.8	0.9	0.75	0.3
Ratio / Flow (m³/s)	1.9	1.6	1.6	1.8	5.6	1.63
Percent Peak Diff.	9.73	-6.67	0.00	-0.41	1.96	-0.80
Percent Volume Diff.	-12.00	-2.54	3.68	3.83	-4.36	-0.99

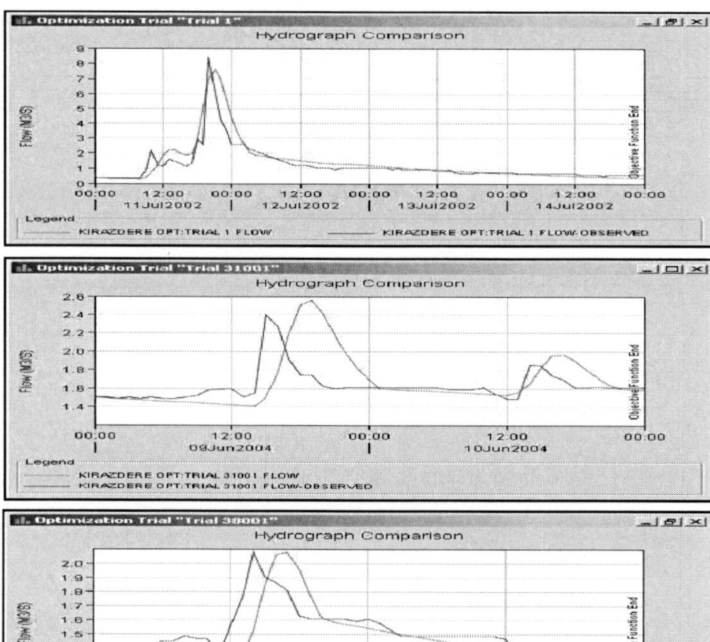

Figure 6: Calibration results between simulated and observed hydrographs

ATMOSPHERIC MODEL INTEGRATION STUDIES
As the event calibration process is continued as one phase of the study, on the other hand numerical weather prediction data are collected and processed. Numerical weather prediction data, Mesoscale Model 5 (MM5), are processed in the supercomputers of State Meteorological Organization (DMI) each day and uploaded through file transfer protocol. The input prediction data to the hydrological model consist of daily average temperature and total precipitation. To test the accuracy of the numerical weather prediction data, the data collected from the 11 automatic ground weather observation stations (AWOS) in the vicinity of the Yuvacık Basin is plotted against the corresponding MM5 temperature and precipitation grid data. The resulting trends seem to match quite accurately although there can be certain offset adjustments needed due to the underlying topographic map mismatch between the AWOS stations and MM5. Daily average temperature and total precipitation between AWOS measurements and one-day ahead predicted MM5 results are presented in Figure 7.

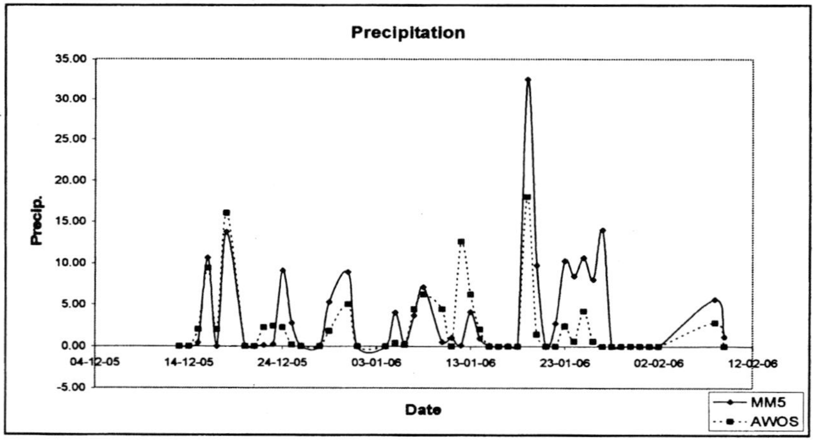

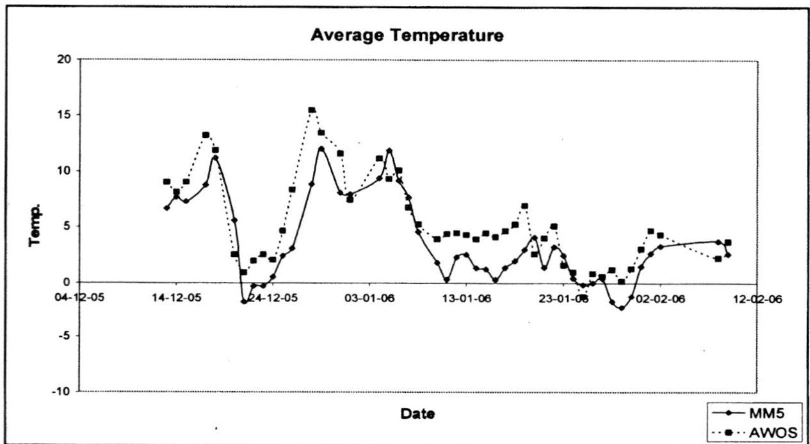

Figure 7: Comparison of MM5 and AWOS average temperature and total precipitation data

DISCUSSION AND RESULTS

The study has been proceeding on schedule (Progress Report I, November 2005). The new instrumentation was placed in time to collect valuable hydro-meteorological data. During the 2006 winter the weather conditions were harsher than average with quite a lot of solid precipitation (snow) being present in the basin. With the construction of snow depth sensors it will be possible to monitor snow data in real-time and hence take precautions especially from high discharges coming from snowmelt.

So far the calibration procedure is being done for rainfall events only and seems to give promising results. As the HEC-HMS program is very recently upgraded to version 3, it also includes snowmelt component using the simpler temperature index method. For those certain selected events that have high discharge values due to snowmelt and rain-on-snow, the next step is to use the new HEC-HMS version 3.1 to calibrate these mixed events in the very near future.

The calibration procedure is to be completed in a few months time (2-3 months) according to the work timeline with the atmospheric-hydrological model integration being the following step. In the end, watershed modelling, when combined with atmospheric forecast data, will allow water control managers to prepare one-day ahead forecasts of hydrological conditions in the watershed. The managers will then use the results to help make reservoir regulation decisions with the knowledge of how those decisions will affect both reservoir operation and the downstream populated and industrial areas of İzmit.

REFERENCES

Fritz J.A., Charley W.J., Davis D.W., Haines J.W. (2002). New water management system begins operation at US projects, *Hydropower and Dams*, Issue 3, United Kingdom.

Progress Report I (2005) Middle East Technical University and Anadolu University, November 2005.

USACE (2005) Hydrological Modelling System, HEC-HMS, User's Manual, Version 3.0.0, December 2005.

WMO (1994). *Guide to Hydrological Practices*, World Meteorological Organization, WMO-No. 168, Geneva, Switzerland.

Water management at Dinorwig pumped-storage power station.

MARK I BAILES, EngD Research Engineer, Cranfield University
OWEN P WILLIAMS, Company Civil Engineer, First Hydro

SYNOPSIS.   Optimum operation of the Dinorwig pumped-storage scheme requires a constant volume of water within its closed reservoir system. Heavy rainfall and the subsequent floods can cause additional or 'excess water' to spill into the closed reservoir system. Whilst the reservoirs are designed to cope safely with the rainfall and subsequent floods in the case of extremely rare events, optimum operation of the pumped-storage scheme can be vulnerable on an annual (or even more frequent) basis to rainfall and the subsequent floods.

This paper describes how 'excess water' is currently managed and then describes a computer model of the system, which was constructed on behalf of the power station operators. The model links hydrology, hydraulics and power station operation. The aim of the model was to increase understanding of the link between upstream catchment conditions, current operational conditions/rules, 'excess water', and downstream catchment conditions, on a day-to-day basis. (rather than extreme event basis). The model was then used to simulate the system with modified upstream conditions and modified operational conditions/rules.

Overall the work has increased understanding of water management at Dinorwig and assessed commercial and environmental implications of different water management strategies.

Translations: Afon = River, Nant = Stream, Llyn = Lake

INTRODUCTION

A *pure* pumped-storage scheme consists of two reservoirs, an upper and a lower one. Inputs are the natural streams that flow into the reservoirs and direct rainfall. Outputs are evaporation, controlled release (e.g. compensation flow) and uncontrolled release (e.g. flow over a spillway) Water falling from the upper reservoir drives turbines in order to generate electrical energy. The water is collected in the lower reservoir and is later pumped back to the upper reservoir. Generating is usually carried out when electricity has high value to the electricity network and pumping when the cost of electrical energy is lower. Thus pumped-storage can provide both economic generating capacity and special support services to the national electricity system. The lower lake of a pumped-storage scheme minimizes the effect of the generating/pumping cycle on the natural river system.

In a *pure* pumped-storage scheme the *total* volume of water in the reservoir system i.e. upper lake, lower lake, shafts and tunnels, must remain constant. This constant volume of water is simply moved between the upper and lower reservoirs during the generating/pumping cycle. If there is more water in the reservoir system than required then, there is said to be positive 'excess water'. If there is less water in the reservoir system than required then there is said to be negative 'excess water' (Jog, 1989).

Both positive and negative 'excess water' can have an economic impact on the optimum operation of a pumped-storage power station.

CASE STUDY: DINORWIG PUMPED-STORAGE POWER STATION

Dinorwig pumped-storage power station, in North Wales, is currently owned and operated by First Hydro Company. First Hydro Company also own and operate Ffestiniog pumped-storage power station. Dinorwig has a generating capacity of 1728 MW (First Hydro Company, 2005). The major constructions to form Dinorwig power station were:
- a dam at Marchlyn Mawr (the upper reservoir) to enlarge an existing lake,
- a dam at Llyn Peris (the lower reservoir) to enlarge an existing lake,
- a shaft and tunnels between Marchlyn Mawr and Llyn Peris to allow flow in either direction
- an underground complex containing the power station.

Downstream of the power station system is a natural lake, Llyn Padarn.

Prior to the construction of the Dinorwig scheme the Afon Hwch, Afon Nant Peris and Afon Dudodyn flowed into Llyn Peris. To reduce the natural inflows into Llyn Peris and hence the build up of 'excess water' in the reservoir system, the following changes were made when the scheme was built:
- Diversion of the Afon Hwch so it now flows into Llyn Padarn. All flow is diverted.
- Diverting the Afon Nant Peris so that it now flows through a Diversion Tunnel into Llyn Padarn. Under low flow conditions all flow in the Afon Nant Peris is diverted. Under high flow conditions spill into Llyn Peris still occurs.
- Diverting the Afon Dudodyn into Afon Nant Peris (upstream of the Diversion Tunnel) using a diversion pipe. Under low flow conditions all

flow in the Afon Dudodyn is diverted. Under high flow conditions spill into Llyn Peris occurs.

Figure 1 shows Dinorwig scheme, including the remaining inputs and outputs to the closed reservoir system, after catchment diversion. Also shown in Figure 1 is Llyn Padarn. The pumped-storage operation results in the level of Llyn Peris only rising higher than Llyn Padarn for a short window each day, if at all.

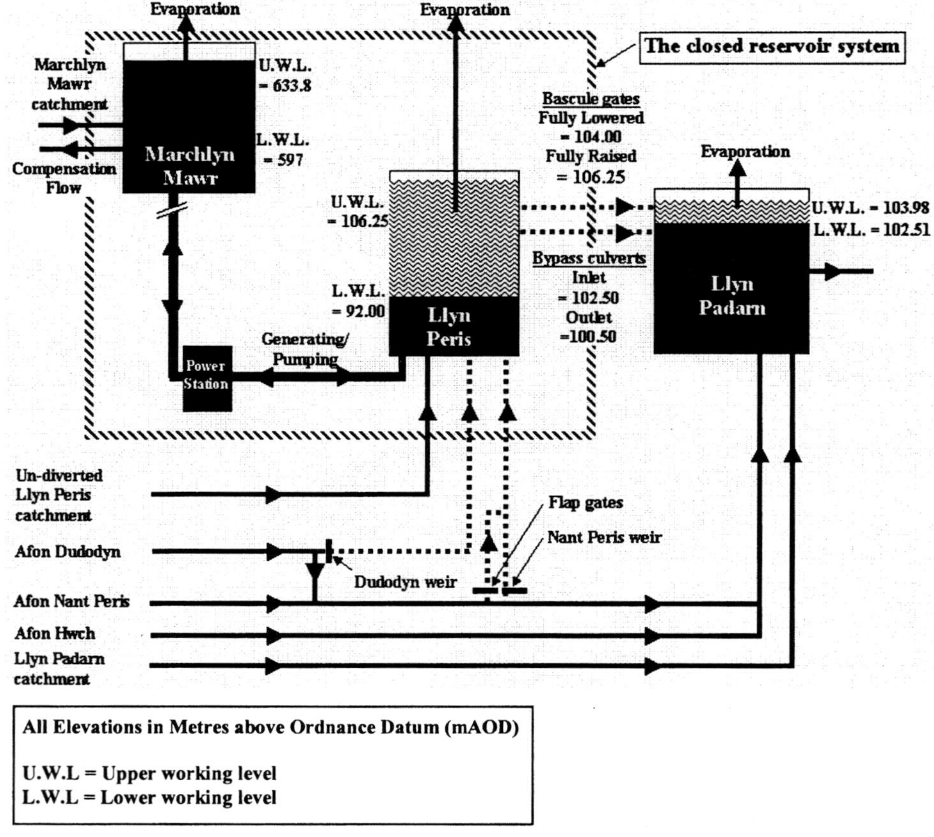

Figure 1. Dinorwig pumped-storage power station.

Although the catchment diversions described above exclude the majority of otherwise natural 'excess water' there will still be a residual inflow into the reservoir system, especially at times of high river flow. Three Bypass culverts and three Bascule gates were therefore incorporated into the lower dam to discharge 'excess water' from Llyn Peris into Llyn Padarn. The Bypass Culverts are either open or closed. The Bascule Gates have six possible positions from fully-lowered to fully-raised.

For a release of 'excess water' to take place:
Llyn Peris level > Level of the Bypass culverts/Bascule gates
AND Llyn Peris level > Llyn Padarn level
AND Llyn Padarn level < a level specified by the Environment Agency

These three conditions constrain when 'excess water' can be released from the reservoir system. Figure 2 shows typical operational data, from the power station, for October 2004. There were only 8 opportunities to discharge 'excess water' in the 31 day period, and the amount of water that could be discharged on each occasion will have been constrained either by the power station generating schedule or Llyn Padarn level.

In 2002, First Hydro Company commissioned an Engineering Doctorate (Engineering and Physical Sciences Research Council, 2005) study to increase their understanding of 'excess water' on an operational basis (rather than extreme event basis, which is already covered under Section 10 of the Reservoirs Act)

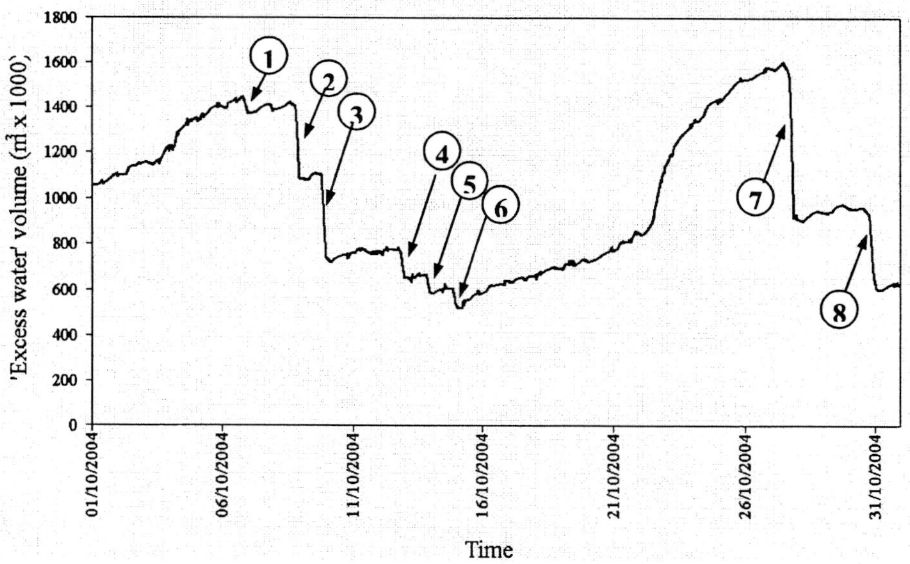

Figure 2. 'Excess water' volume, October 2004

ENGINEERING DOCTORATE PROJECT

The Engineering Doctorate project looked at the Dinorwig scheme on a catchment basis, including both upstream and downstream catchments.

An extensive database of existing hydrological data (rainfall, automatic weather station data, river flow and lake levels) was compiled using data supplied by First Hydro, the Environment Agency, the Environmental

Change Network (ECN) and the Met Office [through British Atmospheric Data Centre, (BADC)].

Data from the automatic weather station allowed open water and catchment evaporation to be estimated, an example of which is shown in Figure 3.

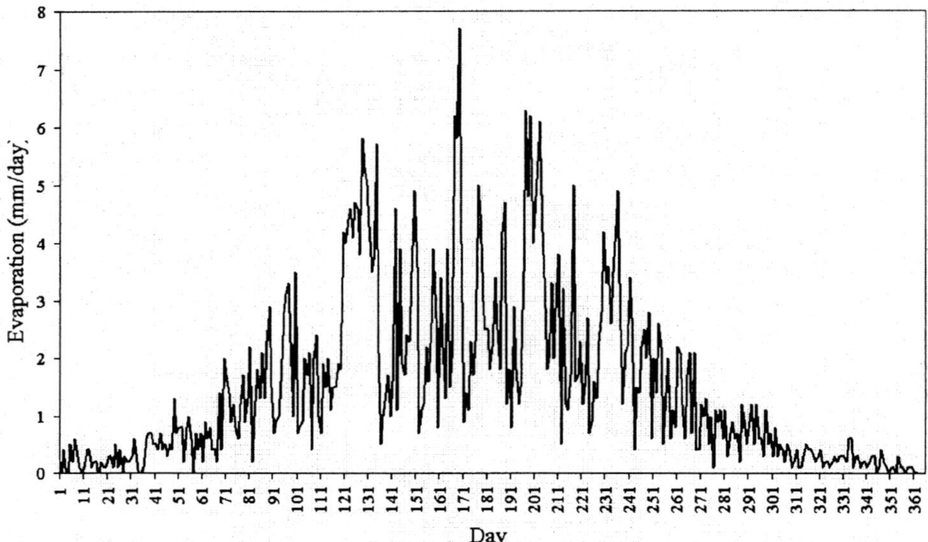

Figure 3. Estimated open water evaporation at Dinorwig, 2000

Data from the rain gauges was first used to construct a spatial rainfall model. This spatial rainfall model was then used as one of the inputs to a hydrological model of each of the catchments upstream and downstream of Dinorwig power station. The hydrological model chosen for this application was the ISIS Probability Distributed Moisture (PDM) model. (Wallingford Software, 2005b). Where flow data, with a natural flow regime, were available, each hydrological model was calibrated and validated against the gauged flow data. Where flow data were not available the hydrological model parameters were estimated from those used in the gauged catchments. Figure 4 shows an example of the flow predicted by the model and that recorded by the Environment Agency at the Nant Peris gauging station.

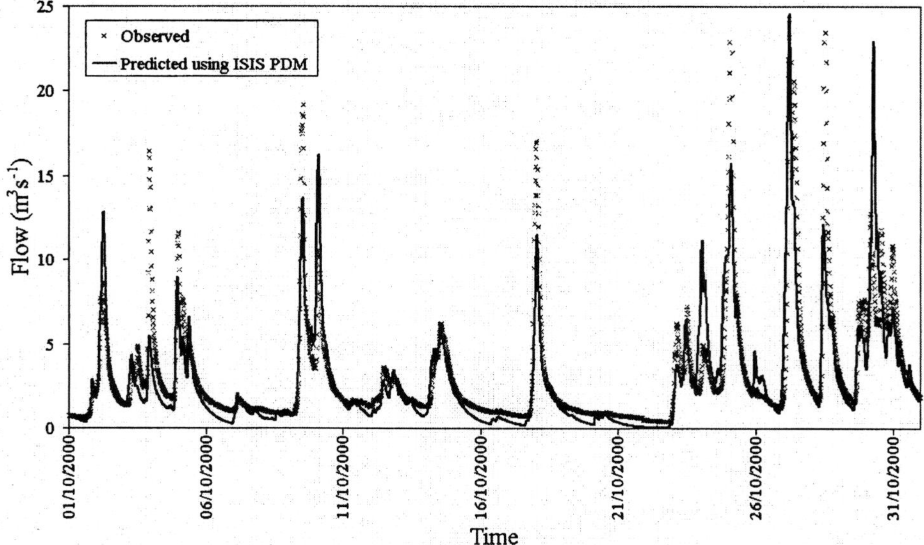

Figure 4. Nant Peris flow at Nant Peris gauging station, October 2000

The catchment flows predicted by each catchment model were then used as the inputs to a hydraulic model. The hydraulic modelling software chosen for this application was ISIS. (Wallingford Software, 2005b). The hydraulic model was designed to simulate both the physical system (including reservoirs, lakes, rivers, diversion structures and discharge structures) and the Power Station operation.

A wide range of data sources was used to construct the hydraulic model. These included Design Reports, As-Constructed Drawings and Topographical Surveys.

To operate the hydraulic model an initial Llyn Padarn level, an initial 'excess water' volume and a generating schedule for the entire simulation period are specified. The hydraulic model will then predict 'excess water' volume and Llyn Padarn level at each time step, based on the volume of water spilling into and released out of the closed reservoir system. The model will automatically open/close the Discharge culverts and lower/raise the Bascule gates according to the operational rules.

The hydraulic model was calibrated against historic 'excess water data' and Llyn Padarn level data. This calibration process highlighted that both the existing Llyn Padarn level–area and level-discharge relationships were inadequate. The Llyn Padarn level-area relationship was improved using NextMap Digital Terrain Model data. (Intermap, 2005) and the Llyn Padarn Level-Discharge relationship was improved by installing flow monitoring instrumentation at the discharge point. Through this work it was possible to improve the performance of the model.

Figures 5a and 5b show the results of a model simulation for October 2000. This was the wettest period on record. Also shown on Figures 5a and 5b is the observed data. Figures 6a and 6b shows the results of a model simulation for the July 2003. This was the driest period on record.

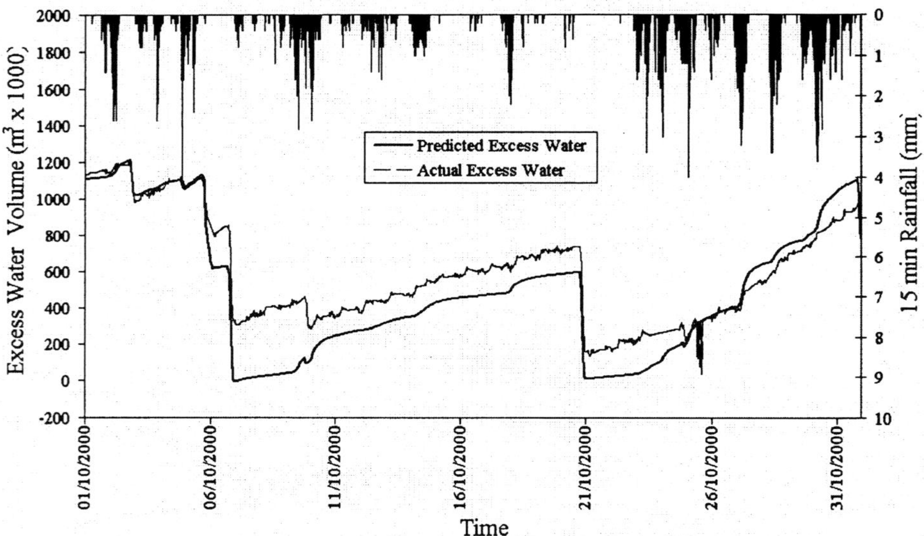

Figure 5a. Results for October 2000. 'Excess water' volume

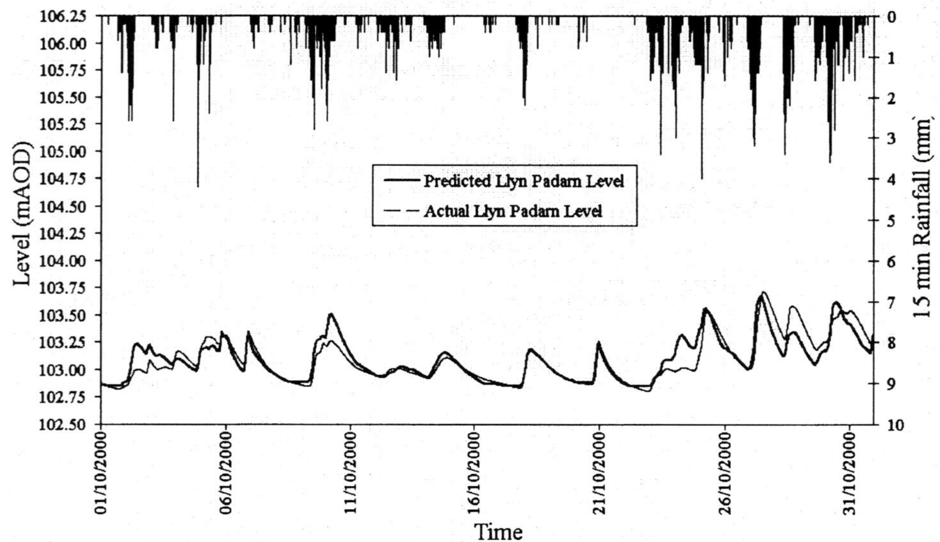

Figure 5b. Results for October 2000. 'Llyn Padarn' level

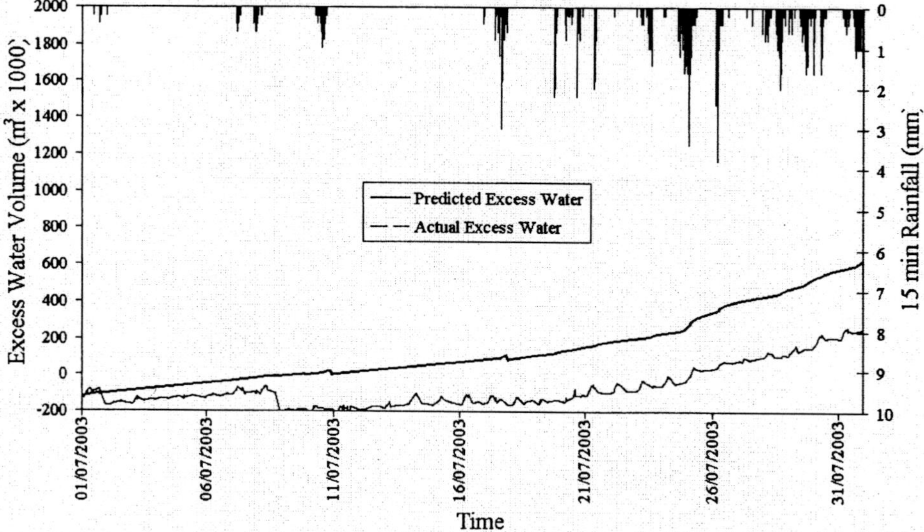

Figure 6a. Results for July 2003. 'Excess water' volume

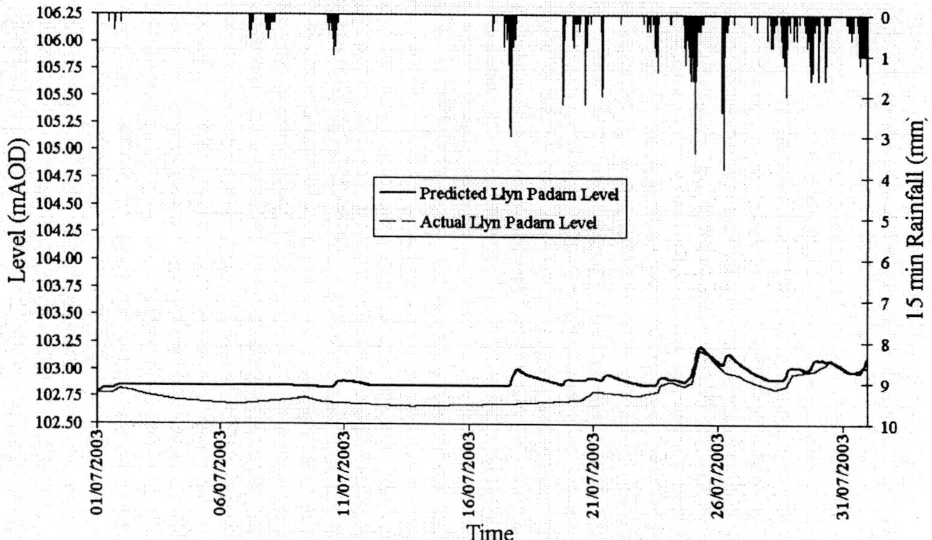

Figure 6b. Results for July 2003. 'Llyn Padarn' level

Figures 5a/5b and 6a/6b showed the model performed reasonably well, when tested during both the wettest and driest periods on record. Further model refinements are planned as additional information becomes available. In the simulations presented it is important to note that any error in predicted 'excess water' will accumulate over the simulation period and that any errors in Llyn Padarn level may also result in 'excess water' been released incorrectly.

It is also possible to use the model to quantify the different sources of 'excess water' in the power station system. Figure 7 shows that during October 2000 most of the 'excess water' came from the direct catchment of Llyn Peris, a smaller proportion came from the direct catchment of Marchlyn Mawr and an even smaller proportion came from spill at the diversion structures. This information can be used to inform investment decisions about future catchment diversion schemes.

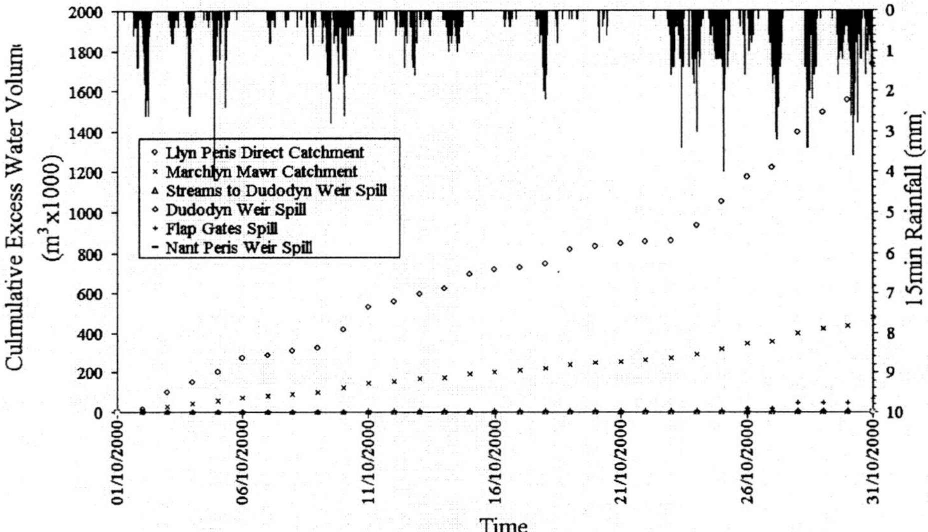

Figure 7. October 2000: Sources of 'Excess water'

Having shown the model to perform well against historic events, the model was then used to explore how the Dinorwig system performed under different operating rules.

Having developed and proved the model concept First Hydro now also have the option to implement the model as a real-time flood forecasting system. One possible software system is Floodworks (Wallingford Software, 2005a) which would use the existing ISIS PDM/ISIS model as the basis for a real-time flood forecasting system. Such as system has already been implemented at Dolgarrog Hydroelectric power station, also in North Wales. (Kalken & Billcliff, 2004).
Implementing a complete FloodWorks model was considered beyond the scope of this work therefore a simple 'excess water' forecasting tool has also been developed, in MS Excel/Visual Basic for applications (VBA) based on the principles of the ISIS PDM/ISIS model. The software is designed to forecast 'excess water' up to 120 hours ahead, based on an

anticipated generating schedule and current catchment conditions. The software allows Control room operators to trial different Culvert openings and Gate positions to ensure that the operational rules are not broken. First Hydro can also trial different generating schedules to see the impact on 'excess water'. Using MS Excel as the interface for the software makes it accessible to Control Room staff at the power station, without the need for detailed training in hydrology and hydraulics. It is also very easy to link this software to First Hydro's existing plant databases. This software is currently undergoing testing within First Hydro Company.

FUTURE WORK

It is now proposed that the existing model will be tested against future rainfall scenarios. It is proposed that a 'Weather generator' will be used to simulate the likely rainfall *patterns* that could be expected in 2020 and 2050, under Low and High emission scenarios. This will allow an assessment to be made of likely impact of climate change on the sustainable operation of the power station.

REFERENCES.

Engineering and Physical Sciences Research Council (2005) *Engineering Doctorate Homepage*. Available at: http://www.epsrc.ac.uk/PostgraduateTraining/EngineeringDoctorates/defaul t.htm. Accessed 2005.

First Hydro Company (2005) *First Hydro Company Home Page*. Available at: www.fhc.co.uk . Accessed 2005.

Intermap (2005) *NextMap*. Available at: http://www.intermap.com/corporate/nextMap.cfm. Accessed 2005.

Jog, M.G. (1989) *Hydro-electric and pumped storage plants*. New Delhi: Wiley.

Kalken, T.V. & Billcliff, A. (2004) *Application of the Flood Watch Flow Forecasting System for the Optimization of Hydro-Power Generation*. Available at: http://www.dhisoftware.com/uc2001/Abstracts_Proceedings/Papers01/123/1 23.htm. Accessed 2004.

Wallingford Software (2005a) *FloodWorks Homepage*. Available at: http://www.wallingfordsoftware.com/products/floodworks/. Accessed 2005.

Wallingford Software (2005b) *ISIS Homepage*. Available at: http://www.wallingfordsoftware.com/products/isis/. Accessed 2005.

Boscastle and North Cornwall Floods, August 2004: Implications for dam engineers.

Dr R. BETTESS, HR Wallingford
V. BAIN, HR Wallingford

SYNOPSIS. On the 16 August 2004 over 180 mm of rainfall in 5 hours caused severe flooding in the River Valency and adjacent catchments. Following the flood event, HR Wallingford, CEH Wallingford and the Met Office jointly carried out a study to understand the flood and to determine the peak discharge. The flood in Boscastle was one of the best recorded extreme flood events in the UK and there was good photographic and trash mark evidence from the flood.

The hydrological and hydraulic simulations of the flood showed that the application of standard Flood Estimation Handbook methods did not reproduce the observed flood characteristics very well. A better simulation was provided by assuming high values of the Percentage Runoff and halving the Time to Peak of the unit hydrograph. This modification to the Time to Peak is more extreme than that recommended by the Flood Studies Report for dam studies. This has implications for the methods that should be used when assessing the Probable Maximum Flood for dams on small, steep catchments. The implications for dam engineers are discussed.

INTRODUCTION

Boscastle entered the UK's flood annals, in dramatic fashion, on 16th August 2004. Prolonged heavy rainfall centred over Otterham, on the edge of Bodmin Moor near the North Cornwall coast, led to severe flooding in a number of river catchments. Those most affected were the River Valency and the Crackington Stream, but flooding and damage also occurred on the River Ottery and the River Neet. Mercifully, no one was killed; but the event scarred the landscape, caused damage to buildings and infrastructure.

To understand the event studies were undertaken on behalf of the Environment Agency by a consortium led by HR Wallingford, with a brief to report early findings within a few weeks and considered conclusions to report early findings within a few weeks and considered conclusions within a few months of the event. The work amounted to assembling forensic evidence, reconstructing the characteristics of the event and deducing its causes, on the basis of best available data and methods. The meteorology of the event was analysed by the Met Office; the hydrology by CEH Wallingford; the hydraulics and geomorphology by HR Wallingford. Halcrow and Royal Haskoning undertook post-flood surveys (in the Valency/Jordan and Crackington Stream catchments, respectively), and they and the Environment Agency assembled witness evidence from local interviews. The project team's final report (HR Wallingford, 2005) contains full details of the analyses undertaken and the conclusions reached. The work pointed to shortcomings in the current procedures used to predict extreme floods on small catchments and so has implications for dam engineers concerned with estimating Probable Maximum Floods on such catchments. This paper reflects on the implications of the findings of the studies for dam engineers. Whilst the floods affected both the Valency/Jordan and Crackington catchments, this paper concentrates on the Valency/Jordan catchment only.

THE CATCHMENT

Figure 1 shows the geographical location of the Valency river and its tributaries above Boscastle; the locations shown refer to points at which inflow hydrographs were derived. The catchment is located on the north coast of Cornwall. The catchment area above Boscastle is approximately 20 km^2. The catchment rises to approximately 300m AOD and the main branch of the River Valency is approximately 7 km long. Thus the slope of the river is steep. There are a number of tributaries which are also steep, and some of them are incised as they approach the main channel. The soils are generally thin over impermeable bedrock. The catchment is predominantly rural with much of the land given over to grassland. There are significant areas of woodland adjacent to the main river and its tributaries.

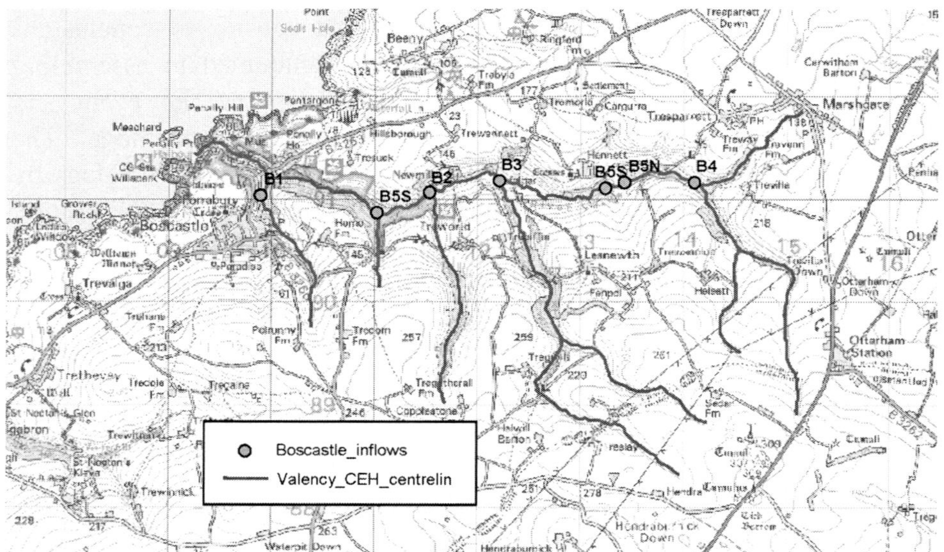

Figure 1 The Valency catchment above Boscastle, showing the river and main tributaries, and points at which inflow hydrographs for hydraulic modelling were calculated.

THE FLOOD EVENT OF 16TH AUGUST 2004, AS WITNESSED
The flooding of Boscastle on 16th August 2004 must be one of the best-recorded extreme flood events in the UK and there is a good photographic record of the event. The evidence indicates that the flood was out of bank for around 5 hours, rising to a peak (from the bankfull stage) in 1.5 hours. As the flood rose, some individuals reported very rapid, short-term rises in water level of 1 to 1.5 metres ("walls of water") in periods of a minute or less.

RECONSTRUCTION OF THE RAINFALL EVENT
The area around Boscastle experienced extreme rainfall accumulations resulting from prolonged intense rain over a four hour period from 13:00 to 17:00 BST (1200-1600GMT) on 16th August 2004. The exact track of the rainfall cells varied slightly during this period, but between the Camel Estuary and Bude the variation was sufficiently small to ensure that the heaviest rain fell into the same coast-facing catchments throughout the period. The intensity of the precipitation was probably enhanced by large-scale uplift associated with larger scale weather troughs.

The Tipping Bucket Rain gauge (TBR) at Lesnewth recorded maximum short period accumulations of 68mm in 1 hour, 123mm in 3 hours, and 152mm in 5 hours. Comparison with the quality controlled check gauge indicates that these should be increased by 20% to 82mm, 148mm & 183mm, respectively, to allow for under-reading by the TBR. The

Lesnewth TBR also recorded a peak rain rate of nearly 300mm/hr at about 16:35 BST (1535 GMT).

Observations of the spatial and temporal pattern of precipitation were well captured by the Cobbacombe and Predannack radars. Maximum values over $4km^2$ pixels differed from those observed by the TBRs due to sampling differences, but the overall pattern was consistent. The highly localised character of the event can be seen clearly in the sharp spatial gradients of the Cobbacombe Cross radar data, in Figure 2. Note the high values around the SW-NE track through Lesnewth and Otterham, and the sharp reductions in rainfall totals away from it.

The FORGEX method (NERC, 1999) was used to assess the probability of occurrence of the observed rainfall. The adjusted, observed maximum one-hour fall at the Lesnewth TBR of 82 mm has an annual probability of occurrence of around 0.13%. The three-hour total, again at Lesnewth, is comparable with the Camelford flood in 1957, and with several events in other parts of the country, most of which were accompanied by large hailstorms. The annual probability of occurrence is about 0.08%. The overall storm has an annual probability of occurrence less than 0.05%, which is larger, that is, less extreme, than that of the 1953 Lynmouth event and the 1955 Martinstown event. It is notable that all three events covered very small areas.

The South West peninsula has been subjected to six extreme rainfall events in the last century, of which three occurred in the decade 1951-60. The point ($1km^2$) probability deduced from an examination of these events indicates a similar annual probability to that derived using the FEH method. Allowing for the sparse observational network, the evidence indicates that an extreme rainfall event will occur somewhere in the South West region once every 20 years, on average.

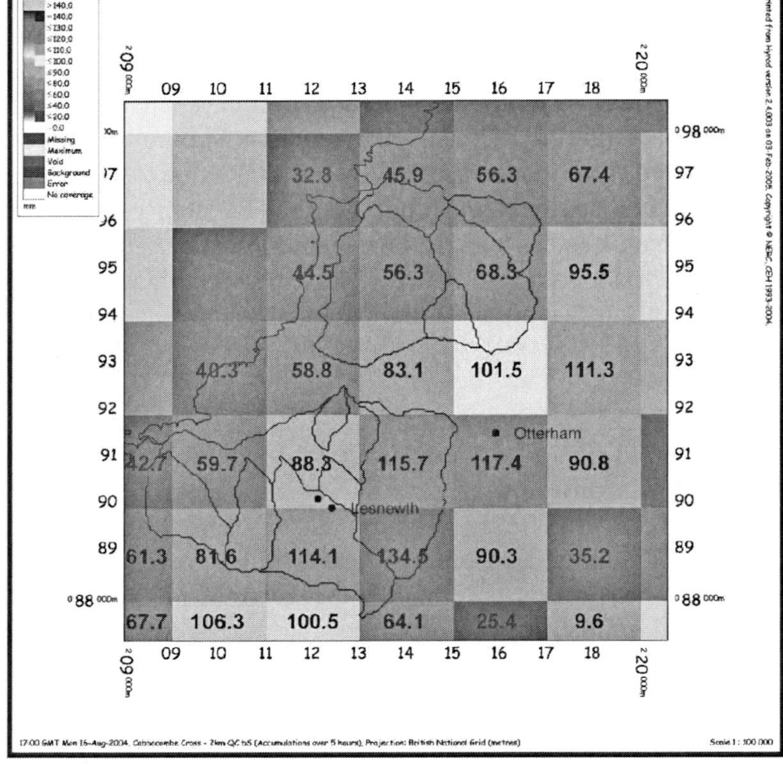

Figure2: 2 km gridded rainfall estimates based on data from the Cobbacombe Cross radar

HYDROLOGICAL & HYDRAULIC MODELLING OF THE FLOOD EVENT

<u>Modelling strategy</u>
Integrated hydrological and hydraulic modelling was undertaken to simulate, and thereby understand, the rainfall-runoff transformation and the development and passage of the resultant flood through the catchment. Using the available rainfall radar data as input, hydrological modelling was used to generate discharge hydrographs for selected sub-catchments of the Valency system. These flows were then routed down the catchment using a hydraulic model to generate discharge and stage hydrographs in Boscastle. Modelled stage hydrographs were compared with wrack mark and eye-witness accounts of flood levels, with the parameters of both the hydrological and hydraulic models being calibrated in reasonable fashion so as to achieve best-possible representation of the characteristics of the flood event by the hydraulic model.

<u>Hydrological modelling</u>
The currently accepted 'best UK methodology' for flood flow estimation is provided in the Flood Estimation Handbook (FEH; Institute of Hydrology 1999). Its focus and methods are geared towards more commonplace floods than those that affected the North Cornwall coast in August 2004, but it remains the only practical tool for modelling flood events on small ungauged catchments in the UK. The flooding at Boscastle from the Valency and Jordan catchments was very severe, and in consequence, difficult to reproduce reliably using FEH methods, as will be clear from what follows.

In the absence of flow records, the required parameters for rainfall-runoff estimation were determined using the standard FEH procedures for ungauged catchments. The required (spatially complete) rainfall data for the Valency catchment was derived from rainfall radar data, normalised to agree broadly with the adjusted rain gauge data. To obtain agreement with the best indications of water levels at given times and places (as obtained from eyewitness accounts and from wrack mark levels), it proved necessary to make a number of adjustments to parameters in the FEH method. To obtain reasonable agreement with best evidence flood levels, the time to peak had to be reduced by 50%, and the percentage runoff had to be adjusted, iteratively. The FEH constant percentage runoff (PR) was replaced by a time -varying PR related to antecedent and developing conditions. PR at the start of the event was calculated using the FEH methodology, but was then increased as the storm proceeded, according to the formula given below, to reflect the progressive wetting of soils and the expansion of the variable contributing area of the catchment:

$$PR_t = PR_{urb} * (1 + 0.8(\sum P_t / P_{TOTAL})$$

where PR_t is the percentage runoff at time t during the storm, PR_{urb} is the FEH design percentage runoff derived from soil and storm rainfall total, $\sum P_t$ is cumulative rainfall from the start of the storm to time t, and P_{TOTAL} is the rainfall total for the entire storm.

The factor of 0.8 was determined empirically, as that needed to generate the necessary gearing factor to increase PR_{urb} from the FEH initial condition to the 85 to 95% values that probably prevailed towards the end of the storm. The high percentage runoff towards the end of the event, coupled with the steep slopes of the catchment, undoubtedly led to high volumes of fast flow running off from increasing areas of the catchment. The destruction of field walls, the under-mining of roads and tracks, and the washing away of fords in the upper parts of the catchment testify to the occurrence of significant, fast-flowing torrents running overland. The departures required from the standard FEH methodology were, in the circumstances, deemed reasonable and understandable.

<u>Hydraulic modelling</u>
Floodwater flows and levels were simulated with an INFOWORKS-RS model of the Valency river system. The model was constructed using post-flood cross sections and structure survey data. No pre-flood data were available. The observed pattern of flow through the streets of Boscastle was represented as a multiple channel arrangement, with flows through and between the various channels being controlled by appropriate spill structure placements and parameters. The model was calibrated, on water level and timing, by reference to observed wrack marks, photographs, video and eye-witness accounts. As noted earlier, the available degrees of freedom in the hydrological and hydraulic phases of flood modelling were co-varied iteratively, to achieve the net best possible (and believable) end result. In the event, it proved necessary both to vary the standard FEH parameters to produce flows of sufficient magnitude, and to model significant blockage of the bridges in order to match water level predictions from the model to the observed profile of peak water levels.

Figure 3 shows the peak level calibration of the final model in the reach above its downstream boundary. The water level effects of the B3263 road bridge at chainage 350m and the smaller bridge at chainage 170m show clearly.

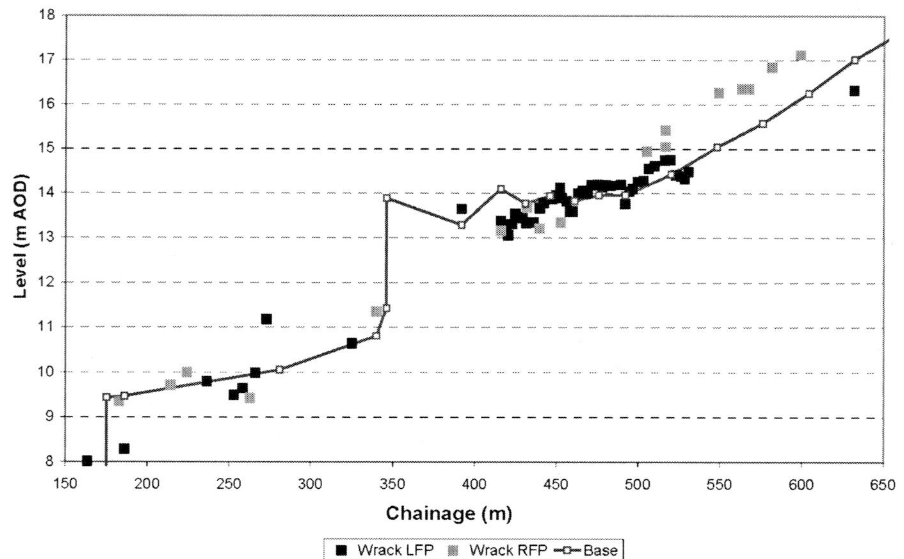

Figure 3 Predicted water levels and observed wrack mark levels in Boscastle. LFP denotes a Left-bank Flood Plain wrack mark. RFP denotes a Right-bank Flood Plain wrack mark.

The predicted peak discharge in the centre of Boscastle was approximately 180 m^3/s. This compares with FEH estimates of the Q_{med} (the median annual flow) at Boscastle of 4 m^3/s and of the 1% annual probability flow from FEH statistical modelling and rainfall-runoff modelling of, respectively, 10.4 m^3/s and of 34.8 m^3/s.

THE EXCEEDANCE PROBABILITY OF THE 2004 FLOOD

The history of flooding in Boscastle and elsewhere in the Valency catchment includes evidence of notable events as long ago as 1824. More recent floods occurred in 1950, 1958 and 1963. A best possible representation of the flood frequency curve of the Valency/Jordan at Boscastle, derived from a combination of FEH statistical and rainfall-runoff methods, supported by historical evidence and considerable judgement, is given in Figure 4. A GEV Type II probability distribution appears to fit the Boscastle data and estimates best. It is clear that the 2004 flood event was a very extreme event. Its estimated annual exceedance probability was 0.30%, the equivalent of a 1 in 350 years return period. The GEV Type II curve indicates that the return period of an event of that magnitude might be as extreme as 1 in 450 years – an annual probability of 0.22%. On the basis of the available data, and recognising the uncertainties involved, it seems reasonable to conclude that the annual probability of a Boscastle-scale flood is around 0.0025.

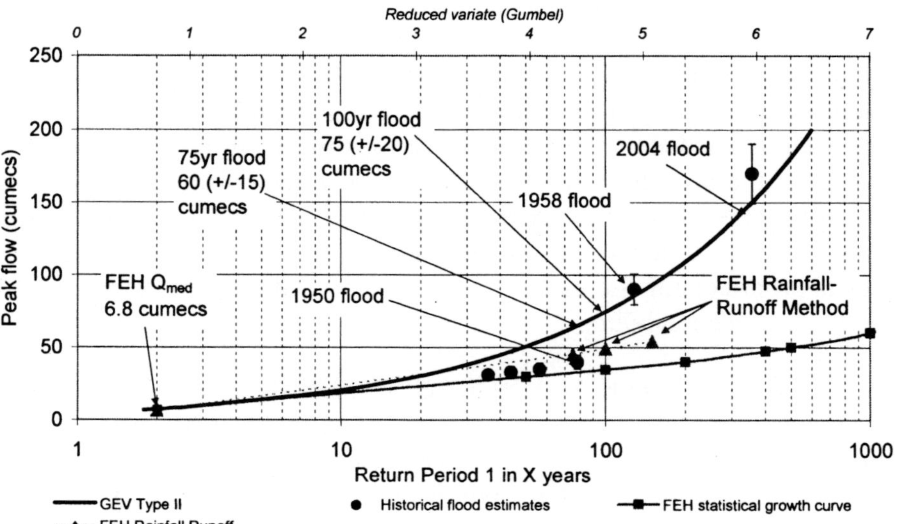

Figure 4 Estimated flood frequency curve for the Valency/Jordan catchment at Boscastle

LIMITATIONS AND UNCERTAINTIES OF DATA, MODELS AND ESTIMATES

The various models used and the various estimates produced with them are based on a set of assumptions and are subject to a range of uncertainties, in both the base data and in the representation of physical processes and conditions within the models.

It is clear from the rainfall radar data that the area of the rainfall event was limited, and that the spatial gradients of rainfall were large. The spatial resolution of the radar is only 2 km, which is coarse in comparison with the spatial gradients of the rainfall. In addition the catchment is near the limit of the area covered by the rainfall radar, and the data from the two rainfall radar stations do not always agree.

Hydrological and hydraulic modelling of such an extreme event is more uncertain still. The FEH method places proper reliance on available hydrological data, but there are little data available from similar catchments within the South West region, and for storms with the rainfall experienced in August 2004. Standard FEH methods had to be varied to simulate the extreme character of the rainfall-runoff processes experienced in the August 2004 event. Thereafter, the hydraulic model had to be constructed using post-event survey data, in the knowledge that wrack marks could not necessarily be relied on as peak water levels in a channel subject to such change as occurred during the flood. The division of flow down the various streets of Boscastle depends upon local features which are difficult to reproduce within a numerical model. The Froude number of the flows through the centre of Boscastle was relatively high, and this introduces numerical uncertainty into the hydrodynamic modelling. A further complication is added by the changes that took place during the event. The blockage of the bridges in Boscastle has already been discussed above. In addition walls and buildings were destroyed during the event. This means that a description of the topography of the floodplain at the start of the event is not appropriate for the end of the event. All these effects add to the uncertainty in the modelling.

EXPLANATION OF LOCAL HYDRAULIC PHENOMENA

Eyewitnesses testified to the occurrence of a number of transient, but significant, rises in water level at various places and times, during the flood. Many observers believe that the rapid increases in water level they observed during the flood event were caused by the rapid and progressive failure of blockage or trash dams in a downstream sequence. The most likely explanations for such local hydraulic phenomena, however, would seem to be blockage of a flow route, with subsequent diversion or failure. Scenario testing with the hydraulic model indicated that rapid blockage of the bridge led to rapid increases in water level upstream of the bridge, and a significant re-distribution of flow into the streets of Boscastle. Thus rapid blockage of the bridge could have led to sudden changes in flow route, and to the observed rapid rises in water level.

ESTIMATION OF THE PROBABLE MAXIMUM FLOOD (PMF)

The Flood Estimation Handbook (FEH) recommends that the estimation of the PMF is determined using the estimated Probable Maximum Precipitation (PMP). The FEH recommendation is then to use a unit hydrograph approach to determine the PMF from the PMP. In carrying out this approach the FEH recommends two modifications to the standard procedure. In determining the unit hydrograph, FEH recommends that the Time to Peak is reduced by one third, that is, the standard value is multiplied by 0.67. This has the impact of increasing the estimated peak discharge by a factor of 1.5 but does not affect the overall volume of water within the PMF. In addition when calculating the PMF, FEH also suggests that an allowance can be made for frozen ground by increasing the standard percentage runoff to 53%.

To get a satisfactory agreement between the modelling and the observations on the Valency a number of changes had to be made to the FEH approach. The Time to Peak was reduced to one half of the value derived from catchment predictors while the Percentage Runoff was increased so that it approached 90% towards the end of the event. If a similar reduction in the Time to Peak was used in estimating the PMF then the peak discharge would be increased by approximately 17% in comparison with the method described in the FEH.

The modelling of the Valency catchment suggested that towards the end of the event the instantaneous Percentage Runoff was high and higher than that which would be estimated for such a catchment using standard FEH methods. If such a Percentage Runoff were used to estimate the PMF then this would have the impact of increasing the overall volume of the flood. The increase in the flood volume depends upon the specific catchment characteristics but could approach 100%.

A study of extreme events in the UK has not shown any detectable variation in the location of extreme rainfall events. This implies that one must assume that a rainfall event of similar severity to the Boscastle event could occur anywhere in the country with the same probability. It has been estimated that the annual probability of a similar rainfall event somewhere in the country is approximately 30%. There is no evidence to suggest that the frequency of such extreme events is changing through time.

The FEH method is based on extensive data sets containing thousands of records and so there is a danger in drawing conclusions from just a single event. The implication of the modelling of the North Cornwall, however, would suggest that the present methods for estimating the PMF when applied to small steep catchments may underestimate the peak discharge and overall volume of the flood. Until such times as these issues are resolved, it would appear to be prudent for dam engineers to be aware that using the presently recommended FEH methods to determine the PMF from the PMP may underestimate both the peak discharge and flood volume.

From the above the issue arises as to how small and how steep a catchment has to be before there is a significant impact on the predicted PMF. The Valency catchment has an area of 20 km^2. Within the FEH database approximately 15% of catchments have an area smaller than 40 km^2. It would seem prudent that special attention is given to any catchment with an area less than 60 km^2. Within the FEH the average slope of the catchment is represented by the parameter DPSBAR which is the average slope of the channel network. The value of DPSBAR for the Valency catchment is 115m/km. This value is exceeded by more than 20% of the catchments within the FEH database. It is tentatively suggested that the values of Time to Peak and Percentage Runoff determined from FEH are modified for any catchment with both an area less than 60 km^2 and a DPSBAR value greater than 110m/km. This is based purely on the experience from the Boscastle event and needs to be confirmed by the collection of additional data of extreme events in small steep catchments.

CONCLUSIONS
Evidence and analyses indicate that the Boscastle flood of 16th August 2004 was unusual in origin, highly localised in extent and extremely rare in occurrence.

The rainfall event of the 16 August 2004 was brought about by an extremely unusual combination of circumstances, none of which are rare but their combination is extremely so. Using FORGEX, it has been estimated that the annual probability of occurrence of the rainfall event is less than 0.05%. The South West peninsula has been subjected to six extreme rainfall events in the last century, of which three occurred in the decade 1951 to 1960. Allowing for the sparse observational network, the evidence indicates that an extreme rainfall event will occur somewhere in the South West region once every 20 years, on average. There is at present no clear-cut evidence to suggest that long-term climate change may be affecting the probability of such extreme events.

The flood event on the Valency was modelled using a combination of hydrological and hydraulic models. A number of changes had to be made to the standard FEH methodology in order to simulate the hydrological processes. The Time to Peak had to be reduced to 50% of that predicted by using standard catchment descriptors. In addition a variable Percentage Runoff regime had to be used, which increased from an initial value determined through the standard soil and storm rainfall total calculated, but which then grew to 95% as the catchment wetted up and its effective contributing area expanded.

Modelling suggested that the rapid blockage of the main bridge in the centre of Boscastle would have led to large and rapid increases in water level upstream as a result of changes to flow paths, which is a likely explanation of the reported rapid increases in water levels that occurred during the rise of the flood. The best estimate of the peak flow of the flood in Boscastle is $180 \text{ m}^3/\text{s}$.

The implication of the modelling of the North Cornwall would suggest that the present methods for estimating the PMF when applied to small steep catchments may underestimate the peak discharge and overall volume of the flood. Until such times as these issues are resolved, it would appear to be prudent for dam engineers to be aware that using the presently recommended FEH methods to determine the PMF from the PMP may underestimate both the peak discharge and flood volume. It is tentatively suggested that the values of Time to Peak and Percentage Runoff are varied from the FEH predicted values for catchments both with areas less then 60 km^2 and DPSBAR values greater than 110 m/km

ACKNOWLEDGEMENTS

The contents of this paper are based on a study carried for the Environment Agency by a team led by HR Wallingford. Other members of the team were the Met Office and CEH Wallingford. Halcrow undertook the post-flood surveys in the Valency and Jordan valleys, and Royal Haskoning undertook those in the Crackington Stream catchment. The contributions of all parties and individuals involved, including those received from the people of Boscastle and Crackington Haven, are gratefully acknowledged.

REFERENCES

HR Wallingford (2005) Flooding in Boscastle and North Cornwall, August 2004. Phase 2 Studies Report. HR Wallingford Report.

Institute of Hydrology (1999) The Flood Estimation Handbook.

Reconstruction of the Znojmo Dam – practical application of hydraulic research

VLASTIMIL STARA, Brno University of Technology, Brno, Czech Rep.
MIROSLAV ŠPANO, Brno University of Technology, Brno, Czech Rep.
JAN ŠULC, Brno University of Technology, Brno, Czech Rep.

SYNOPSIS.   Many dams and waterworks had been built in the Czech Republic during 1950s and 1960s. One of them is the Znojmo Dam. It is an earth dam with a combined spillway block. A nominal discharge of $Q = 355 \text{ m}^3.\text{s}^{-1}$ was considered for designing the safety spillway. Extreme flood events which occurred in the Czech Republic in 1997 and 2002 evoked questions relating to the safety of dams. The nominal discharge of the Znojmo Dam safety spillway was enhanced up to $Q = 610 \text{ m}^3.\text{s}^{-1}$. Therefore, the decision on reconstruction of the Znojmo dam spillway block was made. The following modifications were proposed: sinking the existing spillway crest, replacement of the existing flap gates with tainter gates with flaps, installation of breakwater on the dam crest, and necessary modifications of the stilling basin. With regard to the importance of the dam and the volume of the reconstruction planned, assessment on a hydraulic model was necessary. This paper summarises the results of the hydraulic research and its practical application to the processing reconstruction of the Znojmo Dam. It also describes some differences between the modelled situations and reality.

INTRODUCTION

The Znojmo dam is located on the river Dyje close to the town of Znojmo in southern Moravia. The supervisor of the dam is Povodí Moravy, s. p. (hereinafter referred to as the Client). The Znojmo dam had been built between 1962 and 1966. This rock-filled dam with a watertight core is combined with a spillway block which comprises two sections of safety spillway gated with flap gates, two bottom outlets, two small hydropower stations, and water intakes.

The capacity of the safety spillway was exceeded during the 2002 flood, see Figure 1. A flood peak of $Q = 379 \text{ m}^3.\text{s}^{-1}$ was estimated (Kadeřábková, Krejčí, 2004). A nominal discharge of $Q = 355 \text{ m}^3.\text{s}^{-1}$ was estimated during

Improvements in reservoir construction, operation and maintenance, Thomas Telford, London, 2006, 106–115

the hydraulic research performed in 1960 (Rybnikář, et al, 1960). Therefore, a study Glac, et al 2003) with variants of technical measures for safe management of extreme water passage was made. The study investigated a design flow rate of $Q = 610 \ m^3.s^{-1}$ and a 10 000-year control flood with a peak of $Q = 732 \ m^3.s^{-1}$. The variant called 2b was experimentally assessed on an hydraulic model. It consisted of the following proposed modifications: replacement of the existing flap gates with tainter gates with flaps and sinking the existing spillway crest. These proposals are combined with enhancement of the watertight core using a concrete breakwater. The above-mentioned modifications should ensure enhancement of the dam safety against overtopping as requested by current standards (TNV 75 2935).

Figure 1. Upstream and downstream air view of the Znojmo Dam during the August 2002 flood

PROPOSED MODIFICATIONS OF THE ZNOJMO DAM

The functional block of the Znojmo dam contains a safety spillway, bottom outlets, and water inlets. The existing safety spillway has two 8.70 m wide sections gated with 3.0 m high flap gates. The sections are divided by a 1.80 m wide central pier. To increase the capacity of the Znojmo dam the designer (VODNÍ DÍLA-TBD a.s.) proposed a modification of the existing spillway surface. Its crest was sunk from 223.00 m a.s.l. to 221.80 m a.s.l. A replacement of the existing flap gates with tainter gates with flaps is also included. There are two bottom outlets under the left section and two small hydropower stations (HPS) under the right one. The inlets of bottom outlets and HPS are divided by 3.30 m wide separating piers, see Figure 2. The stilling basin is 20.8 m wide, 33.0 m long, and 3.5 m deep. The necessary modifications of the stilling basin were designed, including increasing the height of the basin walls. A debris-retaining system in front of the spillway was also designed. With regard to the importance of the dam and the volume of the reconstruction planned, assessment on a hydraulic model was necessary. The Client required:

- verification of the modified spillway capacity and a discharge coefficient evaluation,
- assessment of the spillway surface mainly with respect to the negative pressure occurring during extreme flow rates,
- proposal and design of modifications of the stilling basin at a design flow rate of $Q = 610 \text{ m}^3.\text{s}^{-1}$,
- specification of the designed modifications if necessary.

PHYSICAL MODEL AND MODEL TESTS

For the purpose of hydraulic research of the spillway block of the Znojmo Dam a physical model was built in a length scale $M_L = 35$. The Froude similarity was used for modelling due to the dominating gravity forces. The length scale was used with regard to the spatial and capacity limits of the laboratory.

The dam model was installed in a glass-walled channel which is part of the equipment of the new fully automated Laboratory of Hydraulic Research of the Faculty of Civil Engineering, Brno University of Technology. The channel forms a part of the hydraulic circuit maintaining a stabilised rate of flow of up to $Q \sim 150 \text{ l.s}^{-1}$. The model consisted of a spillway block with parts of the dam, a stilling basin, and a 157 m length (in real scale) of the Dyje river bed. There are 31 pressure holes in the spillway surface. The upstream face of the dam was covered with gravel with a particle diameter of 10 to 15 mm. The bottom of the channel downstream of the basin step was covered with the same gravel as corresponds to the real riprap material. A view of the model is shown in Figure 2.

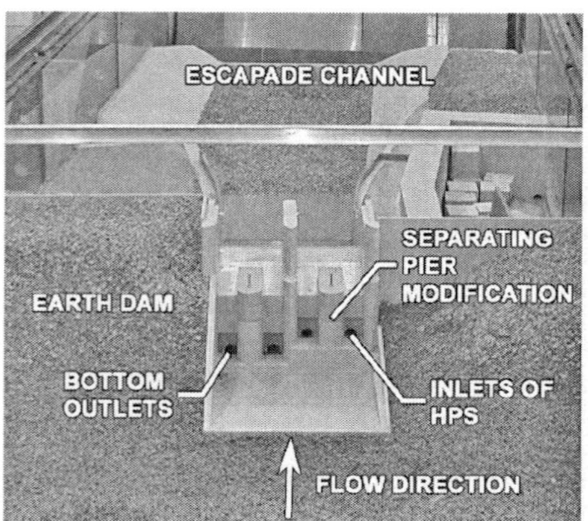

Figure 2. Downstream and upstream view of a model of the Znojmo dam spillway block

RESULTS OF THE HYDRAULIC RESEARCH AND RECONSTRUCTION OF THE ZNOJMO DAM

Reconstruction of the Znojmo dam spillway block has been carried out by the AQUAS vodní díla s.r.o. company (hereinafter referred to as the Constructor). All further values are valid are valid for the real situations.

Tests of modified spillway capacity

Model tests proved that the modified spillway enables safe transfer of a flow rate of $610 \text{ m}^3.\text{s}^{-1}$. However, the capacity is affected by the modification of the separating piers placed in front of the spillway, see Figure 2. The results of the research led the designer to the decision to sink the pier top elevation from 223.00 m a.s.l. to 221.60 m a.s.l. The influence of the tested pier shape modifications was negligible (Stara, et al, 2005)). It was recommended to sink the piers without any shape modifications, see Figure 3.

The edges of the separating piers were lightly damaged during the sinking of their top face. So, the edges will be chamfered after alignment and effacement. No significant difference between modelled and real situations is expectable because the shape modifications had negligible effect on the discharge coefficient of the new spillway surface.

Figure 3. Recommended and applied modification of separating piers

The values of the discharge coefficient of the modified spillway surface are summarised in Table 1.

Table 1. Values of the discharge coefficient (m) as a function of the discharge (Q), valid for final modification

Q [m³.s⁻¹]	51	143	282	606	725
m	0.392	0.372	0.374	0.398	0.401

No values of pressure sufficiently negative to cause any cavitation effect on the spillway surface can occur even if the extreme flow rate is reached (discharge rate $732 \ m^3.s^{-1}$, water level in the reservoir corresponding with the breakwater crest elevation).

A sharp edge between the horizontal and curved parts of the spillway surface appeared at the end of the central pier after dismantling of the flap gates and demolition of the sill where the gates were installed, see Figure 4. This sharp edge was modelled as rounded according to the given plans. The edge could negatively affect the pressure conditions on the spillway surface compared with the modelled situation.

Figure 4. Modelled and realised shape of the spillway surface at the end of central pier

The shape variety of the side pier wall was simplified in the model, see Figure 5. This simplification should not cause significant distortion of the results measured and applied.

Figure 5. Simplification of shape indentation of diffusion at the installation site of the flap gate

<u>Regulation of flow conditions</u>
A model test proved that the contractions at the side piers have no effect on the spillway capacity because the side piers are ahead of the spillway surface. Wakes cause a water level increase of approximately 1.0 to 1.1 m at side piers when the discharge is $Q = 610 \text{ m}^3\text{.s}^{-1}$. An improvement of wake effects, i.e. decreasing the wave crest, was necessary due to the maximum gate opening during floods. Therefore, several tests of flow improvement and wake elimination were performed by using wing-like steel elements attached to the front part of the pier walls. The minimisation of plate dimensions was achieved by taking step-by-step measurements while using different plate shapes and sizes [Stara and Šulc, 2004), see Figure 6.

The stream-leading plates were realised as a concrete body set on the original piers. The grooves of the cofferdam had to be included, see Figure 6. Keeping the shape of the leading face was necessary to ensure good agreement between modelled and real situations.

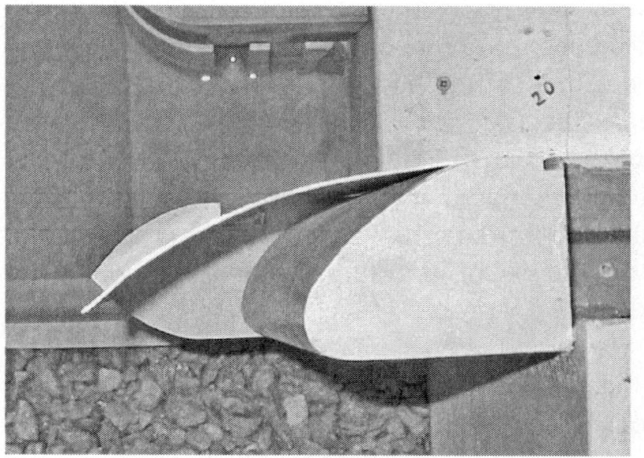

Figure 6. The stream-leading plate set on top of the lower part of the side pier and its realisation by a concrete body with a groove of the cofferdam

The circular trailing end of the central pier was replaced with the rear part of the NACA profile (Ladson, et al., 1996) with the aim to eliminate the wave formed downstream of the pier. The hydraulic efficiency of the NACA profile is shown in Figure 7. This modification will not be implemented.

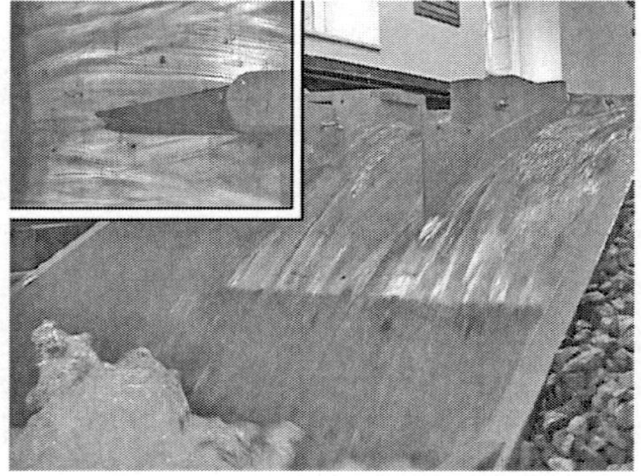

Figure 7. Deformation of the level arising behind the end of the middle pier and its elimination by the NACA profile when $Q = 610 \text{ m}^3.\text{s}^{-1}$

<u>The stilling basin modifications</u>
The first model experiments have shown that the existent stilling basin of the Znojmo Dam (length L = 33.0 m, depth h = 3.5 m) can no longer comply with the requirement for sufficient energy dissipation at a design flow rate of $Q_N = 610 \text{ m}^3.\text{s}^{-1}$. It was necessary to propose its modification. The Client wanted to keep the elevation of the stilling basin bottom at 207.65 m a.s.l., because the bottom is formed by natural rock. Baffle blocks at the bottom of the basin and the chute blocks at the end of the spillway surface with a combination of basin length were proposed and tested basing on preliminary computations (Stara and Šulc, 2004). The criterion for finding the optimum stilling basin modification was the extent of deformations of the bottom of the channel downstream of the sill.

The use of chute blocks had a negligible effect on the total energy dissipation in the basin. The use of chute blocks considerably increased the aeration of the stream entering the space above the basin and spread a more uniform load on its bottom. The use of the chute blocks according to variant 3 was proved as more suitable (Figure 8). A 48.5 m long basin combined with five baffle blocks ensures low deformations of the channel bottom. Baffle blocks may be laid out in one or two rows depending on the chute blocks used, see Figure 8.

The variant with no chute blocks at the end of the spillway surface combined with installation of five baffle blocks laid out in one row at the basin bottom was chosen for realisation, see Figures 8, 9, and 10. A basin length of 48.5 m was also recommended. Due to the spray formed by water impinging on the baffle blocks in the basin, a full concrete barrier about 1.2 m high was added on top of the basin walls. A heavy riprap 15 - 20 m long with concrete filling in an approximately 5 - 7 m long section was recommended as channel bottom protection.

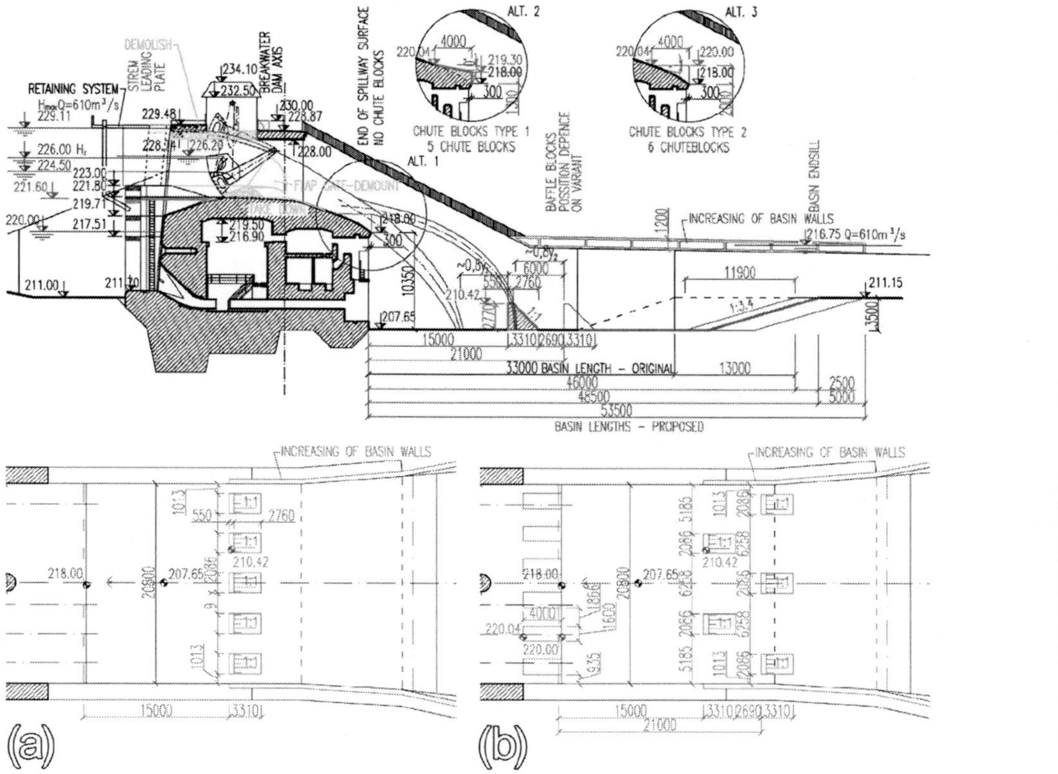

Figure 8. Vertical plane section of operational structure and the possible modifications and ground plan of the optimal layout of baffle blocks: (a) with chute blocks used; (b) with no chute blocks used

During the basin elongation we found a shallow foundation of the basin walls in comparison with the basin bottom elevation. Therefore, additional sinking of the footing bottom with concrete filling and anchoring into the bearing subsoil were performed. Two variants of stabilizing system were proposed, adding conical concrete elements or adding anchored walls, see Figure 9. The second variant was finally realised. The efficiency of the final basin modification was also assessed on a hydraulic model. The results of measurements have shown that the deepest scours occur near the left bank as was expected. The channel bed scours were similar in all cases, of an approximate depth of 1.2 m located some 5 - 7 m far from the basin sill. Thus it is possible to surmise that the final modification of the stilling basin has negligible effect on its efficiency, see Table 2.

Table 2. Comparison of the volume of bottom deformations for particular modifications of basin walls

Variant	Max. scour depth	Position	Max. deposition height	Position
	m	m	m	m
(1) original proposal	1.13	6.0	0.27	10.5
(2) conical elements	1.09	23.0	0.12	33.0
(3) anchored walls	0.93	5.2	0.23	9.0
Note: position means distance from the basin sill				

Figure 9. Upstream view of the basin bottom with five baffle blocks laid in one row and variants of modifications of basin walls: (1) original proposal, (2) conical elements, (3) anchored walls

The reduction of stilling basin diffusiveness by left anchored wall (see Figure 9 (3)) positively affected flow in the basin which resulted in symmetrical channel bed scour behind the basin sill and smaller deformations near the left bank.

Figure 10. Upstream and downstream view of stilling basin of the Znojmo dam; a part of the basin is under the cofferdam and the preparation of fabrication of baffle blocks is in progress

CONCLUSION

The construction and reconstruction of dams represent complex problems which demand technical, technological as well as financial expertise of the designer, the constructor, and the investor. The present paper is focused on the proceeding reconstruction of the Znojmo dam from the point of view of hydraulic research. The differences between modelled and real situations are discussed as well as their possible effect on the accuracy of experimental results. Absolute accuracy between the modelled and the real situation cannot be reached anyway, but it is necessary to ensure good correspondence between both the model and the real structure. Hydraulic research is still a very important, efficient, and economical tool for the proposal and assessment of an appropriate design of both new and reconstructed hydraulic structures.

REFERENCES

Glac, F., Janků, O., Bubeník, M. (2003). VH soustava Vranov – Znojmo, převedení N-letých průtoků, studie. VODNÍ DÍLA - TBD a. s., Brno 2003

Kadeřábková, J., Krejčí, V. (2004). Floods on the Dyje river and reconstruction of the dam Znojmo. *Water Management 12/2004, pp. 380-386*

Ladson, Charles L., Brooks, Cuyler W. Jr., Hill, Acquilla S. (1996). Computer Program to Obtain Ordinates for NACA Airfoils. National Aeronautics and Space Administration, Langley Research Center - Hampton, Virginia, 1996

Rybnikář, J. et al. (1960). Vyrovnávací nádrž na Dyji u Znojma, Výzkumný úkol č. 4200, Závěrečná zpráva, VUT v Brně, Brno, prosinec 1960

Stara, V., Šulc, J. (2004). Hydrotechnický modelový výzkum VD Znojmo. Závěrečná zpráva VUT v Brně, FAST ÚVST-LVV, Brno, 2004

Stara, V., Šulc, J., Špano, M. (2005). Hydraulic research of the spillway block of the Znojmo dam. *Water Management 4/2005, pp. 104-107*

TNV 75 2935 Assessment of dams during floods, Hydroprojekt, a.s., Praha, 2003

3. Internal erosion

Identifying Leakage Paths in Earthen Embankments

VAL KOFOED, Willowstick Technologies, Utah, USA
JERRY MONTGOMERY, Willowstick Technologies, Utah, USA
KEITH GARDINER, United Utilities, Warrington, UK

SYNOPSIS.   The leakage of water from a dam suggests a number of possible scenarios, and none of them are good. At best, the reservoir owner is losing the power-generating potential of the escaping water. At worst, the leakage may be the precursor of dam failure.   Among the technological advances that will enhance the safety and efficiency of hydropower in the future, is a remarkable breakthrough in the area of seepage diagnosis and remediation.

Researchers, working under the auspices of Willowstick Technologies, have developed a system that has been proven to reduce significantly both the time and the expense associated with seepage diagnosis. It thus represents a major step forward in dam safety.

In the new procedure, electrodes are placed strategically upstream and downstream of the dam structure, and the water between them is charged with a low voltage, low amperage, and audio-frequency electrical current. The current creates a distinctive magnetic field that represents the location and character of the water flow occurring between the electrodes. This field can be identified and surveyed from the surface using a specially tuned magnetic receiver.   Through this technique and hardware, investigative teams have accurately diagnosed seepage problems in locations throughout the United States and Canada and in the United Kingdom.

INTRODUCTION

All dams leak to some extent, especially earth embankment dams. Dam owners usually know where the leak is emerging from the embankment and can easily monitor the flow over time to ensure that the leak is not getting larger or carrying fine material. However, the only way to cure a leak successfully is to find the point where the leakage path crosses the "impermeable" barrier and plug it. This is notoriously difficult to do unless there is evidence on the surface such as a depression in the pitching or a vortex in the water; even then the fault in the barrier itself could be some distance away.

What is needed is a technique that will locate leakage paths through, under or around a dam at any depth, at any time of year and through all types of fill and foundation material. Ideally the technique should provide an accurate 3D map of the leak and be able to provide a reasonably degree of certainty about the origin of the leakage water, e.g. is it coming from the reservoir, a spring under the dam or from the natural hillside?

A United Utilities Engineer encountered the AquaTrack™ technique while researching water flows into water abstraction boreholes on the internet. The website also gave an account of leakage tracking through an embankment dam in the United States using the AquaTrack™ technique which appeared to cover all the requirements listed in the preceding paragraph. Contact was made with Willowstick Technologies of Draper, Utah and in November 2005, a contract was signed to trial the technique on several dams that had long standing leaks that had proved impossible to trace and where speculative grouting appeared to be the only answer.

The trial revealed some surprising and unexpected results and will enable decisions to be made about the nature and location of the remedial works with far greater certainty. The leakage maps produced will also help provide answers to two important questions: Is any remedial action required at all?, i.e. is continuing close observation and monitoring sufficient?:and What will be the effect of the proposed remedial action?, i.e. could it make matters worse?

This AquaTrack technology is also known by its technical name of Controlled Source Frequency Domain Magnetics which shall be referred to throughout the remainder of this paper as (CS-FDM).

The Willowstick crews mobilized on two different occasions throughout the winter of 2005-2006. A total of five different test sites were used to determine seepage paths. For the sake of space, only two of these test projects will be described in this paper following a brief introduction of the technology.

INTRODUCTION TO CS-FDM

Although the science behind the CS-FDM technology is rather complex, at its root it relies on a set of basic physical principles. The most important of these principles is known as "transformer theory." Transformer theory holds that two coils, set in close proximity, can be electromagnetically coupled. When the first coil is electrically charged, it emits a magnetic field, which then induces electricity in the second coil. Countless types of electrical transformers utilize this rudimentary principle.

In essence, the CS-FDM apparatus and its associated procedure use aqueous systems to form a virtual electromagnetic transformer. The initial stage of the procedure entails the strategic placement of electrodes into the water above and below the dam structure. Connected on one side by wire and on the other by water, the electrodes form the primary coil of the hypothetical transformer.   The electrodes are then charged with a low voltage, low amperage audio frequency electrical current. (This AC current is harmless to aquatic life). Per the transformer theory, the charged coil conveys a particular magnetic field to the second coil, which is formed by the CS-FDM receiver device. As the current gathers into the channel that seeps through the dam, it emits a magnetic field characteristic of that channel (Biot-Savart law). Thus, when that field is conveyed to the second coil, the receiver can analyze its unique attributes to infer the shape, location, and path of the seepage flow.

To understand further how the CS-FDM apparatus works, it is helpful to consider the strength vectors of the magnetic field produced by an electrical current. The horizontal and vertical vectors reach zero at the center of the current and approach their maximum as they move outward. Therefore, the rates of change of the magnetic field strength in both vertical and horizontal directions can be used to determine the location, width and depth of the conductor in question. Furthermore, the vector known as "the horizontal minimum" can be used to identify the conductor's orientation. With its ability to read and analyze these three components of the emitted magnetic field, the CS-FDM apparatus can offer a complete picture of subsurface channels.

INSTRUMENTATION

The apparatus used to measure the magnetic field induced by the electrical current includes three magnetic sensors oriented in orthogonal directions (x, y, and z) and a Campbell Scientific CR1000 data logger which collects, filters and processes the sensor data. A Global Positioning System (GPS) instrument spatially defines the field measurements, while a Windows-based, Allegro CE handheld computer stores and couples the GPS data with the magnetic field data. All of this equipment is mounted on a surveyor's pole and hand carried to each measuring station.

During the investigation, more than 5,000 readings are taken every 4 seconds at frequencies from 30 Hz to 720 Hz. For quality control, a base station is established within the survey area, and base measurements are taken at the beginning, midpoint and end of each field day. The base data are used to identify any changes in the background magnetic field and/or diurnal drift. The magnetic field measurements collected during the survey are then normalized to compensate for these factors.

To ensure data quality at each measurement station, the Campbell Scientific CR1000 calculates the 380 Hz magnetic field strength (after Fast Fourier Transform, statistical analysis, and stacking of sixteen separate readings) and compares the signal to the background or ambient magnetic field strength at numerous frequencies. These data are compared to pre-determined signal quality criteria and signal-to-noise ratio criteria to establish data legitimacy and repeatability.

QUALITY CONTROL

The CS-FDM procedure calls for the processing and correction of the field data to account for distance from the source electrode, to reduce the impact of antenna interference, and to remove the effects of ambient and shallow subsurface sources of electricity. The processed and corrected data are then used to generate contour maps of the induced magnetic field. Relative changes in the magnitude and/or gradient of the horizontal and vertical fields—rather than the absolute magnitude of the induced field—are used in making interpretations.

The magnetic field observed at the surface, due to subsurface electrical current flow in water, is dominated by a horizontal component; therefore, interpretations of subsurface saturation are based primarily on the horizontal magnetic field readings. Vertical magnetic field gradients can supplement the channel characterization by helping to identify structural edges that influence the hydrology. However, the vertical data also reflect near-surface features more strongly than the horizontal component (including the influence from the antenna and electrodes), so in most cases the vertical data is less constructive in the final interpretation.

Obviously, it is preferred that manmade interferences are known prior to the investigation. If unknown, however, these interferences can often be recognized by their specific signature signals in the data, especially by analyzing the vertical field data in conjunction with the horizontal data. Once recognized, these features can be accounted for, corrected, and/or removed from the final reduced data set. Some of these possible interferences include the following:

- Ground noise from 50 Hz signal (from nearby electrical generating equipment, overhead or buried power lines, any subsurface cathodic protection of pipes, etc).

- Cultural features (buried pipes, steel cased wells, etc).

- Atmospheric noise (diurnal magnetic variations, electrical storms, solar activity and related magnetosphere activity, etc).

These various features are specific to an individual site and can vary significantly.

TEST PROJECTS

The following information highlights the application of the CS-FDM technology on two dams owned by UU referred to as Dam A and Dam B.

The objective or purpose in performing the CS-FDM geophysical investigation was to characterize and delineate areas of greatest groundwater concentrations through, beneath and/or around the earthen structure. By identifying the preferential flow paths between the reservoir and downstream seeps or tailwater, the CS-FDM technology can successfully answer questions about where water, leaking from the reservoir, originates and how it moves through the earthen structure. Understanding the location and extent of seepage moving through an earthen dam is of significant value in engineering a remedy.

<u>Mapping Seepage at Dam A</u>

Dam A was experiencing leakage into the outlet control tunnel, which curves beneath the earthen structure. The tunnel and associated piping and valves control the reservoir levels and flow from the storage reservoir. This leakage was first recorded on a drawing dated 1883 at the same locations as found today. In his Statutory Inspection report of Sept 2003, the Inspecting Engineer recommended that the leaks be grouted up. Before doing so, United Utilities wished to find the leakage path and determine the extent of any possible defects in the embankment which may have been caused over the preceding century.

Also of interest to the client was the origin of a seep flowing from a small diameter pipe located in the downstream groin (mitre) of the left abutment.

Three horizontal dipole antenna/electrode configurations were employed to energize the earthen dam structure for the purpose of conducting the AquaTrack geophysical investigation in an effort to delineate the origin of the seeps and their flow paths through the dam. Two configurations were used to investigate saturation levels and flow paths through the overall earthen embankment. This was done by placing injection electrodes in the reservoir and then placing a return electrode in the tailwater located below the dam. A third configuration targeted the seepage in the left groin by placing an electrode in the reservoir and a return electrode in the very pipe which carried the seepage waters. In all cases, the antenna connecting the injection electrodes with the return electrodes was placed in the shape of a large horseshoe around the area of investigation.

An alternating electrical current, with a specific signature frequency (380 Hertz), was applied to the electrodes. The induced magnetic field, produced from the electrical current, was measured along a series of lines along the embankment. The measurement grid was designed to provide sufficient detail and resolution. Magnetic field readings were collected at 115 measurement stations. These measurement stations were established on lines spaced 25-50 feet apart with measurements taken on each line at roughly 25 feet intervals, resulting in a 25 by 25 foot grid pattern covering the entire earthen embankment.

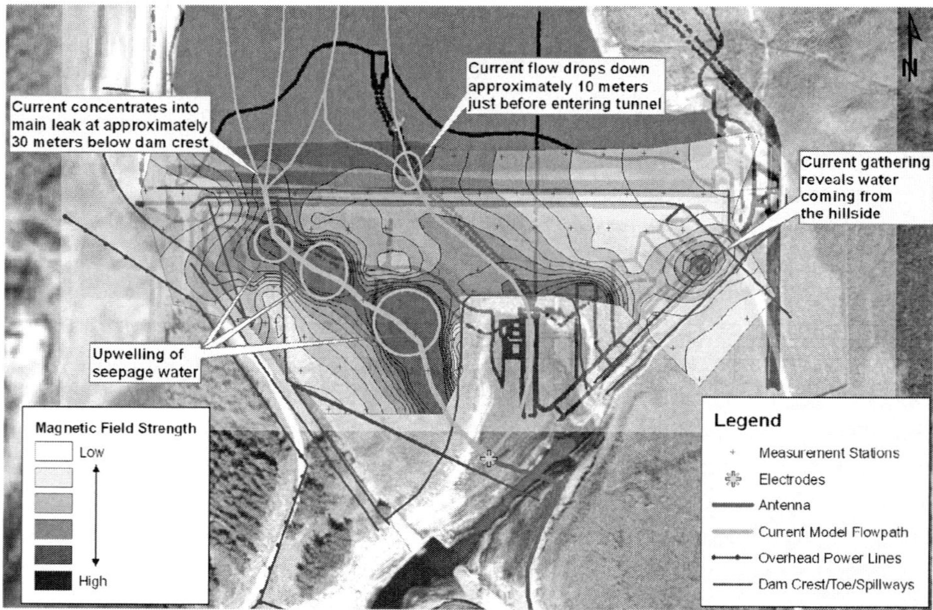

Figure 1 - Dam A Survey Map

The results revealed two seepage paths through and beneath the earthen embankment. Also, the origin of the seep in the left groin was identified. The figure below (Fig. 1 – Dam A Survey Map) provides a summary interpretation of the final results for the earthen embankment at Dam A. The dark grey shading in Figure 1 indicates conductive highs and the light grey shading indicates conductive lows beneath and/or through the dam. In the most general terms, these highs and lows can be viewed as areas of high and low groundwater saturation.

Figure 1 shows two seepage paths. The first seepage path flows around and underneath the concrete cutoff trench of the right abutment between two geologic features consisting of horizontal marine sediments and steeply dipping fractured or brecciated shale. The second identified seepage path flows along the outlet control tunnel were it intersects the clay core of the dam and it finds its way through the tunnel's concrete and stone construction and into the tunnel.

The first noted seep, previously unknown to the client, is believed to be the larger of the two seeps and appears to flow beneath the west abutment's cut-off trench through what is likely a thrust fault contact zone. This contact surface is formed between fairly impermeable horizontal marine sediments (clay and shale) and steeply dipping fractured or brecciated shale. At this interface, between the two formations, water is seeping beneath the dam. It appears that the dam's cutoff wall did not extend deep enough through the fractured or brecciated shale into the impermeable clay/shale to prevent seepage from flowing underneath the dam.

The second noted seep path, which is believed to be the lesser of the two seeps, follows the outlet control tunnel. Water flows along the interface between the exterior concrete and stone tunnel wall and the outer, more porous fill material. As water pipes along this interface and approaches the core of the dam, where earthen materials are suspected of being less porous, water is concentrated and pressured along the interface of the two earthen materials (concrete/stone and clay). Eventually, the water is forced through the tunnel wall through small cracks. These cracks are believed to have been caused by differential settlement of the rigid tunnel conduit straddling the dam's concrete cut-off trench.   A previous survey using temperature probes had not detected any flow through the core itself and this finding was confirmed by AquaTrack.

The seep flowing from the small diameter pipe located in the downstream groin of the east abutment did not originate from the reservoir, but almost certainly originates from the hillside east of the dam. When placing a return electrode in this seep and an injection electrode in the reservoir, virtually no current flow through the ground could be established.

<u>Mapping Seepage at Dam B</u>
Dam B has a clay core linked to a clay blanket and a puddle filled arm trench which cuts off the reservoir from the porous rock in the valley side. Water flowing within the rock from the steep hillside is intercepted by a rubble drain trench and flows through a short adit into the draw off tunnel. A weir has been constructed in the drainage/drawoff tunnel at the exit from the adit to monitor the drainage flow. In recent years flow from the adit has increased considerably and it is suspected that there may be a fault in the clay blanket allowing reservoir water to add to the drainage flow. UU was

desirous to identify the location of any faults in the blanket so that remedial measures could be taken. Original sketches of the dam show the approximate locations of the adit and rubble drain trench, however, the exact length and location of the adit and rubble drain trench were uncertain

A horizontal dipole antenna/electrode configuration was employed to energize the earthen dam structure. This configuration consisted of an

injection electrode in the reservoir at a point upstream from the embankment. A return electrode was located in the drainage tunnel in a weir box at the entrance to the adit in direct contact with the adit leak. An alternating electrical current was applied to the electrodes. The resultant

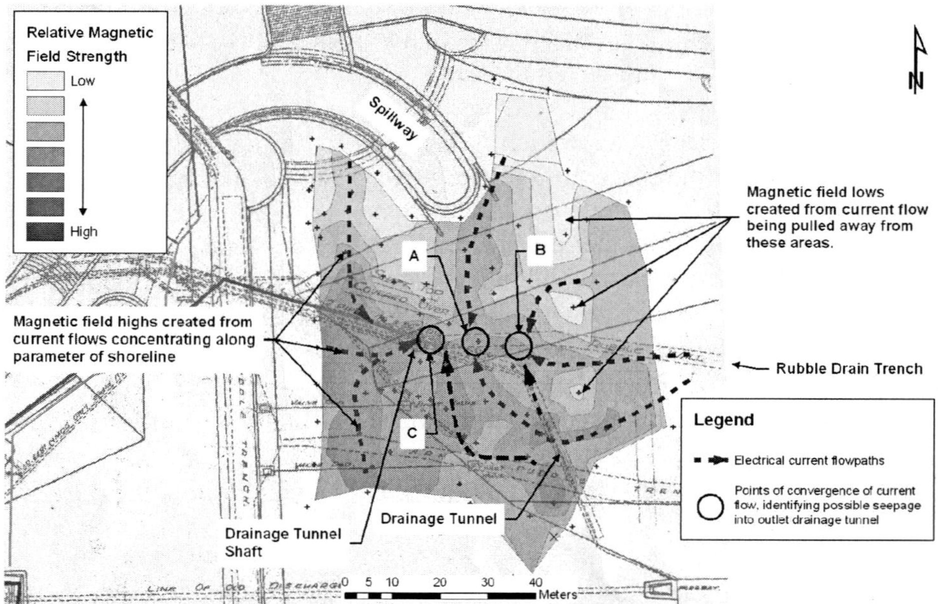

Figure 2 - Dam B Survey

magnetic field was measured and recorded at measurement stations along the embankment and in the reservoir from a boat. A total of 466 measurement stations were established on lines spaced roughly 25 feet apart with measurements taken on each line at roughly 25 foot intervals.

The results revealed the locations of the adit and rubble drain trench beneath the reservoir. The location of water leaking into the adit was also identified. Figure 2 – Dam B Survey shows the results of the adit leak investigation. The darker grey shading in Figure 2 indicates conductive highs. These conductive highs are highlighted and connected with dark grey dashed arrows forming preferential flow paths the electrical current takes as it flows from the injection electrode located in the upper reaches of the reservoir to the return electrode located in the drainage tunnel. The lighter grey shading indicates conductive lows that were observed in the magnetic field data. These high and low conductive areas are created from the signature electrical current introduced and driven through the reservoir.

Five main conductive pathways where observed in the magnetic field data. In all instances, these conductive paths appeared to be influenced by long continuous conductive features known to exist in and/or around the reservoir. These conductive features are noted in Figure 2 and include the shoreline, drainage tunnel, concrete spillway, drainage tunnel shaft and rubble drain trench and/or adit.

The survey indicated that no leakage occurs through the dam itself.

A dashed line labeled "Rubble Drain Trench" in Figure 2 shows a magnetic high following a long conductor that corresponds well with the theorized location of the adit and rubble drain trench.

The survey identified three seepage paths into the outlet drainage tunnel. These paths are noted as points "A", "B" and "C" in Figure 2. Point "A" represents a leak in the adit and is the most dominant of the three leakage points. Point "B" represents a second leak in the adit or possibly the location of where the rubble drain trench connects to the adit. The survey data was not able to determine from the data where the adit ends and the rubble drain trench begins. In either case, current flow is concentrating on two points along the suspected alignment of the adit and rubble drain trench. Therefore, leakage points "A" and "B" are the locations from which water from the reservoir is leaking into the adit and through the weir box into the outlet drainage tunnel. Point "C" represents leakage into a capped shaft, used to construct the tunnel, which co-mingles with the leakage from the adit and is conveyed into the drainage tunnel.

The adit leak investigation was a unique survey. Never before had Willowstick Technologies performed a survey where a water leak was mapped beneath a body of water, in this case the reservoir itself. Theoretic ruminations regarding the ability of CS-FDM to perform this task were proven correct during this survey.

CONCLUSION
The CS-FDM technology provided critical insight as to how seepage was affecting Dams A and B. The client now possesses valuable information that will allow him to improve and optimize repairs, monitoring and management of these important facilities. Willowstick Technologies does not specialize in earthen dam remediation, engineering and construction,

rather they focus their expertise on groundwater characterization, mapping and modeling.

The information provided by the CS-FDM survey should also be compared with known information of the site to further characterize and substantiate subsurface conditions impacting the earthen embankment. Willowstick remains committed to assisting United Utilities with whatever effort is required to fully understand the information provided.

The diagnostic investigations which took place at Dams A & B provide substantial evidence regarding the efficacy of this new water-mapping technology. CS-FDM's particular utilization of basic scientific principles—including transformer theory, current gathering and Biot-Savart law—has resulted in an elegant and efficient method for identifying seepage points in earthen dam structures.

Is internal erosion detectable?

J. DORNSTÄDTER, GTC Kappelmeyer GmbH, Karlsruhe, Germany
D. DUTTON, British Waterways, Watford, UK
A. FABRITIUS, GTC Kappelmeyer GmbH, Karlsruhe, Germany
P. HEIDINGER, GTC Kappelmeyer GmbH, Karlsruhe, Germany

SYNOPSIS.   Three possible methods of estimating seepage flow velocity, using the well-proven ground temperature sensing technique, are presented for use as a means of detecting and monitoring the progress of internal erosion within an embankment dam.

INTRODUCTION

Embankment dams account for nearly 80% of the dams in the UK and, of these, about 75% were built before 1940, thus predating modern geotechnical engineering. There have been a number of catastrophic failures in which lives have been lost; the best known being Bilberry (which failed in 1852 with the loss of 81 lives) and Dale Dyke (which failed in 1864 with the loss of 245 lives). Overtopping and internal erosion have been the two main causes of embankment dam failure in the past; the risk of catastrophic overtopping during floods has now largely been eliminated with improvements that have been made in the methods of hydrological analysis over the last 75 years, leaving internal erosion as the most likely cause of dam failure in this country in the future.

INTERNAL EROSION

The processes of initiation and progression of internal erosion are not, as yet, well understood. Modern dam design practice includes the use of zoned fills and filters to combat internal erosion but, for existing, old dams, the emphasis has to be on surveillance and monitoring for the early detection of progressive internal erosion.

The following case study, involving the embanked River Rhine, demonstrates a possible way to detect internal erosion. Over a period of several years, ground temperatures at different depths in the river embankment were measured, together with the temperature of the river water. Initially, there was almost no phase shift between the water temperature and the ground temperature (Fig. 1) as would normally have been expected *(Armbruster et al 1992)*. After a slurry trench wall had been built, a phase shift and a difference in amplitude of the temperatures occurred and increased over time, indicating that Darcy-velocities had decreased in the embankment. As a consequence of constructing the slurry trench, all indications of seepage - and therefore internal erosion - ceased. Inverting the argument, one can conclude that a reduction in phase shift between water and ground temperatures is caused by increased seepage, itself the consequence of internal erosion. Therefore, ground temperature measurements can help to detect internal erosion.

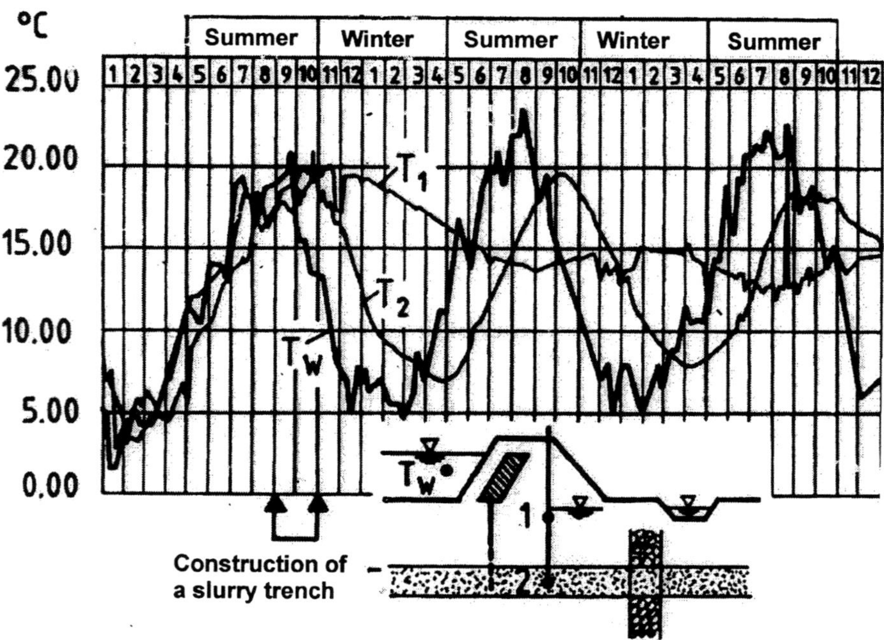

Figure 1: The building of a slurry trench in an embankment leads to a phase shift in the temperature between the water and the ground at the positions T1 and T2.

GROUND TEMPERATURE MEASUREMENT

The technique exploits the differing seasonal variations in temperature which occur within the ground and within surface water bodies. In non-percolated ground, the temperature at any depth is a function of the temporal variation in surface temperature and the thermal conductivity of the soil.

The flow of fluid from a surface water source through the soil will alter the non-percolated ground temperature distribution, creating temperature anomalies. Measurable temperature disturbances will be created by Darcy-velocities as low as 10^{-7} m/s to 10^{-6} m/s.

To measure ground temperatures at depth, a series of small-diameter, thread-coupled hollow steel tubes are installed vertically at intervals along the line of the embankment; in the case of homogeneous dams, and those with a watertight element within the embankment, the probes are sited close to the downstream edge of the crest. The tubes are driven by hand-portable rammers; depending on the nature of the ground, the distance between probes is generally 10m or 20m, but may be reduced to 5m for particular investigations, and probe depths of 20-30m can be achieved. For short-term or limited-extent investigations, ground temperatures are measured by lowering a string of temperature sensors into each tube; the tubes are withdrawn from the ground at the end of the investigation. For long-term monitoring, temperature sensors can be permanently installed within the tubes, and the cables run to a nearby data-reader unit for remote interrogation.

Because of the hand-portable nature of all the equipment, access is not normally a problem and no heavy vehicles are required on the embankment. It is quite possible to carry out all site operations at a 150m long, 15m high dam in one working day.

METHODS OF COMPUTATION OF DARCY-VELOCITY

To detect internal erosion (and the corresponding Darcy-velocity) with ground temperature measurements over short time periods, or with only one set of measurements, the following methods have been used: the phase shift approach, the determination of Darcy-velocities by scaling and the use of apparent thermal conductivity.

A. Phase shift approach

This approach has two essential requirements. The first is a period of frequent measurements of the reservoir water temperature and the ground temperatures within the dam; the second requirement is that the monitoring takes place at a time of year when the water temperature is not static and, during the period of monitoring, there must be a clear temperature turning point, either a maximum or a minimum.

The method of analysis requires first that a leakage investigation is undertaken to locate the percolated and non-percolated sections of the dam, using the ground temperature sensing technique developed and patented by GTC Kappelmeyer GmbH (Dornstädter 1997). Knowledge of the ground temperatures at the non-percolated sections of the dam allows the temperature changes due to seepage to be identified.

At the percolated sections, variations in the reservoir water temperature will, after a time interval, be reflected in the ground temperatures. Establishing this phase shift between the reservoir water temperature and the ground temperature at each point of leakage, and knowing the length of the flow path from the water/dam interface to the position of the ground temperature sounding, enables the Darcy-velocities to be calculated. For leakage at shallow depth below the crest of the dam, the assumption of a horizontal flow path may provide a sufficiently accurate estimate of the flow velocity but for leakage at greater depth, or for a more accurate estimate of the flow velocity, it would be necessary to install, and monitor, a second temperature probe on the percolated section, at a known distance from the initial leakage investigation probe. Under these circumstances the frequency of temperature monitoring may have to be increased to ensure that the phase shift is accurately measured.

B. Temperature and Spacial Difference Scaling approach

Under certain circumstances, this method offers a fast and easy estimation of the Darcy-velocity distribution within a dam. The approach is based on a single measurement of the ground temperature distribution in the dam and the flow rates of all issues from the dam, together with a knowledge of the geometry of the dam. The measurement of ground temperatures in the dam must show at least one non-percolated section in order that the temperature anomalies due to seepage can be identified. The method of analysis also assumes that:
- the water temperature is constant throughout the full depth of the reservoir and that it has not varied significantly for a period of time (~ 14 days) prior to the ground temperature investigation,
- the dam foundation is impermeable and the measured issue flows represent the total leakage through the dam

The leakage potential at every point (node) in the dam where the ground temperature is measured, and the Darcy-velocity at these nodes can then be determined by scaling.

The average Darcy-velocity through the dam is by definition:

$$v_{\bar{d}} = \frac{Q}{A} \tag{1}$$

where

$v_{\bar{d}}$:   average Darcy-velocity [m/s]

Q :   total leakage flow rate [m^3/s]

A :   longitudinal profile area of dam [m^2]

This average Darcy-velocity represents an evenly distributed seepage velocity, as if through an isotropic and homogenous dam. In practice the water will flow in some places more than in others. Where there is little or no leakage, the ground temperatures in the dam will approach, or be identical to, those at the non-percolated sections of the dam but, at places with a high seepage flow, the ground temperatures will be closer to the temperature of the reservoir water.
An area of flow is assigned to every measured temperature node. This assigned area covers, both vertically and horizontally, the half distance to the next temperature node. Thus, if all the temperature nodes except one display a non-percolated ground temperature, then all seepage water is assumed to flow through the assigned area of that one particular node. If more nodes with disturbed temperatures exist, the flow must be distributed to the areas assigned to them. This flow distribution has to be done in accordance with the scale of the temperature disturbance at each node. The sum of all flows leads to the total leakage flow:

$$Q = \sum_{n=1}^{m} v_n \cdot A_n \tag{2}$$

where

m :   number of nodes

v_n :   Darcy-velocity at node n [m/s]

A_n : assigned area of node n [m^2]

A scaling factor can be implemented in the following way:

$$v_n = f_n \cdot v_{ref} \tag{3}$$

where

f_n :   scaling factor for node n

v_{ref} :   reference Darcy-velocity [m/s]

The reference Darcy-velocity is not necessarily a physical value; it is more a calibration value, which can directly be determined out of equations (2) and (3):

$$v_{ref} = \frac{Q}{\sum_{n=1}^{m} f_n \cdot A_n} \qquad (4)$$

With the assumptions shown above, the mathematical description of the temperature field inside the dam can be simplified from the unsteady (Bear 1972, Huyakorn and Pinder 1983) to the general steady heat conduction and convection equation of an isotropic and homogenous porous media. The steady state equation has the following form:

$$0 = A' + b \cdot v_d \cdot grad\ T \qquad (5)$$

where

A' :   conductive heat transport [W/m^3]

b :   constant factor [J/K/m^3]

v_d :   Darcy-velocity [m/s]

Neglecting the conductive heat transport[1], a scaling of the Darcy-velocities must be done according to *grad T* . The gradient of the temperature is defined as the quotient of the temperature difference and the spatial difference. Therefore, to fully implement the gradient of the temperature, we have to split the scaling factor f_n into two sub-scaling factors:

$$f_n = f_{\Delta T} \cdot f_{\Delta X} \qquad (6)$$

where

$f_{\Delta T}$:   scaling factor for temperature difference

$f_{\Delta X}$:   scaling factor for spatial difference (geometry factor)

The scaling factor for the temperature difference can be found by comparing the water/ground temperature variance. A leakage potential at any depth d below the dam crest level is represented by:

$$\Delta T_{(NW)d} = T_W - T_{Nd} \qquad (7)$$

where

T_W :   reservoir water temperature [°C]

T_{Nd} :   non-percolated ground temperature at depth d [°C]

---

[1] This can be done at first order, because the amount of heat transported by conduction is generally far less than the amount transported by convection. The amount is the same at a Darcy velocity of about 1.E-7m/s.

The measured ground temperature anomaly of a node is represented by:

$$\Delta T_{(Gd)n} = T_n - T_{Nd}$$ (8)

where

T_n: ground temperature at node n [°C]

Then, the ratio of the measured ground temperature anomaly to the leakage potential is the scaling factor for the temperature difference:

$$f_{\Delta T} = \frac{\Delta T_{(Gd)n}}{\Delta T_{(NW)d}} = \frac{T_n - T_{Nd}}{T_W - T_{Nd}}$$ (9)

This scaling factor $f_{\Delta T}$ ranges from 0 (the measured temperature equals the non-percolated ground temperature - no leakage) to 1 (the measured temperature equals the water temperature - maximum leakage). Where, due to inherent physical limitations of the method of investigation, $f_{\Delta T}$ is found to be less than 0 or greater than 1, its value is limited to 0 or 1 respectively.

The scaling factor for the spatial difference can be found by consideration of the geometry of the dam; the longer the seepage flow path, the greater the significance of the measured ground temperature anomaly. The scaling factor again ranges from 0 to 1 and is calculated from:

$$f_{\Delta X} = \frac{d_n}{d_{max}}$$ (10)

where

d_n: flow path length to node n [m]

d_{max}:   maximum flow path length to any node [m]

Now all the necessary equations are established. For an analysis, first all scaling factors of the nodes are calculated and added, weighted by their assigned area (Equation 4). Thereby, the calibration Darcy-velocity v_{ref} is determined. Then, the Darcy-velocity for each node can be specified, according to equation 3.

C. Flow velocity determination by the use of heat conduction

C.1. Approximate solution for the determination of thermal conductivity

It is possible to establish the differential equation describing unsteady conductive heat flow from a long, cylindrical, ideal heat source embedded within a homogenous matrix. (*Blackwell, 1954* and *Jaeger, 1956*), but the complexity of the equation makes it of limited practical use. A simpler form of the equation, requiring only the direct determination of the thermal conductivity provides an approximate solution, although this is only valid for a certain time frame. The error involved is no more than ± 10% (*Sattel, 1982*).

for $\dfrac{\kappa\, t}{a^2} \gg 1$ follows:

$$T_i(t) = A \ln\!\left(\frac{t}{t_0}\right) + B \tag{1}$$

where

$$A = \frac{q_L}{4\pi\lambda}$$

and

$$B = \text{constant}$$

with:

$T_i(t)$:   temperature of the heat source after time t [°C]
λ: thermal conductivity of the homogenous matrix [W/m/K]
q_L: heat capacity per unit length of the cylindrical heat source [W/m]
κ: thermal diffusivity of the homogenous matrix [m²/s]
t: time since the start of the heat-pulses [s]
t_0:   time unit (1 sec.)
a: radius of the cylindrical heat source [m]

In equation (1) – for a semi-logarithmic display of the temperature – the gradient of the curve is inversely proportional to the thermal conductivity of the surrounding matrix and this simple relationship is used to determine the conductive heat flow. The constant shift B depends on the heat transfer coefficient at the heat source/matrix interface and is not taken into account in the evaluation of the conductive heat flow *(Dornstädter 1987)*.

To use this method to investigate seepage through an earth dam, a linear heat source is installed at locations within the embankment. Where seepage is occurring, heat will be transported away from the heat source by convection as well as conduction, and the assumptions which underlie equation (1) are no longer valid. The heat flow is no longer solely dependent on the material surrounding the heat source, but is also influenced by the seepage flow; the value of λ derived under these circumstances using the approximate solution is referred to as the apparent thermal conductivity.

The apparent thermal conductivity can also be determined by numerical modelling although, unfortunately, there is no exact analytical solution because the cylindrical symmetry of the purely conductive heat flow is corrupted by the seepage flow and complicates the heat flow equation. In practice there are two different approaches which enable the influence and the quantitative value of the seepage flow velocity (Darcy velocity) to be determined.

C.1.1. Péclet number analysis

The Péclet number analysis is a one dimensional approach to determine the Darcy velocity. The Péclet number P_e describes the ratio between the convective and conductive heat flow. This ratio must be determined by using the approximation solution (equation (1)). Sections with high apparent thermal conductivity can be identified as percolated layers (convective and conductive heat transport) and sections with low apparent thermal conductivity as layers without seepage flow (pure conductive heat flow). If the same thermal conduction is assumed for the ground matrix in all the layers, the thermal conductivity λ_{cond} in the layer without seepage flow is proportional to the conductive heat transport, and in the layer with seepage flow the apparent thermal conductivity $\lambda_{cond+conv}$ is proportional to the sum of the convective and conductive heat transport. Therefore, the Péclet number can be determined by:

$$P_e = \frac{\lambda_{cond+conv.} - \lambda_{cond.}}{\lambda_{cond}} = \frac{\lambda_{cond+conv}}{\lambda_{cond}} - 1 \quad (2)$$

and the Darcy velocity can be calculated by using the definition of the Péclet number (*Zschocke, 2003*):

$$P_e = \frac{q_a}{q_c} = \frac{\rho\, c_p\, v_f\, \Delta T}{\lambda\left(\dfrac{\Delta T}{l}\right)} \quad (3)$$

$$v_f = \frac{P_e\, \lambda}{l\, \rho\, c_p} \quad (4)$$

$$v_f = \frac{\lambda_{cond+conv} - \lambda_{cond}}{l \, \rho \, c_p} \qquad (5)$$

with:

q_a:	convective (advective) heat flow [W/m^2]
q_c:	conductive heat flow [W/m^2]
ρ:	density of the fluid [kg/m^3]
c_p:	specific heat capacity of the fluid at constant pressure [J/kg/K]
v_f:	flow velocity of the fluid [m/s]
ΔT:	temperature difference [K]
$\lambda = \lambda_{cond}$:	thermal conductivity of the ground [W/m/K]
l:	characteristic length [m]

The Péclet number analysis is a fast and direct method of determining the seepage velocity by the use of the heat pulse method. The accuracy of the results, though, is strongly dependent on the thermal conductivity of the ground material. Using incorrect thermal conductivities for the ground material may mean that the conductive part of the heat flow is not determined correctly, with errors of up to 100% or more. Nevertheless the method can be used for a quick assessment of a change in the seepage flow velocity with respect to previous measurements, thus helping to determine the onset of internal erosion.

C.1.2 FD Modelling

The second method of determining the Darcy velocity uses FD- (or FE-) modelling, involving the following steps:

- measuring the heat-up curve
- simulation of the heat-up curve with a finite difference (or finite element) program for various seepage velocities
- comparison of the simulated curves with the measured ones from which conclusions about the Darcy velocity can be drawn..

This approach enables a fairly exact determination of the Darcy velocity in the ground, even if the parameters necessary for the modelling (permeability, porosity, thermal conductivity and specific heat capacity) are not known exactly. The accuracy which can be achieved with this method is ± 25% (*Heidinger, 1998*). The disadvantage of this method lies in the considerable effort needed, and power required, to provide all the data and conditions for the model.

CASE STUDIES

Cam Loch

Cam Loch is at the head of a cascade of reservoirs feeding the Crinan Canal in Scotland. The loch is retained by two dams, the more westerly one of which is about 60m long and has a maximum height of 7.5m. A small issue of water has been known to occur at the toe for many years. A ground investigation carried out at the dam in 1989 did not locate a clay core but showed that the embankment consists mainly of peat on a rock foundation; a rockfill berm, with a drain at its toe, was constructed on the downstream face during remedial works undertaken in 1990.

A temperature sounding leakage investigation was carried out in March 2005 when the reservoir water was still at a typical winter temperature of 3.5 °C. The measured temperatures are shown on the longitudinal profile of the dam in Figure 2. Issue flows at the dam toe totalled 0.6 l/s and the investigation indicated temperature anomalies partway along the dam, down to a depth of 2 m below crest (1 m below TWL), and also adjacent to the left abutment.

Analysing the temperature data by means of the Temperature and Spatial Difference Scaling method outlined above gave a Darcy-velocity distribution shown on the longitudinal profile of the dam in Figure 3.

River Rhine embankment

This investigation was conducted to check the impermeability of a newly installed sealing wall. Temperature measurements using the Heat-Pulse method were conducted along a several hundred metre long section of the river embankment. These results, combined with the Peclet number analysis, enabled Darcy-velocities of a possible seepage flow through the embankment to be determined.

At three points within the investigated section, temperature soundings were performed on the upstream and the downstream side of the sealing element. The soundings were positioned at 1m distance from the sealing element and reached 1m below it.

At one of the soundings, the results showed a slight seepage flow – in the range of 10^{-6} m/s. The results are displayed in Figures 4, 5 and 6.

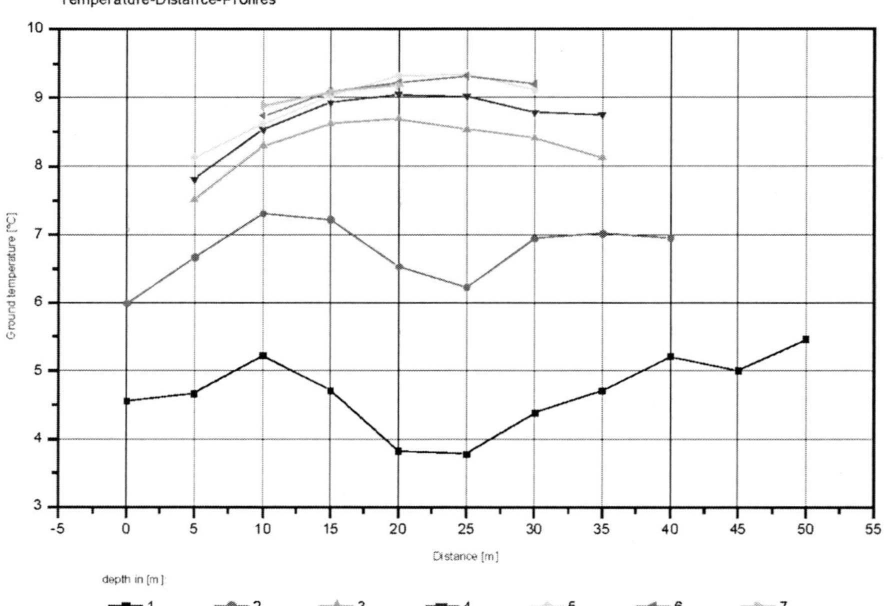

Figure 2: The temperature-distance profile recorded during the temperature sounding leakage investigation.

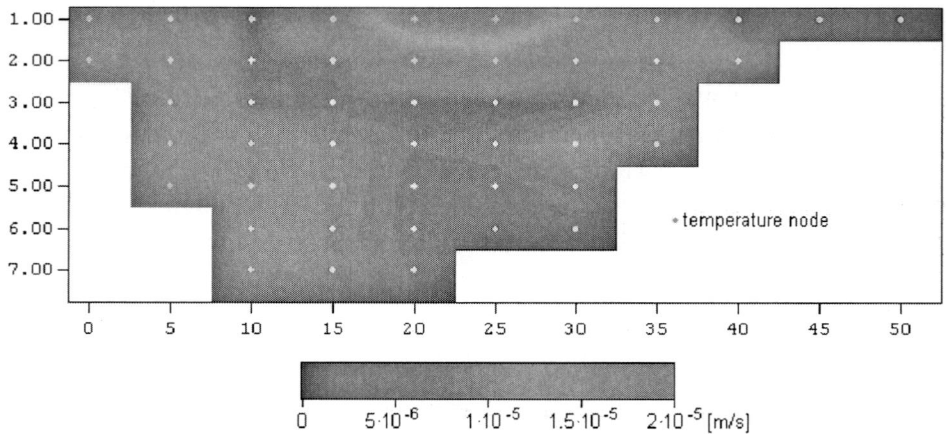

Figure 3: Darcy-velocity distribution. The maximum flow velocity was calculated to be $1.87*10^{-5}$ m/s.

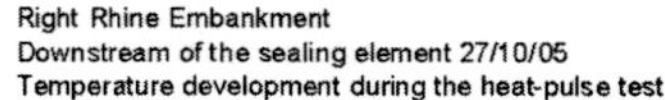

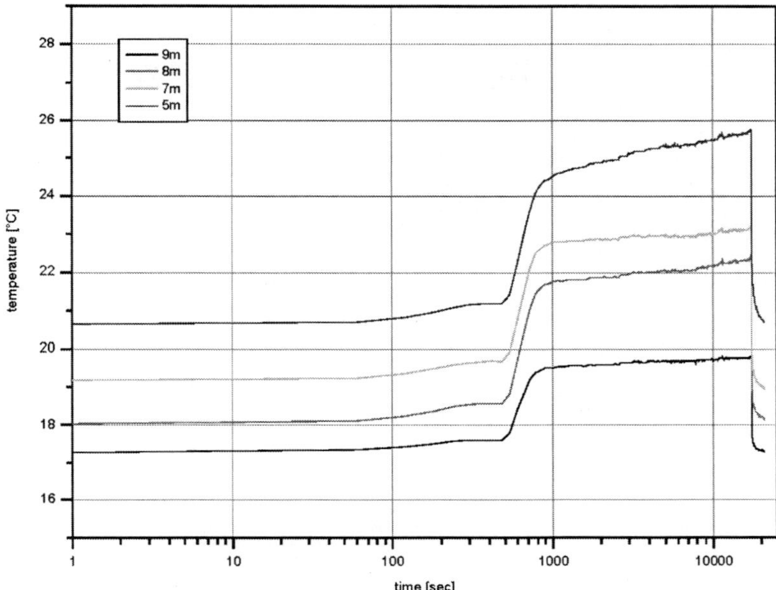

Figure 4: Temperature development during the heat-pulse test shown on a logarithmic time scale.

CONCLUSIONS

Seepage through embankment dams can be detected by means of ground temperature measurements. Ground temperature measurements taken over a period of time or, under certain conditions, during a single investigation, can provide an initial estimate of the Darcy-velocity of the seepage flow. This allows an early diagnosis of internal erosion, to be confirmed if necessary by further investigation. If, during a second investigation, an increase in the Darcy-velocity is detected with the same hydraulic loading conditions, progression of internal erosion is indicated.

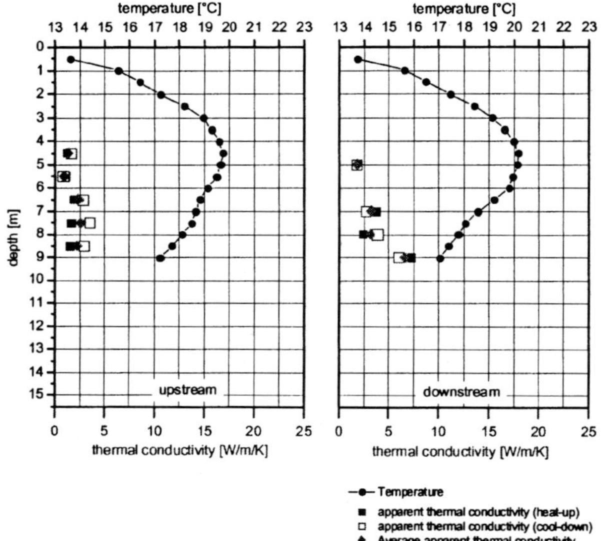

Figure 5: Thermal conductivity upstream and downstream of the sealing element.

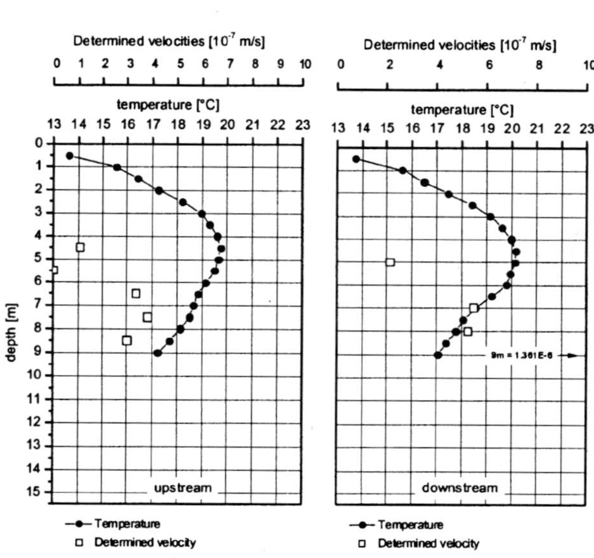

Figure 6: Darcy velocities on both sides of the sealing element.

REFERENCES

Armbruster H.; Dornstädter J., Kappelmeyer O. and Tröger L. (1992) *Detection of seepage and flow phenomena by temperature measurements in soil.* Tracer Hydrology - Proc. of the 6th Int. Symp. on Water Tracing. Eds.: H.Hötzl&A.Werner. A.A.Balkema Verlag, Rotterdam.

Bear J., (1961) *On the tensor form of dispersion.* JGR, v. 66, no. 4, p 1185-1197.

Blackwell J.H., (1954) *A transient-flow method for determination of thermal constants of insulating materials in bulk, Part 1* – Theory. J. Phys., 25: 137-144

Dornstädter J., (1987) *Das Temperaturangleich-Verfahren (TAV). Diplomarbeit;* Geophysikalischen Institut der Universität Karlsruhe

Heidinger P., (1998) *Bestimmung der Sickerwasserfließgeschwindigkeit durch Wärmezufuhr.* Diplomarbeit, Geophysikalisches Institut der Universität Karlsruhe

Jaeger J.C, (1956) *Conduction of heat in an infinite region bounded internally by a circular cylinder of a perfect conductor.* Aust. J. Phys.,9: 167-179

Huyakorn P.S. and Pinder G.F., (1983) *Computational methods in subsurface flow.* New York, Academic Press, 473p.

Sattel G., (1982) *In-situ Bestimmung thermischer Gesteinsparameter aus ihrem Zusammenhang mit Kompressionswellengeschwindigkeit und Dichte.* Dissertation, Geophysikalisches Institut der Universität Karlsruhe 1982

Zschocke A., (2003) *Software für die Anwendung der Pécletzahlanalyse auf Temperaturlogs in einem geschichteten Untergrund.* GGA-Institut, Hannover

Leakage investigations at Lower Carno dam

A. ROWLAND, Black & Veatch, Redhill, UK
A. POWELL, Black & Veatch, Haywards Heath, UK

SYNOPSIS. Lower Carno reservoir is situated 3 km north of Ebbw Vale. The 27 m high dam, completed in 1911 to the designs of GF Deacon, has a central puddle clay core supported by shoulders of clay and stony material. The dam has had a history of leakage and a succession of remedial works has been carried out.

In January 2005 operations staff, who visit the reservoir on every working day, notified the supervising engineer of increased flows in the toe drains. After investigations to trace the source of the flows were unsuccessful and the rate of leakage increased rapidly, it was decided to empty the reservoir.

This paper describes the history of the reservoir, indicators of the leakage that were observed, the conclusions reached to date on the pathway of the leakage and the ground investigation being undertaken to confirm potential remedial options.

DESCRIPTION OF RESERVOIR

Introduction

The reservoir was completed in 1911 and is impounded by an earthfill embankment dam built across the headwaters of the River Ebbw. The capacity of the reservoir is 800,000 m^3 and the direct catchment area of the reservoir is 5.39 km^2. The catchment is generally moorland. The town of Ebbw Vale is a short distance downstream and the dam is Category A.

On 20th January 2005 operations staff notified the supervising engineer of increased and turbid water flows in the toe drains. The previous inspecting engineer of the reservoir was consulted and it was decided to empty the reservoir. During further investigations more significant flows started to emerge from the area at the downstream toe of the dam. The flow was estimated to be 200 l/s with a significant (1%) silt content.

Improvements in reservoir construction, operation and maintenance, Thomas Telford, London, 2006, 144–153

Black & Veatch were retained by Dwr Cymru (Welsh Water) to carry out a literature review and visual inspection of the dam to determine the potential cause of the leakage event. The study identified possible mechanisms for the leakage and recommended intrusive investigations to provide more information for establishing the leakage path and for the design of remedial works.

<u>Dam details</u>
The dam, which has a maximum height of 27 m and is 180 m long is formed by an earth embankment with a puddle clay core. The crest level is nominally 394.8 mODN. Figure 1 shows a typical section of the dam.

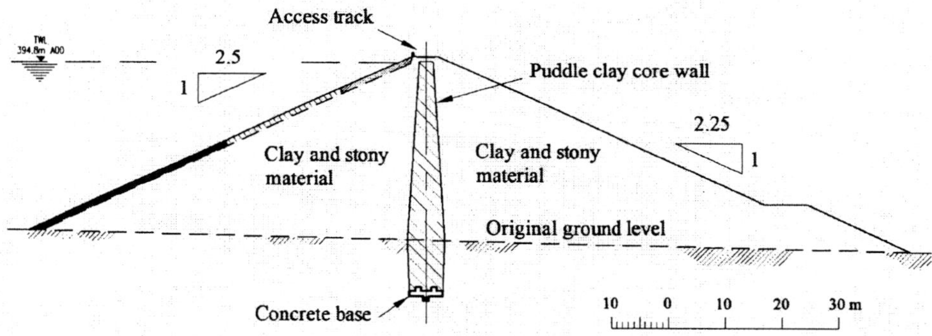

Figure 1. Typical cross section of Lower Carno Dam

The majority of the dam is founded on the Lower Coal Measures of the Westaphalian Sequence which consist mainly of mudstones and siltstones. Glacial Till overlies the Coal Measures but has been eroded away in the valley, which has alluvium along the river course. No faults underlie the reservoir basin or dam.

The puddle clay core is supported by shoulders of clay and stony material. The shoulder fill is of variable composition. Generally it is a heterogeneous firm sandy gravelly clay and mudstone/siltstone gravel with some cobbles. Zones of stonier material adjacent to the core are thought to be present.

There are no drainage blankets or filter zones in the dam but the River Ebbw was filled with a stone product along its course under the embankment.

The clay core is extended down into the foundation as a cut-off trench which has a concrete key at the base. The trench cut-off extends 78 m into the right abutment and 43 m into the left abutment.

The upstream face steepens from the bottom to the top and averages 1 in 2.5. The slope of the downstream face also increases from bottom to top and averages 1 in 2.25. The downstream slope is mainly grassed.

<u>Instrumentation</u>
The locations of all instrumentation on the dam are shown on Figure 2. These installations include 24 standpipe piezometers, toe drain flow measurement, 12 wave wall settlement measurement points, 8 crest settlement points and reservoir level measurement.

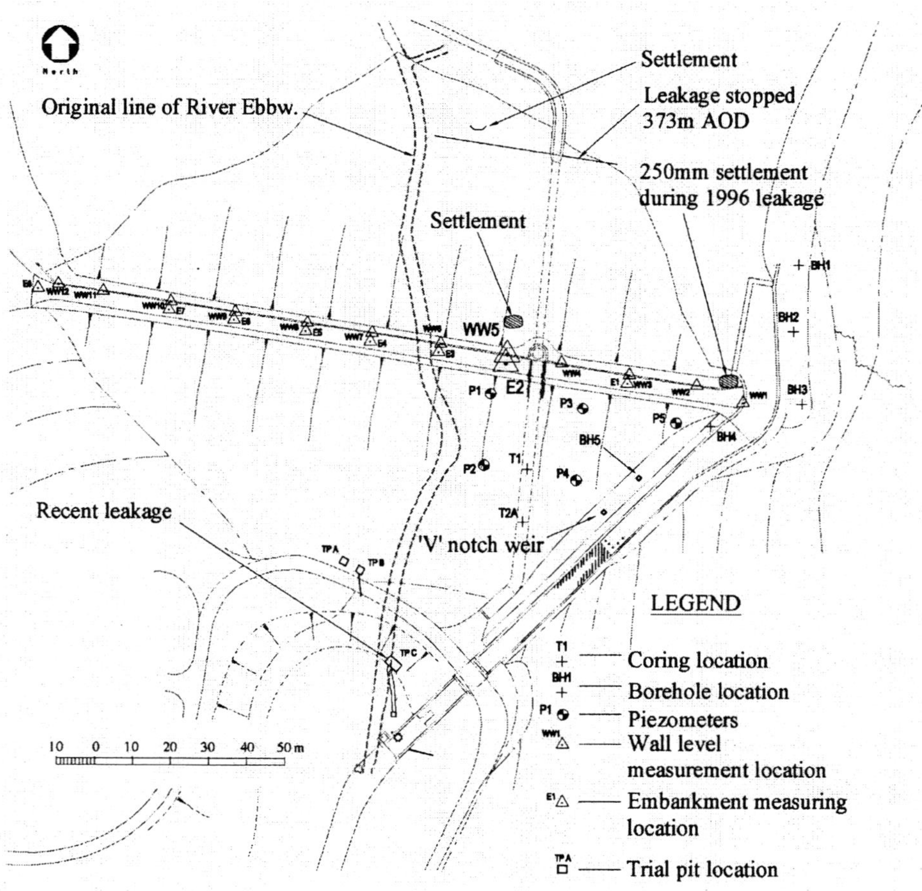

Figure 2. General arrangement of Lower Carno Dam

<u>Draw-off arrangements</u>
There is a concrete culvert under the upstream shoulder, with a reinforced concrete approach channel and a 'dry' concrete culvert under the downstream shoulder, the latter being sealed by a brick and concrete plug

where it passes through the puddle clay core. The section of the culvert passing through the puddle clay core trench is supported on (and enclosed within) a concrete foundation cast into the base of the cut-off trench.

A 450 mm diameter scour main and a 300 mm diameter supply main were laid in the dry culvert, with control valves at the base of what was originally a dry shaft located just upstream of the puddle clay core. These valves were operated from the top of the shaft. This shaft has been filled with water to reservoir level since the outlet arrangements were modified in 1933, after serious leakage started to take place into the shaft. At that time the original supply outlet main was permanently sealed and a completely new draw-off system constructed through the hillside upstream of the right abutment of the dam. The original 450 mm scour pipe through the culvert remains in use.

HISTORY OF LEAKAGE PROBLEMS

There have been many modifications and remedial works undertaken, many to reduce rates of leakage. These are described below:

1911- On first filling in 1911 a flow of 9 l/s was observed leaking through the fissured rock and coal seams which form the left abutment of the dam. The designer decided not to attempt to seal these leakage paths unless the maximum leakage exceeded the statutory compensation flow of 10.8 l/s. Within 18 months the leakage rate had fallen to 2.5 l/s.

1915- A few years after the completion of construction in 1911, unsuccessful attempts were made to reduce the rate of leakage in the left abutment. The designer of the dam concluded that the rate of leakage, which was less than the specified compensation water, caused no threat to the integrity of the dam or the safety of the reservoir. It is not recorded what these remedial works included but injection grouting may have been used.

1933- Strong leakage from the upstream shoulder into the valve shaft via voids in the base of the tower was identified during a diver's inspection. The voids had been formed by disintegration of the concrete in the side walls of the shaft. The diver also noted that significant outflows from the tower were visible. After prolonged attempts to locate the source of leakage and to seal the flow by grouting, it was decided to abandon the valve shaft. The new outlet works were constructed on the right side of the reservoir and the bottom of the old valve shaft was sealed with a concrete and brick plug.

1987- A small amount of leakage into the culvert 1.5 m downstream of the plug on the line of the puddle clay core was identified in 1985. This leakage was sealed by pressure grouting.

1996- A significant leakage emerged from a point about two-thirds of the way up the left abutment of the embankment in April 1996. The source of the water was traced to the underside of the new concrete section of the sloping spillway channel where water was seen to emerge from the stony backfill placed in this area, the soil matrix evidently having been washed away. It was also noted that about 250 mm of settlement of a small area of stone pitching laid on backfill immediately adjacent to the new overflow weir had occurred. The valve shaft was drained to determine that the issue of water was not caused by leakage out of the shaft and thus not through or under the body of the dam embankment. To prove conclusively that the flow was not linked to the valve shaft a tracer test was carried out. No dye was found in the leakage water. Additionally, the water level in the shaft was varied to determine any variation in the leakage flow. None was evident. Conductivity testing of the reservoir water and leakage water confirmed that it was the same water. The flow of water was estimated as 5 l/s. When the reservoir level had dropped to about 8.3 m below TWL the leakage ceased. Pressure relief holes were drilled at an inclination of $30°$ to the horizontal through the spillway channel floor to release some of this water, and the remainder was piped down the lower part of the mitre to rejoin the original leakage water in a chamber near the downstream toe.

SUMMARY OF THE RECENT LEAKAGE EVENT

On 20th January 2005 site operatives informed the Supervising Engineer of increased and discoloured water flows in the toe drains. Investigations were carried out to locate the source of these flows, but without success. Contractors were employed to clean out the manholes and V-notch chambers but the source of water was not found. As a precaution, flow from Upper Carno was diverted past the reservoir using the adit and the scour was opened to lower the water level.

During the week ending 6th February 2005, two drains (one being the discharge from the 1996 installed V-notch and the other being an unidentified drain leading towards the right abutment) had lifted to the west of the stilling basin. Flows were seen discharging from this area (marked on Figure 2) over the ground and into the stilling basin. The previous inspecting, engineer was informed and, on a site visit on 8th February, instructed that the unidentified drain should be traced back towards the reservoir.

During the investigation more significant flows started to emerge from the area to the west of the stilling basin. A channel was cut to contain the flow. The flow was estimated to be approximately 200 l/s with a significant (1%) silt content. The rate of lowering of the reservoir level was increased and the leakage at the toe of the dam substantially stopped with the water level at 373 mODN, 20 m below top water level.

<u>Instrumentation response</u>
Prior to the discovery of the leakage, neither the boreholes in the left abutment nor the V-notch flow showed any abnormal behaviour. All of them have a damped response to reservoir level rather than to seasonal fluctuation in ground water levels. It is concluded that there is a small but not insignificant amount of leakage in the area of the left abutment. It is probable that this seepage is not connected to the severe leakage at the toe of the dam.

There was no visible crest settlement prior to drawing down the reservoir. Settlement occurred suddenly and quickly during the course of emptying the reservoir but this settlement is not associated with a rapid drawdown failure. The most recent settlement measurements indicate that the dam has settled by about 250 mm since 1990 at monitoring points E2 and WW5, as shown on Figure 3. Increased settlement is noted along nearly the whole length of the dam and is significant over a length of 70 m. The width of the settlement trough is as would be expected for ground loss at the depth of the culvert or base of the cut-off trench.

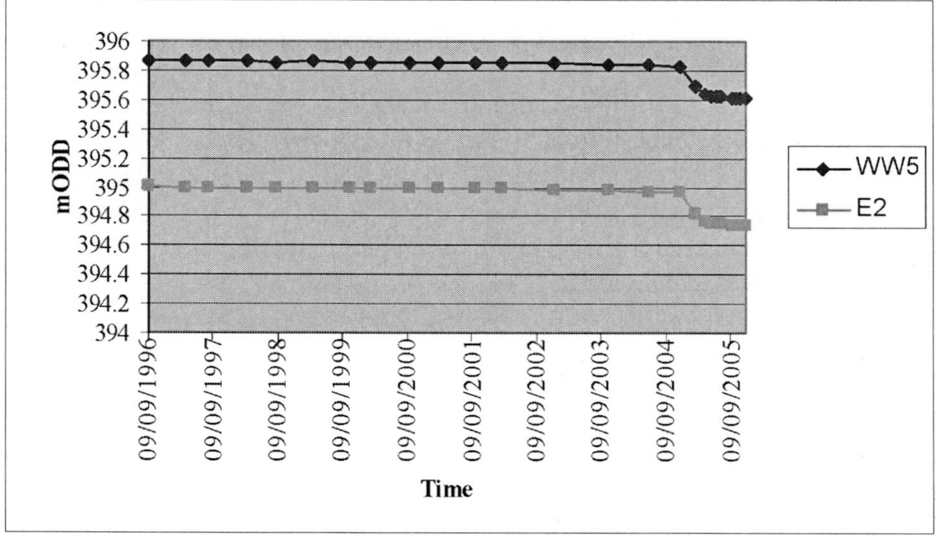

Figure 3. Settlement measuring points E2 and WW5 1996 to 2005

The crest pin monitoring records show that, during the period 1990 to 1996, crest pin E2 and wave wall pin WW5, both close to the line of the draw-off culvert, settled at an average rate of 5.3 mm and 5.8 mm a year respectively. Between 1996 and 2002, the average rates of settlement reduced to 3.1 mm and 2.4 mm a year. Inspection of the plots of crest level against time show that the readings taken in December 2002 suggest an increase in the rate of settlement. Subsequent readings show this trend continuing and the average rates of settlement between February 2002 and November 2004 were 6.2 mm and 8.0 mm a year for the crest and the wave wall. These represent a slight increase over the long term average. Adjacent pins show a similar pattern of settlement. The amount of annual reservoir draw down has not changed significantly and it is possible that this small change in settlement rate was an early indication of the leakage problem.

PRELIMINARY ENGINEERING ASSESSMENT

Lower Carno Reservoir has a long history of leakage events. From the initial filling of the reservoir through to the present day, significant seepage has occurred. However, it is not uncommon for old dams to experience a number of seepage events during their lifetime. This section reviews the likely cause of the leakage scenarios and anticipated future investigations required to determine necessary remedial repair. It is worth noting that mining related subsidence has been ruled out as no deep mining is catalogued in the vicinity of the reservoir.

Shallow mining related voids

Potential cause of leakage: Prior to 1901 a trial coal adit was excavated into the left abutment of the dam at a level of approximately 385 mOD just upstream of the core. The coal deposits outcrop upstream of the dam and seem relatively fissile. Additionally, the leakage through the left abutment does seem to cease once the reservoir level drops below 384 mOD. The recent leakage flow at the toe of the embankment did not significantly reduce until the reservoir water level was at 373 mOD. Thus the leakage through the left abutment is unlikely to be related to the flow path associated with the recent leakage incident.

Further investigation and remedial option: The investigation is designed to assess the need and lateral extent of any positive cut-off to minimise the flow through the left abutment to ensure that further problems do not arise in this area in the future. In the left abutment rotary coring and packer testing will be carried out to determine the presence of any significant leakage pathways in the abutment.

Leakage through the concrete support to the culvert in the core trench

Potential cause of leakage: In September 1932 leakage into the culvert downstream of the core increased significantly. A diver was sent down into the valve shaft and reported strong leakage into the shaft from the upstream side and heavy seepage out of the shaft to the downstream side. The seepage was occurring through voids in the concrete sides of the shaft and existing culvert plug. In 1933 a grouting operation was carried out to reduce the flow into the culvert. Water samples taken in 1953 from within the culvert indicated that the sediment in water discharging to the culvert was practically completely soluble in acid, indicating that it was not a clay-based product. In 1987 a further grouting exercise was carried out to minimise inflow.

It is possible that the concrete below the culvert in the puddle clay core trench has also deteriorated with time resulting in a seepage path developing below the culvert. The seepage could possibly have entered the Alluvial Gravel downstream of the core or gravel filled former river course, eventually emerging at the embankment toe.

Further investigation and remedial option: The proposed investigation will include core sampling of the concrete surround to the culvert. Visual examination of the samples will allow the level of deterioration of the concrete to be determined. Additionally, packer testing will be carried out within the concrete via rotary drillholes to determine the permeability of the in-situ concrete. Where the concrete is found to have deteriorated significantly a grouting exercise from the surface and possibly from the culvert could be undertaken.

Leakage through the puddle clay core

Potential cause of leakage: The puddle clay core sits on the concrete surround to the culvert. The sides of the surround are quite steep, between 30-60°. Prior to remedial works in 1988, comprising enlargement of the upper part of the spillway, restoring the crest level of the dam and constructing a wave wall, it was noted that the crest of the dam had settled approximately 600 mm in the centre of the valley since construction. Where these relatively large settlements occur adjacent to a rigid inclusion, such as a concrete culvert or plinth, fracturing of the puddle clay core adjacent to or at the concrete interface may occur. This would result in the opening of a leakage path through the core. Hydraulic fracture may also have caused the initial leakage path but significant internal erosion has occurred since resulting in the crest settlement and milky discharge observed.

Further investigation and remedial option: Fully cased light cable tool percussion boreholes will be sunk from the crest of the dam through the core. Permeability tests will be carried out at close intervals to determine the presence of leakage paths or voids in the core. Additionally, near continuous undisturbed sampling will be carried out to allow visual examination of the puddle clay. It is also expected that the casing and/or boring tools will drop on locating voids and this will be noted by drillers and the consultant's geotechnical engineer supervising the investigations. Consolidation grouting may be required to fill voids in the shoulder fill and core. A positive cut-off trench could then be excavated through the puddle clay core over the length and depth of core affected by any fracturing.

PROPOSED GROUND INVESTIGATION

At the time of writing this paper, the ground investigation works are programmed to start April 2006. They are designed to determine which of the hypothesized scenarios, defined above, is the most likely cause of the recent leakage event. Between 10 and 15 light cable tool percussion boreholes will be sunk through the puddle clay core and 10 may then be extended by rotary means to allow the concrete culvert surround and dam foundation materials to be sampled and tested.

The upstream and downstream shoulder fill is more problematic to investigate. A section of the reinforced concrete wave wall will require removal to allow staging to be constructed over the upstream face. Additionally, inclined rotary holes may be drilled from the crest of the dam. The presence of a temporary operational 500 mm diameter raw water main on the downstream edge prohibits access onto any staged exploratory holes from the adjacent crest area over the downstream fill. However, a light cable tool percussion rig may be able to be winched up the slope. The extent of any grouting exercise would be determined as part of the grouting works.

The foundation is not thought to be a likely cause of the recent leakage event. However, to confirm that this is the case, rotary coring and packer testing of the foundation will be carried out to confirm the depth of any cut-off wall and confirm the absence of high permeability layers or zones.

CONCLUSION

In retrospect the evidence of the embankment crest settlement readings could infer that the leakage began before August 2002, although the actual changes were very small and there were no other indications of problems in the piezometer or drainage monitoring readings. The leakage accelerated in January 2005 apparently causing considerable internal erosion of the dam and wash-out of fines from the fill.

The design of the dam comprising a central puddle clay core supported by shoulders of clay and stony material is typical of the Heads of the Valleys reservoirs and there was nothing different in the design of Lower Carno to suggest that it would be more vulnerable to leakage and internal erosion than other dams of the type. However, there is some evidence that stones picked from the core may have been placed close to the core rather than in the outer part of the shoulders as planned. It is not possible to determine whether this has contributed to the history of leakage problems. At present more weight is being given to the proximity of the leakage path to the outlet culvert.

The position of the resulting settlement trough which developed at the dam crest and the position of the emergence of the leakage suggests that the erosion was largely confined to the core and probably occurred close to the western side of the outlet culvert. The proposed ground investigations will determine whether a suitable remedial option can be designed to allow the reservoir to return to operation.   Discontinuance of the reservoir has not been ruled out.

Once again the importance of regular, vigilant and effective surveillance and of prompt communication with those qualified and empowered to take appropriate and immediate action has been proven. Delay in detecting the leakage and taking immediate action could have resulted in a major disaster.

Comparison of methods used to determine the probability of failure due to internal erosion in embankment dams.

M. EDDLESTON, MWH, Warrington, UK
I. C. CARTER, MWH, High Wycombe, UK

SYNOPSIS.   Quantitative risk assessment is a routine part of periodic safety reviews in Australia and the United States. Many dam owners have been using these techniques to produce Portfolio Risk Assessments of their dams for a number of years. The recent publication of the DEFRA Interim Guide to Quantitative Risk Assessments has increased awareness of such techniques in the UK. This paper describes the application of two methods used for quantitative risk assessments of internal erosion used in Australia and the United States to six pilot dams and compares the results with the DEFRA method of assessing the probability of failure due to internal erosion.

BACKGROUND

Interest in risk assessment of UK reservoirs followed the Report on the Water Industry by the House of Lords Select Committee on Science and Technology in December 1982 which recommended that *"research should be carried out into risk associated with reservoirs and the methodology for quantitative risk assessment, and as a result of that in the light of this research a wider spectrum of safety criteria should be introduced to take into account of the different degrees of risk in individual reservoirs."* This resulted in a number of feasibility studies being undertaken by a variety of organisations (Clifton et al. 1985a & 1985b, Clark and Tyler, 1987, Water Research Centre 1987, and Cullen, 1990). The last in this series of reports (Cullen 1990) concluded that *"in the light of present knowledge probabilistic risk assessment is not yet a suitable tool for inspection work"*.

In 2002 a DEFRA Research Contract was undertaken and published (DEFRA, 2002) aiming to *"Propose and demonstrate an Integrated System which provides a framework for decision making by Panel Engineers on the annual probabilities of occurrence, consequences and tolerability of all the various threats to safety"*. This publication followed on from a draft report by DEFRA and Babtie giving Guidance on the Stability Upgrades to Embankment Dams (Anon, 2002). DEFRA have recently published an Interim Guide to Quantitative Risk Assessment to UK Reservoirs (Brown and Gosden, 2004, 2005). In a letter to Panel Engineers in July 2004 DEFRA state that *"The Interim Guide is a tool for the management of reservoir safety enabling a screening level assessment to be made to inform decision-making by dam professionals on the annual probability of occurrence of reservoir failure, the consequences and the tolerability of that risk."* and that *"The Guide is intended to assist both Panel Engineers, particularly those who design and inspect reservoirs, and reservoir owners. It is anticipated that risk assessments undertaken using the Guide would be carried out where concern over the safety of a reservoir has arisen, to inform inspections of high hazard reservoirs, and in connection with portfolio risk assessment to rank risks and prioritise expenditure on safety improvement works."*

A major UK dam owner has developed a Portfolio Risk Assessment (PRA), using Quantitative Risk Assessment (QRA) techniques to rank dams within the Portfolio. The QRA techniques for internal erosion were based on those developed at the University of Stanford (McCann et al., 1985) and the University of New South Wales (Foster et al., 1998, 2000a). The output is considered more as a ranking tool rather than reliable absolute values for each embankment.

Following the introduction of DEFRA Interim Guide the dam owner commissioned a study to compare the methods used in their PRA (based on those of the University of Stanford and the University of New South Wales) and those used in the DEFRA Interim Guide as part of the development of a methodology to investigate internal erosion in its stock of embankment dams. This was based on desk study assessments of six embankment dams selected as being relatively high on the PRA probability list and with a variety of perceived problems. The methodology was reviewed by a panel of independent experienced dam engineers. This paper summarizes the results of the comparisons of the different methods of assessment undertaken on the six pilot desk studies to give a better understanding of why the different methods at times gave differing results and to see if more weight can be given to any particular method.

The methods of the University of Stanford (UoS) and DEFRA Interim Guide are based upon historical data of dam safety events related to internal erosion, from which estimates of probabilities of failure have been derived for different dam conditions. The University of Stanford method (McCann, et al. 1985) uses a Bayesian probabilistic model to evaluate the probability of failure (PoF) for four modes of failure: piping, slope stability, piping associated with outlet works and foundation and miscellaneous failures. The model draws on US dam data and practices to estimate the frequency of failure for a number of dam conditions. The DEFRA Interim Guide uses data from the BRE/KBR National Dam Database (NDD), which classifies historical problems and incidents. Both methods require an engineering assessment of the current dam condition to be performed and scored on a scale of 1 (best condition embankment) to 10 (distressed embankment). Based on the current condition score, a probability of failure is 'read off' from the base data.

In the case of the DEFRA Interim Guide method, the 'base data' is the BRE/KBR National Dam Database, with typical probabilities of failure being adjusted to take account of the individual characteristics of the embankment under assessment.

The University of New South Wales (UNSW) method (Foster et al. 2000a and 2000b)uses data from reported dam incidents throughout the world (ICOLD, 1995) to make assessments of the likelihood of failure of embankment dams by three methods; piping through the embankment, piping through the foundation and piping from the embankment into the foundation. These historically determined probabilities of failure are then adjusted using weighting factors to take account of the characteristics of the dam under assessment, considering attributes such as core properties, compaction, conduits and foundation geology and also the performance of the dam in terms of seepage observations and pore water pressures. The frequency of monitoring and surveillance is also incorporated into the assessments. Each of these attributes is assessed as above or below a median probability of failure, representing a more likely and less likely condition to initiate failure. Appropriate weightings applied to give an overall probability of failure.

COMPARISON OF METHODS USED TO CALCULATE PROBABILITY OF FAILURE DUE TO INTERNAL EROSION.

General Comparison

The DEFRA Research Contract (2002b) produced a comparison of the various methods of calculating probability of failure by equating the maximum and minimum weighting in the University of New South Wales Method (UNSW) to the best and worst condition scores used in the University of Stanford (UOS) and DEFRA methods used at the time of publication (i.e. before the publication of the DEFRA Interim Guide). The comparison has been updated to include the new Guide as shown in Figure 1.

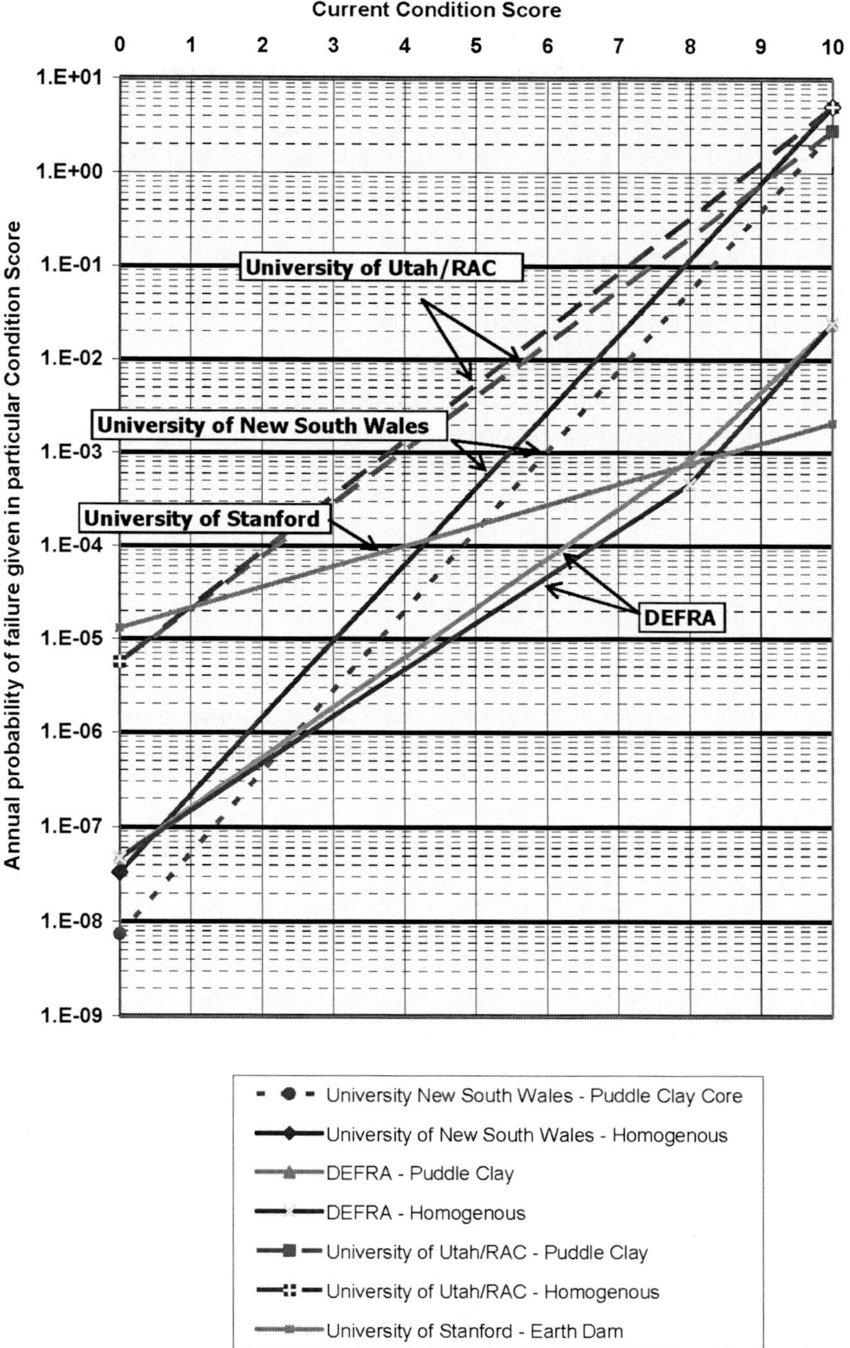

Figure 1 - Comparison of methods of evaluating Probability of Failure and Dam Condition Scores

A number of observations can be made from the above plot including:

- The wide variations in estimated PoFs for both good and poor condition dams by the various methods
- The UoS method shows the least degree of variation in PoF for differing Dam Condition Scores
- The UNSW Worst Case Probability of Failure is greater than one
- The UNSW method gives the lowest probability of failure for best condition dams
- The PoFs for best condition dams for the UoS and DEFRA methods are similar
- The fact that even for Condition Score 10 dams the PoF is still only 1 in 50 reflects the relatively high number of emergency draw downs due to internal erosion which have been reported since 1975, whilst there have been no corresponding reported failures due to internal erosion in the same period. It also assumes that similar surveillance and intervention would continue in future
-

The value of PoF determined by any of the methods is dependant on a consistent evaluation of a Dam Condition Score. Experience in using the methods has shown that, for certain dams, the assessment of condition scores by the different methods gives widely differing values, but the calculated probabilities are similar. Conversely similar condition scores assessed by the different methods can give wide variations in calculated PoFs.

The application of all the methods and interpretation of the resulting PoFs requires considerable engineering judgement.

It is also of note that the incidence of failure of puddle clay core dams in the UK is greater than that for homogenous dams. The reverse is the case for the data from world dams used in the UNSW Method.

Comparison based on application to Pilot Desk Studies

The results of assessment of Probability of Failure and Dam Condition are presented in Table 1 and plotted in Figure 2.

The dams selected for study consisted of three traditional "Pennine" dams with puddle clay cores and three controlled seepage or "Canal" dams.

Looking at the estimated overall Probabilities of Failure it can be seen that there is a wide spread of results. The largest difference noted is for Dam 2 where the highest PoF assessed by the UNSW method was 34 times greater than that calculated by the DEFRA method. The differences ranged from between 1.5 and 34 with a mean variation of 9 i.e. nearly an order of magnitude different. In at least one case (DEFRA PoF 9×10^{-5} for Dam 1) the method gave a result contradicting the judgement of the Review Panel (and others) following review of the data and a site visit, the UOS and UNSW methods do however indicate a problem.

As would be expected Dam Condition Scores upon which the estimated PoFs are based, show a similar variation with none of the three methods giving consistently high or low values. This illustrates the very approximate nature of the derived values and the need to treat them in conjunction with engineering judgement rather than applying them as absolute values.

Table 1 - Assessed Probabilities of Failure and Dam Condition Scores for Pilot Dams

OVERALL SUMMARY OF PROBABILITY OF FAILURE	Dam 1	Dam 2	Dam 3 (with filters)	Dam 3 (without filters)	Dam 4	Dam 5	Dam 6 (Before Grouting)	Dam 6 (After Grouting)
UOS								
Piping	9.5E-05	9.5E-05	2.5E-05	3.8E-05	1.0E-05	1.2E-05	6.9E-04	1.1E-05
Slope Stability	9.2E-06	5.9E-06	5.9E-06	5.9E-06	2.8E-05	2.1E-05	1.1E-06	1.1E-06
Piping/Outlet Works	6.6E-06	1.1E-05	8.0E-07	8.0E-07	7.9E-07	1.6E-06	1.2E-04	2.7E-05
Foundations and Misc	1.1E-05	3.6E-05	5.1E-05	5.1E-05	1.1E-05	1.1E-05	1.0E-05	1.0E-05
Total	1.2E-04	1.5E-04	8.3E-05	9.6E-05	2.6E-05	4.6E-05	8.2E-04	4.9E-05
UNSW								
Embankment Piping	7.4E-04	4.5E-03	5.8E-07	3.0E-04	8.3E-05	3.3E-04	3.6E-03	4.3E-04
Foundation Piping	6.2E-05	4.7E-05	2.1E-05	2.1E-05	2.0E-06	6.1E-06	1.4E-04	2.8E-05
Embankment to foundation	2.1E-04	3.1E-05	0.0E+00	1.0E-07	6.0E-07	3.1E-05	9.1E-05	1.8E-05
Total	8.1E-04	4.5E-03	2.1E-05	3.2E-04	8.6E-05	3.7E-04	3.8E-03	4.7E-04
DEFRA								
Embankment	8.90E-05	9.00E-05	8.6E-05	8.6E-05	1.7E-05	1.7E-05	1.0E-04	1.00E-04
Surface Structures	2.60E-06	-	1.0E-07	1.0E-07	1.0E-06	1.0E-07	4.2E-06	1.10E-04
Buried Structures	-	4.20E-05	9.4E-06	9.4E-06	6.3E-07	1.7E-05	1.0E-02	1.20E-05
Total	9.16E-05	1.32E-04	9.55E-05	9.55E-05	1.86E-05	3.41E-05	1.01E-02	2.22E-04
Max	8.1E-04	4.5E-03	9.6E-05	3.2E-04	8.6E-05	3.7E-04	1.0E-02	4.7E-04
Min	9.2E-05	1.3E-04	2.1E-05	9.6E-05	1.9E-05	3.4E-05	8.2E-04	4.9E-05
Max/Min	8.8	34.3	4.5	3.3	4.6	10.9	12.3	9.6

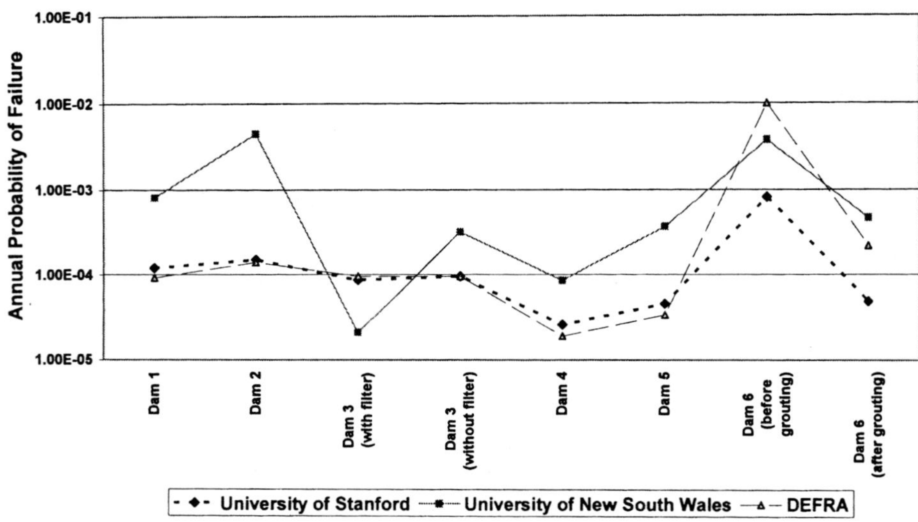

Figure 2 – Plot of Assessed Probabilities of Failure and Dam Condition Scores for Pilot Dams

Comparing the results where known influences occur the following can be highlighted:

Seepage carrying fines

Where seepage carrying fines is noted (as for Dam 6, before grouting) this is allocated a relatively high weighting in all methods and whilst the range of PoFs vary by an order of magnitude they are all generally in the $>10^{-4}$ within the HSE intolerable range. (N. B. General guidance issued by the Health and Safety Executive (2001, 2004) for industries imposing hazards on the public, suggests that annual probabilities of failure of less than 10^{-6} (1 in 1,000,000), are broadly acceptable. Probabilities between 10^{-6} and 10^{-4} (1 in 10,000), requires action to reduce risk to as low as reasonably practicable (ALARP) and probabilities above 10^{-4} are intolerable and require remedial action at any cost).

Conduits through embankment

In the cases where conduits pass through the embankment Dams 2, 5 and 6) an appropriately high weighting is picked up by the UNSW method ($W_{E(con)}$), in the buried structured part of the DEFRA appurtenant works assessment and in the Piping/outlet section of the UoS method. Whilst the PoF difference between the three methods for the case of Dam 2 vary by the highest order (34 times difference) the results all fall within the HSE Intolerable range indicating a potential problem. Similar PoFs are recorded for the influence of a conduit for Dams 5 and 6 by the DEFRA and UNSW methods. However the wide range of condition scores (4-7) and associated probabilities covered under a single description by the UoS method for pipes considered in "neutral" condition makes there assessment very much open to the interpretation of individual assessors.

Filters

The addition of stabilising berms, which may have been designed as filters, at the Dam 3 embankment allowed an evaluation of the effect of including a weighted filter. The effects of filters slightly reduces the PoFs estimated by the UoS method and by an order of magnitude with UNSW method. Surprisingly there is no apparent beneficial effect with the DEFRA method and only a very small benefit with the UOS method. Consideration of the method of evaluation (DEFRA Sheets 4.3 and 4.4) indicate that whilst the effect of filters is taken into effect in establishing the Intrinsic Dam Condition this is only used in the adjustment of the Best Condition Dam Anchor Point. This has little effect on dams with discernible defects. In the assessment of the Current Dam Condition Score used in the assessment of the PoF the effect of filters is related to seepage and only has a major effect when there is a "large amount of uncontrolled seepage discharging i.e. not discharging into a filtered drainage system". This is related to the calculation of a Seepage Index (Charles, 1993). This is clearly an area requiring further consideration in relation to determining the beneficial effects of remedial works using the ALARP methods recommended in the Guide.

Effectiveness of Remedial Works (Grouting)

Dam 6 had previously identified defects that were remediated by grouting. This allowed for the evaluation of PoF before and after the remedial works. Results before the remedial works were implemented by all methods indicated PoFs within the HSE Intolerable range. The DEFRA method gave a value of 1 x 10^{-2} indicating Condition Score 10 equating to a "Serious incident involving emergency action or drawdown" in accordance with the NDD Classification (see Table 2). Results estimated after grouting for the DEFRA and UNNSW remain in the intolerable range indicating that there are still potential problems within the embankment that still need to be addressed.

ADVANTAGES AND DISADVANTAGES OF THE METHODS

The advantages and disadvantages of the methods investigated to determine the PoFs against internal erosion may be summarised as follows:

University of Stanford	
Advantages	Disadvantages
<ul><li>easy to use</li><li>largely based on visual evidence</li><li>can provide quick assessment for ranking a number of dams</li><li>covers a number of potential failure mechanisms</li><li>covers embankment and concrete dams</li></ul>	<ul><li>evaluation scale difficult to interpret and sensitive to choice of Dam Condition Score</li><li>some descriptions for condition scores are not appropriate for large dams</li><li>does not distinguish between types of earth embankments</li><li>based on US data and practice</li></ul>

University of New South Wales	
Advantages	Disadvantages
<ul><li>easy to use and apply</li><li>large range of factors considered</li><li>based on world dam incident data</li></ul>	<ul><li>wide coverage makes interpretation to UK conditions difficult for some factors</li><li>does not give a Dam Condition Score</li></ul>

DEFRA Interim Guide	
Advantages	Disadvantages
<ul><li>based on UK National Dam Database data</li><li>rigorous means of determining Dam Condition Score</li></ul>	<ul><li>anchors to Category 3 NDD incidents only (incidents leading to work). It does not anchor to incidents leading to the involvement of an Inspecting Engineer</li><li>some double accounting of contributory scores (e.g. inspections)</li><li>zero anchor point adjustment does not have a noticeable effect on poor condition dams (e.g. absence or presence of filters, conduits)</li><li>sensitive to scoring of Dam Condition</li></ul>

CONCLUSION

The methods used to provide estimates of probability of failure due to internal erosion are useful values for initial screening purposes. It must be stressed that the calculated probabilities should not be taken as absolute values but to be used in conjunction with engineering judgement, particularly on issues not directly taken into account in the assessment (e.g. downstream consequences). Currently all three methods are used by the authors to make Probability of Failure assessments, the findings of which will assist in refining the current DEFRA Interim Guide as it develops into a national standard. Before being used in ranking and (potentially) decisions on remedial works, the authors recommend that the process be reviewed by an All Reservoirs Panel Engineer (under the Reservoirs Act 1975) working with the team undertaking the data gathering and inputting to the process which should include experienced Geotechnical Engineers, Dan Engineers Supervising Engineers and Reservoir Keepers.

REFERENCES

ANON (2002b) *DEFRA Research Contract "Reservoir Safety-Floods and Reservoir Safety Integration.* August 2002.

Anon (2002b) DEFRA/Babtie, *Guidance on stability upgrades to older embankment dams.* Babtie Ref. BWA20075/JWF/MM.

Brown, A. and Gosden J. (2004) *Interim Guide to Quantitative Risk assessments for UK Reservoirs.* KBR/DEFRA. Thomas Telford, London.

Brown, A. and Carter, I. C. (2004)*European Symposium on Internal Erosion. Dams and Reservoirs* Sept 2004.

Brown, A. and Gosden J. (2005) *An example of the application of the Interim Guide to QRA.* Dams and Reservoirs July 2005, 11-12.

Brown, A. and Bridle, R. (2005) IREX/EDF *Workshop on internal erosion and piping of dams and foundations,* 25-27 April 2005, Aussois, France. Dams and Reservoirs July 2005, 17-18.

Charles, J. A. (1993) *Embankment dams and their foundations: safety evaluation for static loading. Proc. International Workshop on Dam Safety Evaluation,* Grindlewald. Vol. 4. Sutton, Surrey. Dam Engineering, 47-75.

Clake, P. B. and Tyler, M. J. *A probabilistic Risk Assessment of an earth embankment dam* (1987) Report No. HS/RANN/2/29/WP1. Safety an Reliability Directorate, UKAEA, Warrington.

Clifton, J. L., Holden, P. L., Kennard, M. F., Phillips, D. W. and Webber, D. M. (1995a) *A feasibilty study into probabilistic risk assessment for Reservoirs.* Report No. HS/RANN/2/12/WP1. Safety an Reliability Directorate, UKAEA, Warrington.

Clifton, J. L., Holden, P. L., Kennard, M. F., Phillips, D. W. and Webber, D. M. (1995b) *A feasibilty study into probabilistic risk assessment for Reservoirs.* Report No. ER 188. Water research Centre, Swindon.

Cullen, N. (1990) risk Assessment of Earth Dam Reservoirs. Report No. DoE. 002-SW/03. WRC, Swindon.

Foster, M., Fell, R and Spannagle M. (1998) *Risk assessment - estimating the probability of failure of embankment dams by piping.* ANCOLD 98 Conference on dams.

Foster, M., Fell, R. and Spannagle M. (2000a) *A method for assessing the likelihood of failure of embankment dams by piping.* Can. Geotech. J. 37:1025-1061.

Foster, M., Fell, R and Spannagle M. (2000b) *Statistics of embankment dam failures and accidents.* Canadian Geotechnical Journal. Vol. 37 No 5, 1000- 1024.

Health and Safety Executive (2001*) Reducing Risks: Protecting People.*

Health and Safety Executive (2004) *Guidance on "as low as reasonably practicable" (ALARP) decisions in control of major accident hazards (COMAH).*

International Commission on Large Dams (1995). *Dam failure statistical analysis* ICOLD Bulletin 99.

McCann, M. W, Franzinia J. B, Kavazanjian E & . Shah H. C. (1985) *Preliminary safety evaluation of existing dams.* John A. Blume Earthquake Engineering Centre, Dept of Civil Eng., Stanford University US FEMA. Contract EMW-c-0458. Vol. I (Report No 69) Methodology, Volume II (Report No 70)- User manual.

Water Research Centre (1987) *A probabilistic risk assessment of an earth embankment dam.* WRC, Swindon.

In search of the perfect geotextile/geocomposite filter for retro-fitting old embankment dams

M. EDDLESTON, MWH, Warrington, UK
H. TAYLOR, United Utilities, Watrrington,UK
N. ROBINSON, TerraConsult, St. Helens, UK
A. BINNS, TerraConsult, St Helens, UK

SYNOPSIS.   Increased use of quantitative risk assessment techniques to determine the probability of failure of the various threats posed to reservoir embankments has highlighted the potential risk of internal erosion occurring in many older embankment dams. In dams where an internal erosion threat has been identified retrofit filters are commonly incorporated into the downstream shoulder of the dam. The variability of the materials used in the construction of the dams makes conventional filter designs difficult. Locally sourcing and placing of consistent quality filter materials on slopes also poses difficulties. This paper investigates the potential of designing using geotextile filters to retrofit old embankment dams identified as being vulnerable to internal erosion.

INTRODUCTION

In recent years the concepts of reservoir risk assessment have advanced considerably (ICE, 1996, Hughes et al, 2000, Brown and Gosden, 2002, 2004, 2005, Brown and Carter, 2004, Brown and Bridle, 2005). These assessments have identified inadequate overflow capacities of many dams. As a result those dams which were below the current standard have had their spillways upgraded to acceptable levels. This has left the internal behaviour of dams designed and built before the development of soil mechanics and modern compaction plant as a major cause of concern (Charles, 1998, 2001, Vaughan, 2000a). This concern is enhanced by the number of internal erosion incidents which occur each year (e.g. Gardiner et al, 2004, Bridle, 2004). Internal erosion is the damaging process by which seepage flow through the body of an embankment dam carries away fine particles of soil. The flow paths may progressively enlarge resulting in the material carried away becoming larger, increasing in quantity and eventually leading to breaching of the dam. The final stages can be very rapid. Internal erosion can be contained by 'filters' which are sized to retain the soil particles eroded from the soil to be protected (the 'base soil') while allowing water to pass through. Such filters are designed to prevent erosion of the watertight element of the dam and prevents the development of erosion 'pipes'.

DESIGN OF GRANULAR FILTERS

Vaughan and Bridle (2004) have recently published an update on methods of filter design to resist internal erosion. They recognise that most have been developed in relation to coarse granular materials. The application of these methods results in the design of successively coarser layers, each of which is sized so that grains or particles from the adjoining layer will not pass through its neighbour. Such an approach was used in the construction of graded filters and weighted stability berms used in the upgrading of dams in Northern Ireland (Cooper, 1987).

Filter designs are based on the principle that the pore spaces between particles should be just small enough to prevent the passage of the smallest of the protected grains and rely on a 'self-filtering' effect whereby some the protected material is allowed into the filter to make it effective.

In dams, the element most vulnerable to erosion is the waterproofing element, the core, typically clay. This poses special problems in filter design because using traditional rules to design filters to protect cohesive soils usually leads to filters of sizes which are themselves likely to be cohesive. These would be capable of keeping cracks open like the core they are intended to protect. Clearly, this offers no effective protection to vulnerable cohesive clay cores and it is generally accepted that different design principles should be applied.

Vaughan and Bridle (2004) recognise that the filter design principles used for granular soils do not apply to the protection of clay cores. They identify two approaches to the design of filters to resist internal erosion of clay cores The first is known as the 'critical filter' approach developed by USDA Soil Conservation Service (1986) and Sherrard & Dunnigan (1989) and quoted in ICOLD (1994). This is based on the 'no erosion' test where samples of base soil and prospective filter were tested by passing water under pressure through a small diameter hole in the base soil into the filter. Various filter gradings were proposed for different core materials based on the results of conventional particle size distribution tests undertaken on dispersed and flocculated soil.

Vaughan et al 1970; Vaughan & Soares, 1982 and Vaughan, 2000b have proposed an alternative design method for filters for clay cores known as the 'perfect' filter method. This is based upon an assessment of the finest material which might be eroded from the walls of a crack in the core. Vaughan and Soares, 1982 determined the size of particle retained by a given filter experimentally by preparing different sizes of particle and passing them in dilute suspension through the filter. They discovered that this was usually clay flocs of around 10 μm (0.01 mm) particle size.

The filter grading which is required to retain these flocs is based on the finer sizes present, usually the 15% size. A comparison of critical and perfect filter designs has been undertaken by Vaughan (2000b). Vaughan and Bridle (2004) note that the permeabilities of filters retaining clays flocs are low and their drainage capacity is therefore limited. If filters are protecting fills that include permeable layers that may allow substantial quantities of seepage to pass, it may be necessary to provide a coarser drainage filter downstream of them to allow the seepage to escape freely.

DESIGN OF GEOTEXTILE FILTERS

Geotextile filters are manufactured in a variety of forms either single geotextile membranes or as a prefabricated geocomposite comprising two filter geotextiles bonded either side of the water-carrying core. For ease of reference the following terms will be used in subsequent sections when referring to the components of the filter.

Filter   a geotextile filter or geocomposite prefabricated filter.

Geotextile   a single geotextile or the geotextile bonded on either side of the water-carrying core.

Core   the water-carrying core of a geocomposite between the geotextiles outer layers which are bonded to it.

A variety of methods for designing geotextile filters have been summarised by ICOLD (1986), Christopher et al (1991) and Giroud (1982, 1988,1996, 1998). As with granular filters, the methods of design of geotextile filters are predominantly based on filtration rules for granular materials. The main requirements are to provide efficient filtration without clogging or piping and adequate flow capacity under the design loads to provide the maximum anticipated seepage during the design life.

Vaughan (2001) states that geotextile filter design involves the use of maximum openings in lieu of d_{15} size or equivalent in granular filter design.

A variety of other criteria have been produced by others examples of which are detailed in Table 1.

Table 1 - Published Retention Criteria

Geotextile Retention Criteria		
Source	Criterion	Remarks
Rankilor (1991)	$O_{50}/D_{85} \leq 1$	Nonwovens & soils with $0.02 < D_{85} < 0.25$
	$O_{15}/D_{15} \leq 1$	Nonwovens & soils with $D_{85} 0.25$
Giroud (1982, 1984, 1992)	$O_{95}/D_{85} \leq [(9-18)/Cu]$	Dependent on Cu of soil and assumes fines will migrate for large/low Cu soils
French Committee on Geotextiles and Geomembranes (CFGG, 1986)	$O_f / D_{85} \leq 0.38 - 1.25$	Dependent on soil type, compaction, and hydraulic conditions of the site
Fischer, Christopher and Holtz (1990)	$O_{50}/D_{85} \leq 0.8$ $O_{50}/D_{15} \leq 1.8 - 7.0$ $O_{50}/D_{50} \leq 0.8 - 2.0$	Based on geotextile pore size distribution and dependent on Cu of the soil
Terram Design Guide (Anon, 1996)	Piping Limit - Minimum Aperture Opening Size (AOS) less than or equal to 120µm	
American Standard on Geotextile in Highway Applications (AASHTO, 1999)	Maximum AOS, ASTM D-4491 - 200µm	

where:
O_{95} = 95% opening size of geotextile
O_{50} = 50% opening size of geotextile
D_{85} = grain size in millimetres of 15 percent finer by weight
$Cu = D_{60}/D_{10}$

Giroud (1984) points out that such criteria are probably too restrictive for highly cohesive soils. However, cohesive soils are not usually problematic, and Giroud's retention criterion handles worst case (i.e., cohesionless) conditions. ICOLD (1996) has published a list of known applications of geotextiles in embankment dams from twelve countries. Nominal Pore Size of Geotextile Filters used range from 29 -180 µm.

For long term performance, the filtering system development between a geotextile and an adjacent stable soil performs the filtering function for fine particle movement. From a design standpoint, there needs to be as many possible paths per unit area available for water to pass through the soil/geotextile system. The adjacent soil will comprise a percentage of these paths once the filtering mechanism is developed. From a conservative design approach, the more paths available per unit area the more efficient the filtration system will be over time.

In addition to designing a geotextile with sufficient soil retention and filtration properties to control piping, some other important design objectives must be fulfilled. These include:

- sufficient drainage capacity and the ability to dissipate excess hydrostatic pore pressure
- long term effectiveness (ability to resist clogging)
- stability of geosynthetic soil filter systems laid on slopes
- survivability and durability of geotextiles

<u>Sufficient drainage capability</u>

Darcy's equation is the foundation for determining sufficient water passage capability of a soil/geotextile filter system. The adjacent soil's hydraulic properties govern the initial and long term filtration behaviour of a soil/geotextile filter system.

In geotextile design, the adequacy of water passage, rather than Darcy's coefficient of permeability, is considered the important design parameter. Water passage normal to the plane of the geotextile (permitivitivity, Ψ) and in-plane flow along the line of the geotextile (transmissivity, θ) need to be considered.

Permittivity is defined as:

The volumetric flow rate of water per unit cross-section area, per unit head, under laminar flow conditions, in the normal direction through the geotextile.

$$q = \Delta h \, A \, K_n/t = \Delta h \, A\Psi$$

A geotextile's ultimate permittivity, Ψ_{ult}, is determined using ASTM D-4491 (2004). This permittivity result is often used to develop a geotextile permeability value which is dependent on geotextile thickness. De Berardino (1992) Bhatia (1998), Rollin, Mlynarek and Bolduc (1990), and others have commented on the complete lack of a role that thickness plays in soil/geotextile filtration design. Koerner (1990) also uses permittivity for design.

Using Darcy's Law, Koerner has developed the following required soil permittivity:

$$\Psi_r = q / \Delta h\, A$$

where:
Ψ_r = required permittivity
q = flow rate
Δh = head loss
A = area of geotextile

Flow net analysis can be used to determine the required permittivity (Cedergren, 1989).

A factor of safety applied to the required permittivity as follows:
$$\Psi_g \geq F.S.\ \Psi_r \quad (\text{Koerner, 1990})$$

where:
Ψ_g = design allowable geotextile permittivity
F.S. = Factor of safety
Ψ_r = required permittivity

The factor of safety is based on experience. A factor of safety of 5 is recommended for geotextile erosion control structures. A number of similar criteria based on permittivity have also been published a summary of which are given in Christopher and Fischer, 1991.

The transmissivity, in-plane flow within the geotextile, is defined as the in plane permeability times the geotextile thickness.

Transmissivity, $\theta = k_p t$

Thus applying Darcy's law:

$$q = k_p\, i\, A = \underline{\theta}\, i\, w\, t = \theta\, w\, i$$
$$\ \ t$$

where
q = flow rate
k_p = In plane flow rate
t = thickness of geotextile
w = width of geotextile

When a granular drain or filter layer is replaced by a geosynthetic layer, it is often assumed that the two have the same hydraulic transmissivity. In the United States, this approach is often mandated by regulations for the case of leachate collection layers used in landfills. This is true only in the case of confined flow (i.e. if the liquid collection layer is completely filled with liquid). In reality, liquid collection layers should be designed for unconfined flow, (Giroud et al. 2000). To be equivalent under the unconfined flow condition, the geosynthetic liquid collection layer should have a greater hydraulic transmissivity than the granular liquid collection layer (Giroud et al. 2000).

$$\theta_{geocomposite} = E\,\theta_{graular\,drain}$$

$$E = 1/0.88\,[\,1+ (t_{soildrain}/\,0.88L)\,.\,(\cos\beta/\tan\beta)]$$

E	- Equivalency factor
$t_{soildrain}$	- Thickness of soil drain(m).
L	- Slope length (m)
β	- Slope angle
$\theta_{soildrain}$	- Soil drainage transmissivity m^2/s
$\theta_{geocomposite}$	- Geosynthetic drainage transmissivity m^2/s

A typical granular filter thickness of 300mm with a permeability of 1×10^{-4} m/s equates to a geotextile in plane flow rate that requires either a cuspated or geonet drainage core to provide the sufficient drainage capacity. The high water carrying capacity of the drainage core allows for economic provision of drainage capacity in the geocomposite filter.

<u>Long term effectiveness (ability to resist clogging)</u>

The ultimate long term performance concern is potential clogging of the geotextile. A number of additional criteria have been proposed to resist clogging as detailed in Christopher and Fischer, 1991.

Giroud (1996) maintains that an effective filter must be able to retain larger soil particles whilst, at the same time, it must allow very fine particles close to the geotextile to pass which would otherwise block the geotextile. So the geotextile effectively supports the formation of a natural, stable grain structure within the adjacent soil to produce a self filtering effect. To pass through a nonwoven geotextile, a particle must pass between fibres. Giroud (1998) modelled the filtering effects of geotextiles, defining the term constriction, as a passage contained between three or more fibres which are nearly, but not necessarily exactly, in the same plane. The size is defined as the diameter of the sphere which can just pass through the constriction. A soil particle that travels in a nonwoven geotextile follows a certain path until it meets a constriction which is smaller than it is. The level at which a particle is stopped depends on the filtration path. If the particle is not stopped by a constriction, it passes through the geotextile.

Giroud demonstrated that in order that the filter is neither too permeable for the soil nor too prone to clogging, the geotextile must possess an optimum number of constrictions m where:

$$m = \frac{\text{Geotextile Thickness}}{\text{Fibre Diameter}} \times \sqrt{(1 - \text{porosity})}$$

Geotextile fibre diameters vary from 25-50 μm for fine to coarse fibres.

Giroud (1996) produced a performance model of a geotextile in contact with soil which shows that the number of constrictions between the filaments which a soil particle must pass during its way through the geotextile is a significant characteristic. He demonstrated that the optimum number of constrictions for geotextile filters is between 25 and 40. The range of 25 to 40 constrictions guarantees the homogeneity of the geotextile opening size, and the retention and stabilisation of the grain structure at the surface of the geotextile, resulting in the quick formation of a stable, natural filter without internal clogging. At less than 25 constrictions (typical for thin heat-bonded geotextiles) the retention capacity under turbulent flow conditions is no longer guaranteed, and as a consequence the homogeneity of the geotextile filter is too low. At more than 40 constrictions (typical for thick, single-layer geotextiles) soil particles can clog the geotextile (so-called "internal clogging).

Stability of geotextile soil filter systems laid on slopes

The stability of a geotextile filter system is often controlled by the shear strength between the various interfaces, i.e. geotextile/soil and geotextile/geocomposite interface shear strengths. The importance of interface shear strength was illustrated by the slope failure in Phase IA of Landfill B-19 at Kettleman Hills in the USA, which instigated a major investigation carried out by the University of California at Berkeley (Seed et al., 1988). It has also played an important role in a number of UK failures (Jones and Dixon, 2003). Various authors have produced simplified methods of establishing the stability of geotextiles on slopes (Giroud et al., 1995 and Koerner and Hwu, 1991). The analysis can be undertaken on conventional slope stability programmes. It is important to obtain data on the geotextile interface shear strength from shear testing to BS6906:1991. Some typical results are quoted by Jones and Dixon, 2003.

A further consideration is the potential to build up tension forces due to unbalanced friction forces above and below the geotextile liner (Koerner and Hwu, 1991 and Bourdeau et al. 1993). The situation is shown in Figure 1.

Figure 1 – Frictional Forces on Geotextiles placed on slopes

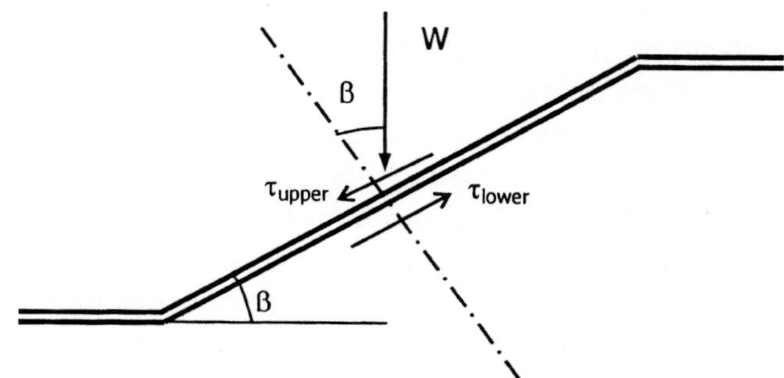

Three different conditions need to be considered as follows:

$\tau_{upper} = \tau_{lower}$ The geotextile goes into a state of pure shear which should not be of great concern for most geotextiles

$\tau_{upper} < \tau_{lower}$ The geotextile goes into a state of pure shear up to a magnitude of τ_{upper} and the balance of $\tau_{lower} - \tau_{upper}$ will not be mobilised

$\tau_{upper} > \tau_{lower}$ The geotextile goes into a state of pure shear up to a magnitude of τ_{lower} and the balance of $\tau_{upper} - \tau_{lower}$ will not be mobilised

Tension will occur in a situation when a material with high interface friction (e.g. sand and gravel) is placed above a geotextile and a material with low interface friction (e.g. a soft clay) is placed beneath the geotextile. The tension developed in this case is given by the equation (Bourdeau et al., 1993)

$$T = \left\{ \left[\frac{\alpha_{upper} - \alpha_{lower}}{F} \right] + \gamma\, H\, \text{Cos}\beta \left[\frac{\tan\delta_{upper} - \tan\delta_{lower}}{F} \right] \right\} L$$

where:

T   = Tension force in geotextile
α_{upper} = adhesion between geotextile and upper soil
α_{lower} = adhesion between geotextile and lower soil
γ = soil density
H   = slope height
β = slope angle
δ_{upper} = friction angle between geotextile and upper slope
δ_{lower} = friction angle between geotextile and upper slope
L   = length of geotextile

Placing the geotextile to limit tension stresses developing during laying of geotextiles will also need to be specified.

<u>Survivability and Durability of Geotextile</u>

Giroud et al. (1998) recognised that a geotextile with adequate and homogenous filtering properties can be specified. However a geotextile will also need appropriate mechanical properties to resist mechanical damage and deformation when being placed. To achieve this, Giroud considered a two-layer non-woven needle punched geotextile; a filter layer constructed with fine fibres to give the required filter properties and a protective layer providing the required mechanical properties to protect the filter layer from construction activities. Such products are now commercially available from a number of suppliers. Indeed multi-layered composites can be constructed by several manufacturers using lamination or needle-punching with project specific elements. Composites may also comprise a geo-net drain or cuspated plastic sheet (fin drain) with a geotextile bonded on one or both sides to provide filtration, separation and/or protection.

One of the first applications of two-layer filter geotextiles in embankment dams was on the Valcros dam in France (Artieres and Tcherniavsky, 2002). The dam incorporated some sampling points whereby the long term performance of the geotextiles could be monitored. These allowed visual monitoring of the soil particle retention mechanism. The observed particles retained inside the first filament were consistent with Giroud's (1996) constriction concept to achieve filtering requirements (Anon, 2001)

Durability of geotextiles over time is an important consideration for long term effectiveness. Polypropylenes and polyesters are long chain polymers in which structural degradation can be initiated by heat, light and chemical/biological action. The resistance properties are therefore important in specifying a filter and method statements must limit the exposure to ultra violet light during storage and installation of materials to avoid early deterioration.

Koerner (2005) provides a methodology to define as-manufactured properties of geosynthetics for design purposes. Strength should be reduced using the equation:

$$T_{allow} = T_{ult} \left[\frac{1}{RF_{ID} \times RF_{CR} \times RF_{CBD} \times RF_{SM}} \right]$$

The Reduction Factors (RF) are for installation damage (ID), creep (CR), chemical and biological degradation (CBD) and seams (if appropriate)

(SM). Koerner (2005) lists many common uses with recommended reduction factors. These numerical values are site specific and material specific and can be changed dependent upon the risk assessment of the construction methods and anticipated service within the project.

Allowable flow rates also require a range of reduction factors:

$$q_{allow} = q_{ult} \left[\frac{1}{RF_{SCB} \times RF_{CR} \times RF_{IV} \times RF_{CC} \times RF_{BC}} \right]$$

Here the Reduction Factors are for soil clogging and blinding (SCB), creep into void space (CP), intrusion into voids (IV), chemical clogging (CC) and biological clogging (BC). Again appropriate ranges are provided by Koerner (2005).

CONCLUSIONS

Design methods are available to produce a perfect filter geotextile, however due consideration also needs to be given to providing sufficient drainage capacity, the ability to resist clogging, the stability of geotextile/geocomposite soil filter systems laid on slopes and the survivability and durability of geotextiles. Geotextile/geocomposite filter systems are more economical and simpler to lay than granular filters leading to quicker and simplified construction. They also require less site quality control compared to that required to achieve a properly graded granular filter. However they do require adequate design and specification, based on an understanding of the specific issues relating to geosynthetics to guarantee long term effectiveness.

ACKNOWLEDGMENTS

The advice and assistance of David Young, ABG, Doug Langton of Geofabrics and Peter Assinder of Polyfelt Geosynthetics in producing this paper has been greatly appreciated.

REFERENCES

AASHTO M288-99 - Geotextile Specification for Highway Applications

Anon (1982) A Design Primer: Geotextiles and Related Materials, Industrial Fabrics Association International, St. Paul, MN, pp. A-2, A-4.

Anon (1996) Designing for subsurface drainage. Terram Geotextiles.

Anon (2001) Filtration Guidelines - A Technical Manual for the use of geotextiles in hydraulic engineering applications. Polyfelt Group.

American Society for Testing and Materials D4491-99a (2004) Standard Test Methods for Water Permeability of Geotextiles by Permittivity.

Artieres, O. Tcherniavsky J.G. (2002) Geotextile filtration systems for dams - 30 years of improvement 7th International Conference on Geosynthetics, 969-974.

Bhatia, S. K., Smith, J.L., Ridgway, J., and Hawkins, W., Are Porosity and O_{95} Important in the Retention and Particulate Clogging Behavior of Geotextiles Geosynthetics '99, Boston, Volume II, pp 813 – 832, 1999.

Bhatia, S. K., and Moraille, J., Comparative Behavior of Granular vs. Geotextile Filters, Geo-Congress, ASCE, Geotechnical Special Publication No.78, pp.1-29, 1998.

Bridle R. C. (2003). Melton Mowbray flood storage dam. Dams & Reservoirs, Journal of the British Dam Society, Vol 13 No 1.

Bridle R. C. (2004) Lessons Learnt from a dam incident. Long-term benefits and performance of dams, 585-596, Thomas Telford, London.

British Standards Institution (1991) BS 6906: Part 8. Methods of test for geotextiles. Determination of sand-geotextile frictional behaviour by direct shear. BSI, London.

Bourdeau, P. L., Ludlow, S. J. & Simpson, B. E. (1993). Stability of Soil-Covered Geosynthetic-Lined Slopes: A Parametric Study, pp. 1511-1521 In Proc. Geosynthetics '93 Vol. 3, Vancouver, Canada, North American Geosynthetics Society.

Brown, A. and Gosden J. (2004) Outline strategy for the management of internal erosion in embankment dams. Dams and Reservoirs March 2004, pp13-20.

Brown, A. and Gosden J. (2004) Interim guide to quantitative risk assessment for UK reservoirs. KBR/DEFRA ; Thomas Telford, London.

Brown, A. and Carter, I. C. (2004)European Symposium on Internal Erosion. Dams and Reservoirs Sept 2004.

Brown, A. and Gosden J. (2005) An example of the application of the Interim Guide to QRA. Dams and Reservoirs July 2005, 11-12.

Brown, A. and Bridle, R. (2005) IREX/EDF Workshop on internal erosion and piping of dams and foundations, 25-27 April 2005, Aussois, France. Dams and Reservoirs July 2005, 17-18.

Cedergren, H. R., Seepage, Drainage, and Flow Nets, John Wiley & Sons, New York, 1989, pp. 151, 154, 156, 165.

CFGG (1986) Recommendations for the use of geotextiles in drainage and filtration systems. French Committee of Geotextiles and Geomembranes.

Charles J. A. (1998). Internal erosion in European embankment dams Progress report of Working Group on internal erosion in embankment dams. Dam Safety (ed L Berga). Proceedings of International Symposium on New Trends and Guidelines on Dam Safety, Barcelona, vol 2, pp 1567-1576. Balkema, Rotterdam.

Charles J. A (2001). Internal erosion in European embankment dams. Proceedings of ICOLD Europeam Symposium, Geiranger, Norway, Supplementary volume, pp 19-27. [Reprinted in Proceedings of 12th British Dam Society Conference, Dublin, September 2002, pp 378-393. Thomas Telford, London.

Christopher, B. R. and Fischer, G. R., (1991) Geotextile Filtration Principles, Practices and Problems, 5th GRI Seminar on Geosynthetics in Filtration, Drainage and Erosion Control, Philadelphia, P, pp. 6-9.

Christopher, B. R. and Holtz, R. D. (1985), Geotextile Engineering Manual, U.S. Federal Highway Administration, Washington, D. C., FHWA-TS-86/203, 1044 pp.

Christopher, Barry R. and Fischer, Gregory R., "Geotextile Filtration Principles, Practices and Problems", 5th GRI Seminar on Geosynthetics in Filtration, Drainage and Erosion Control, Philadelphia, PA, 1991, pp. 6-9.

Cooper G. A. (1987) The reservoir safety programme in Northern Ireland. IWEM Journal August 1 (1) 39-51.

DeBerardino, S. J., (1992) Permeability or Permittivity? What Should Be Specified?, Geotechnical Fabrics Report, Industrial Fabrics Association International, St. Paul, MN, Nov. 1992, pp. 852.

Dixon N and Jones D R V. (2003) Stability of Landfill Lining Systems: Report No. 2 : Guidance. Environment R&D Technical Report P1-385/TR2

Fischer, G. R., Christopher, B. R., and Holtz, Richard D., (1990) Filter Criteria Based on Pore Size Distribution, Fourth International Conference on Geotextiles, Geomembranes and Related Products, A.A. Balkema Publishers, Brookfield, VT, pp. 289-294.

Gardiner K. D, Hughes A. K and Brown A. (2004) Lessons from an incident at Upper Rivington Reservoir – January 2002. Dams and Reservoirs.

Giroud, J. P., (1982) Filter Criteria for Geotextiles, Second International Conference on Geotextiles, Las Vegas, NV, pp. 103-108.

Giroud, J. P., Delmas, P. and Artieres, O. (1998), Theoretical basis for the Development of a Two-Layer Geotextile Filter, Proc. 6th ICG, Atlanta, IFAI, pp. 1037-1044.

Giroud, J. P., (1984) Geotextiles and Geomembranes: Definitions, Properties and Design, Industrial Fabrics Association International, pp. 37-38.

Giroud, J P, Williams N. D, Pelte, T. and Beech J. F., Stability of Geosynthetic-Soil Layered Systems on Slopes, Geosynthetics International, Vol. 2, No. 6, 1995, pp. 1115-1148.

Giroud, J. P., (1988) Review of Geotextile Filter Criteria, First Indian Geotextiles Conference on Reinforced Soil and Geotextiles, India, pp. 1-6.

Giroud, J. P., Zhao, A., and Bonaparte, R., 2000, The Myth of Hydraulic Transmissivity Equivalency Between Geosynthetic and Granular Liquid Collection Layers, Geosynthetics International, Special Issue on Liquid Collection Systems, Vol. 7, Nos. 4-6, pp. 381-401.

Hughes, A, Hewlett, H, Samuels, P. G, Morris, M, Sayers, Moffat, I, Harding, A and Tedd, P (2000). Risk Management for UK Reservoirs. Construction Industry Research and Information Association, CIRIA C542.

Hughes A. K, Lovenbury H and Owen E (2001). Audenshaw Reservoir. Dams & Reservoirs, Journal of the British Dam Society, Vol 11 No 1.

Institution of Civil Engineers, Floods and reservoir safety: 3rd Edition. Thomas Telford, London.

ICOLD Bulletin No. 55 (1986) Geotextiles as Filters and Transitions in Fill Dams.

International Commission on Large Dams (1994). Embankment dams - granular filters and drains. Bulletin 95. ICOLD, Paris.

Jones D R V and Dixon N. (2003) Stability of Landfill Lining Systems: Report No. 1 (2003) Literature Review Environment R&D Technical Report P1-385/TR1

Koerner, R. M., (2005) Designing with Geosynthetics, 5th Edition, Prentice-Hall, Englewood Cliffs, NJ.

Koerner, R. M. and Hwu, B.-L., (1991) Stability and Tension Considerations Regarding Cover Soils on Geomembrane Lined Slopes, Jour. Geotex. And Geomemb., Vol. 10, No. 4, pp. 335-355.

Rankilor, P. R. (1981) Membranes in Ground Engineering, Wiley, 377 pp.

Rollin, A. L., Mlynarek, J., and Bolduc, G., (1990) Study of Significance of Physical and Hydraulic Properties of Geotextiles Used as Envelopes in Subsurface Drainage Systems", Fourth International Conference on Geotextiles, Geomembranes and Related Products, A.A. Balkema Publishers, Brookfield, VT, pp. 363.

Seed, R. B., Mitchell, J. K. & Seed, H.B. (1988). Slope Stability Failure Investigation: Landfill Unit B-19, Phase 1-A, Kettleman Hills, California, Report No. UCB/GT/88-01, Dept. of Civil Engineering, University of California, Berkeley, July 1988. 96pp.

Sherrard J L and Dunnigan L P (1990). Critical filters for impervious soils. ASCE Journal of Geotechnical Engineering Division, January, Vol 115 (GT7) pp 927-947.

Standard Specification for Geotextiles: AASHTO M288-90, Federal Highway Administration, pp. 689-692.

USDA Soil Conservation Service (1986). Guide for determining the gradation of sand and gravel filters. Soil Mechanics Note No 1 210-VI. United States Dept of Agriculture, Soil Conservation Service, Engineering Division, Washington.

Vaughan P. R., Kluth D. J, Leonard M. W and Pradoura H. H. M.(1970). Cracking and erosion of the rolled clay core of Balderhead dam and the remedial works adopted for its repair. Transactions of 10th International Congress on Large Dams, Montreal, Vol 1, pp 73-93.

Vaughan P. R. & Soares H. F. (1982). Design of filters for clay cores of dams. ASCE Journal of Geotechnical Engineering Division, January, Vol 108 (GT1) pp 17-31.

Vaughan P. R. (2000a). Internal erosion of dams – assessment of risks. Filters and drainage in geotechnical and environmental engineering, proceedings of the third international conference - Geofilters 2000, Wolski W & Mlynarek J, eds. Balkema Rotterdam. ISBN 90 5809 146 5

Vaughan P. R. (2000b). Filter design for dam cores of clay, a retrospect. Filters and drainage in geotechnical and environmental engineering, Proceedings of the Third International Conference - Geofilters 2000, Wolski W & Mlynarek J, eds. Balkema Rotterdam. ISBN 90 5809 146 5

Vaughan P. R. (2001) Imperial College – Earthworks Course Notes. Lecture - 16 Internal erosion filters and drains.

Vaughan, P. R. and Bridle R. C. (2004) An update on perfect filters. Long-term benefits and performance of dams, 516-531, Thomas Telford, London.

4. Planning and design

Wave surcharge on long narrow reservoirs- a reality check

ANDREW KIRBY, Mott MacDonald, Cambridge
KENNY DEMPSTER, Scottish and Southern Energy, Perth

SYNOPSIS.   Safety reviews carried out for Scottish and Southern Energy
had initially identified five hydroelectric dams where the wave surcharge
allowance required was significantly greater than the wave freeboard
available. These initial reviews suggested that significant dam heightening,
of up to 3 m in one case, would be required to provide adequate wave
freeboard.

The dams in question impound long narrow lochs, up to 26 km in length.
Basic wave generation formulae suggest that the full length of the loch is
available for wave generation resulting in very large significant wave
heights. A reality check on this simplified approach was required. Wave
analysis methods from the maritime sector were used to model wave
generation on the reservoirs. To reduce survey costs old and new
technologies were combined by turning 100 year old bathymetric survey
into a digital bathymetric model of the loch beds. The modelling output
confirmed that significant engineering judgement is required to assess the
influence of loch shape on wave generation to determine an appropriate
fetch length.

INTRODUCTION

As part of Scottish and Southern Energy's ongoing safety assessments of its
stock of dams a series of initial assessments had been carried out to check
the freeboard of its dams. One such assessment of wave surcharge had
identified four hydroelectric dams in the Highlands with potentially
significant freeboard deficits of up to 3 m. Such a large deficit was a
concern both in terms of dam safety and the potentially high cost of
improvement works. In order to develop a programme of improvements it
was considered that a review was required to confirm the need and scale of
the works. This paper describes the detailed wave surcharge assessment
carried out at the four dams: Lairg dam on Loch Shin, Fannich, Glascarnoch

and Loch Mhor. A fifth dam, Little Loch Shin, was also included following concerns over wave surcharge following a Statutory Inspection.

The purpose of the detailed assessment was to check the initial assessment, review current wave surcharge literature and determine whether site specific conditions influence the applicability of the initial estimates. Where dams were still found to have a freeboard deficit a risk analysis was to be carried out and measures to reduce the risk of damage or failure proposed.

ISSUES TO BE CONSIDERED

Wave Height Prediction

Up to the late 1940's wave height prediction for fetches under 23 miles was carried out using Stevenson's formula:

$$H_w = 1.5F^{0.5} + 2.5 - F^{0.25}$$

where H_w is the wave height from crest to trough (in feet), and F is the fetch in nautical miles. Research in the United States during the 1950s and early 1960s led to the publication in 1962 of what has come to be known as the SMB/Saville method. This is the method used in the 1978 edition of Floods and Reservoir Safety (FRS). The equations can be expressed as:

$$\hat{H} = 0.0026\left(\hat{F}\right)^{0.47} \text{ and } \hat{T}_m = 0.46\left(\hat{F}\right)^{0.28}$$

where $\hat{H} = \dfrac{gH_s}{U^2}$, $\hat{T}_m = \dfrac{gT_m}{U}$, $\hat{F} = \dfrac{gF}{U^2}$

Saville et al found that using a simple fetch length along the wind direction overestimated wave heights where the fetch width was small compared with the fetch length (i.e. in long narrow reservoirs). After assessing a number of different methods to adjust the fetch length the concept of an effective fetch was developed, familiar to users of the FRS 1978 edition, where the length of radials is measured over a range from 45 degrees either side of a central radial and then their component in the central radial direction is summed to give:

$$F = \frac{\sum x_i \cos\alpha_i}{\sum \cos\alpha_i}$$

In the early 1970's a wave measurement programme was carried out in the North Sea as part of the Joint North Sea Wave Project (JONSWAP). The following equations were developed from this work:

$$\hat{H} = 0.00178\left(\hat{F}\right)^{0.5} \text{ and } \hat{T}_p = 0.352\left(\hat{F}\right)^{0.3}$$

Donelan introduced the concept that the fetch length should be measured along the wave direction rather than the wind direction, thus the wind speed used should be the component of the wind speed acting in the wave direction. This became known as the Donelan/Jonswap method and is that used in the latest edition of FRS (1996).

Preference for the use of this method over the SMB/Saville method in FRS appears to be at least partly due to work by HR Wallingford in the late 1980s and in the 1990s comparing wave prediction methods against actual wind and wave measurements at Megget and Glascarnoch reservoirs (Owen, 1987 and Owen and Steele 1988). The work showed that none of the methods gave particularly good agreement with observed data for all wind speeds and directions. It did show that the SMB/Saville methods underestimated wave heights at long narrow reservoirs like Glascarnoch (length to width ratio of 13.5), while giving a good match at shorter reservoirs like Megget (length to width ratio of 7). The Donelan/Jonswap method gave a better match to observed data at Glascarnoch, though over-estimated wave heights at Megget. The conclusion from this work was that the Donelan/Jonswap method should be used in preference to the SMB/Saville method as it was conservative.

<u>Fetch length for irregular shaped reservoirs</u>

At two of the five dams (Lairg Dam and Fannich Dam) the large freeboard deficit could be accounted for by the long fetch length used for estimating the wave surcharge.

Lairg Dam lies at the southern end of Loch Shin. Loch Shin has a total length along its longitudinal axis of around 27km and thus has potentially a very long fetch length. However it is relatively narrow with its width varying between 650m and 2000m. The current edition of Floods and Reservoir Safety (1996) introduced the concept that wind may change direction as it passes down steep sided gently curved reservoir such that changes in direction of the longitudinal axis of a reservoir (up to about 50 degrees) should be ignored when determining fetch length. This advice appears based on information in Herbert et al (1995) and Yarde, Banyard and Allsop (1996). It should be noted that the advice is based on anecdotal evidence rather than observed data. The implications of applying this "bent fetch" approach to Loch Shin is significant increasing the fetch by a factor of 10 from about 3km to nearly 30km. As a consequence the estimated significant wave height is large.

A similar situation occurs at Fannich Dam which has a curved "banana shaped" reservoir, with a total length along the central axis of around 12km but a width of only between 600m and 1400m.

A key issue was therefore whether waves would propagate along such long narrow curved reservoirs.

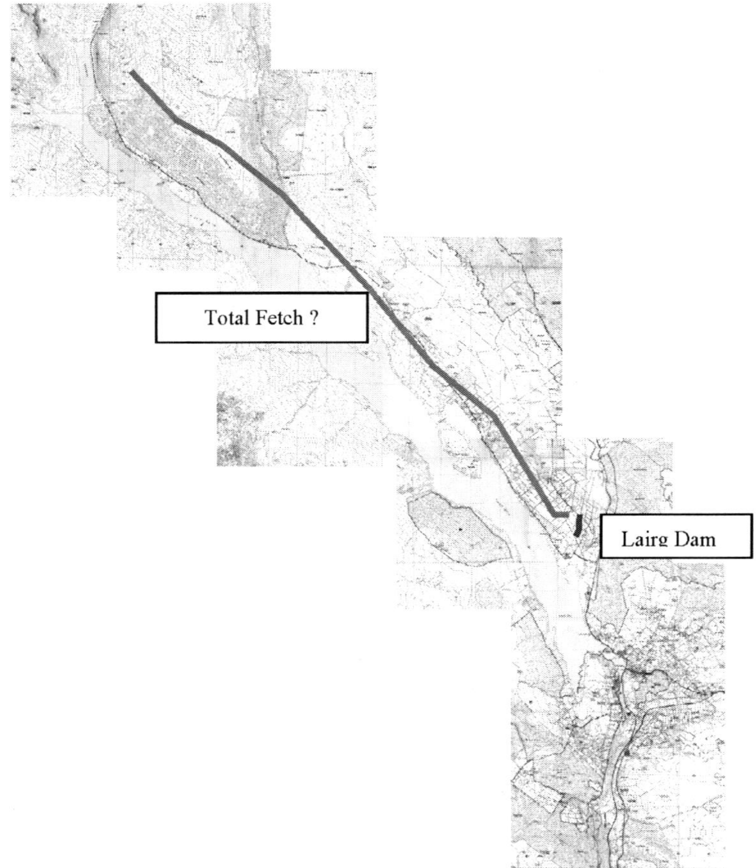

Figure 1. Loch Shin – fetch along longitudinal reservoir axis

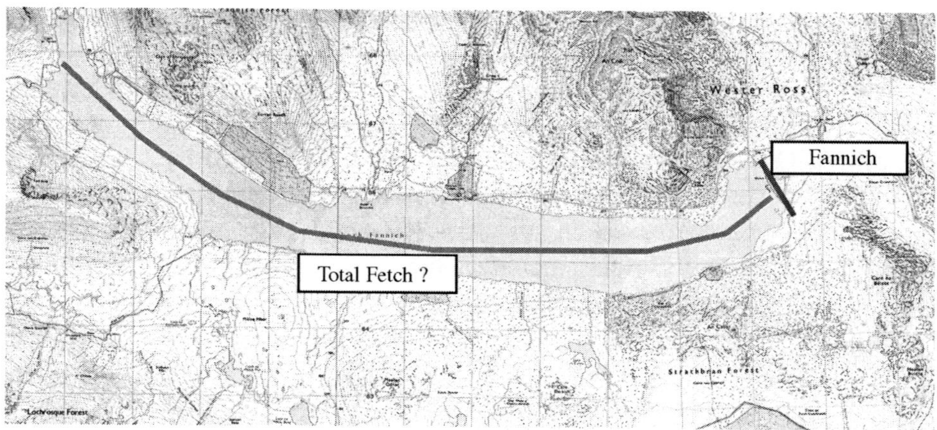

Figure 2. Fannich reservoir - fetch along longitudinal reservoir axis

<u>Wave run-up and overtopping</u>

FRS (1996) uses the concept of a wave surcharge to determine dam freeboard but acknowledges the approach has limitations. These include that it does not cater for composite dam sections, such as an embankment with a wave wall, also no estimation is made of the quantity of water overtopping. It suggests the use of a numerical approach to assess in detail wave carry over and surcharge at existing dams, giving the example of the program SWALLOW, developed by HR Wallingford. SWALLOW uses the method developed by Owen for wave overtopping of sloping embankments: Design of seawalls allowing for wave overtopping (1980). In the early 1990s the method was expanded to cover vertical walls and recurved wave return walls (Herbert, 1993 and Owen and Steele, 1993), however SWALLOW was not updated to cover these wall types. Further research in the 1990s refined the formulae used in Owen (1980). The methods for different types of seawall were brought together in a single volume by the publication by the Environment Agency of Wave Overtopping of Seawalls Design and Assessment Manual (Besley, 1999). This is currently the reference that covers the widest range of embankment types. The methods were developed from series of model tests, some backed up with observed data.

Yarde, Banyard and Allsop (1996) compared the run-up method with the overtopping method and found it to be conservative compared with the overtopping method. They justified the adoption of the overtopping method as a logical and better supported approach, consistent with other practice and easily understood.

The overtopping method should only be used within the range of application covered by the model testing. Where conditions exist that are outside those modelled then modelling of particular conditions is required. This can be using physical models but also numerical techniques are coming to the fore e.g. AMAZON, a 2D wave modelling software developed by Manchester Metropolitan University's Centre for Mathematical Modelling and Flow Analysis (CMMFA). Research on assessing wave overtopping in coastal conditions is ongoing and includes the development of a European overtopping Manual. Some of this research will be applicable to reservoir conditions.

<u>Tolerable overtopping discharges</u>

The overtopping method described above estimates the mean and peak overtopping discharges. To determine a wave freeboard a safe or allowable overtopping discharge needs to be assigned to a dam. This will depend on the type of construction of the dam and the type of access required along the crest. Criteria for safe overtopping discharges were investigated by Simm (1991) and Franco et al (1994). Criteria are summarised in Yarde, Banyard and Allsop (1996) and the Environment Agency's Wave Overtopping of Seawalls Design and Assessment Manual (1999). These safe allowable overtopping discharges primarily relate to sea defence type embankments and therefore may need some consideration if applied to dams.

Some of the dams that were assessed were rockfill which may be able to withstand higher volumes of overtopping without damage than conventional sea defence embankments. However research on flow over rockfill dams has primarily concentrated on two areas: breaching due to overtopping and the use of rockfill spillways as a low cost alternative to concrete structures. However research tends to concentrate on steady uniform flow and could not be directly applied directly to wave overtopping.

<u>Wind Setup</u>

Wind setup (also called wind tide) is the rise in stillwater level caused by wind stress on the surface of a body of water. USBR (1981) and USACE (1997) both present the same estimation method, which in SI unit is:

$$S = \frac{V^2 F}{4850D}$$

It is not mentioned in FRS (1996), probably because reservoirs in the UK are generally relatively small so the effect is considered insignificant. The magnitude of wind set up was verified to confirm that this assumption held true for the long reservoirs being considered.

Wind funnelling

Some of the dams being analysed were at the end of steep sided valleys where wind funnelling may be an influence. Derivation of a design wind speed takes account of the effect of altitude on wind speed but does not generally take account of the effect of topography. The effect, wind funnelling, is more usually associated with building design for the determination of wind loadings.

Information on wind funnelling along valleys is sparse in building codes. BRE Digest 346 (1989) gives a method for calculating the effect on winds of valleys perpendicular to the wind direction. However for winds parallel to the valley axis it states "wind blowing along the axis of a valley is not significantly changed".

The other sector where wind funnelling has been researched is in forestry where the design of plantations requires an assessment of the risk of windthrow (loss of trees by wind damage) which is influenced by topographic effects. The most relevant reference found was by Ruel (1998) where a physically model test of mountainous terrain was used to evaluate wind speed increases in valleys and on hilltops. The model by its nature is site specific, but appears not dissimilar to Highland conditions. Results of the model test suggested wind speed increases of 25 – 40 % for wind directions parallel to valley axes and increases of 80 – 120 % over hilltops.

Wind funnelling will be specific to a particular site. The literature shows that topography can cause variations in local wind speeds but does not provide conclusive evidence that wind funnelling over a long fetch will occur causing increased wave heights.

Waves in Shallow Water

Incoming wave are transformed as they pass from deep water to shallower near-shore waters. The transformation is by refraction, shoaling, diffraction and breaking. Refraction is the change in wave velocity as waves propagate in varying depths. As a result waves change direction so that they tend towards the direction normal to the local bed contours. Shoaling is the change in wave height as waves propagate through varying water depths. This may have an effect, particularly where the bathymetry in front of the dam is high and the wave heights are large. Diffraction is the transformation of waves by obstacles such as headlands or breakwaters. For all but simple obstacle shapes diffraction is complex to analyse requiring numerical modelling. Diffraction can both attenuate and increase wave heights. Wave breaking before waves reach the dams could result in significant wave energy dissipation, thus reducing the hydraulic loading on the structures. Wave breaking is dependent on depth and steepness. The CIRIA/CUR manual: Manual on the use of rock in coastal and shoreline engineering. (1991) provides details of analysis of the above wave transformation conditions.

<u>Outline methodology for analyses</u>

Based on the above the methodology shown in the figure below was adopted to consistently assess each dam.

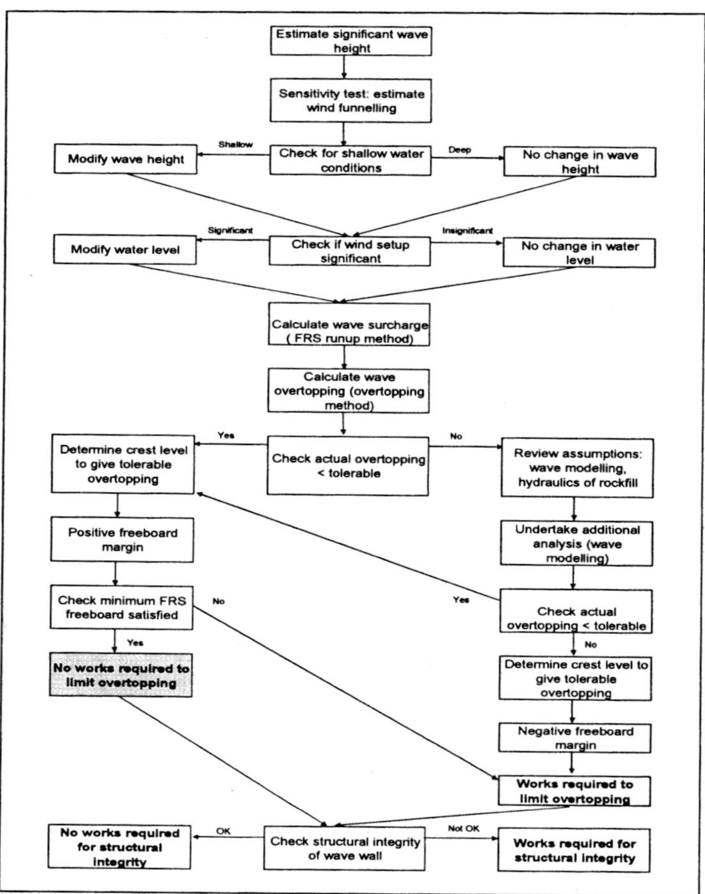

Figure 3. Wave surcharge assessment methodology

WAVE MODELLING

To review the applicability of using the total reservoir length as the fetch length at Loch Shin and Loch Fannich wave modelling was carried out using MIKE21 using the "NSW" module. MIKE 21 NSW is a spectral wind-wave model, which describes the propagation, growth and decay of short-period waves in nearshore areas and are therefore reasonably applicable to reservoir conditions. The model includes the effects of refraction and shoaling due to varying depth, wave generation due to wind and energy dissipation due to bottom friction and wave breaking. The module has some limitations in that it does not include diffraction, reflection or wave-wave interaction. Other modules are available that include these factors, but are more normally used for modelling wave effects in harbours or around breakwaters.

A problem with developing wave models for relatively small scale studies is the cost of collecting bathymetric data, which can easily be significantly more costly than the study itself. We were fortunate in being able to overcome this problem using data from an unlikely source. Between 1897 and 1909 bathymetric surveys were carried out of all the freshwater lochs of Scotland (Murray, 1897-1909). This gives bathymetric data as contours at 50 foot intervals and spot depths along survey lines at about 250 m spacings. Copies of these surveys were held by SSE as well as the National Library of Scotland. Though the survey is old it gives very good detailed coverage of the lochs which would be costly to obtain by re-surveying the reservoir.

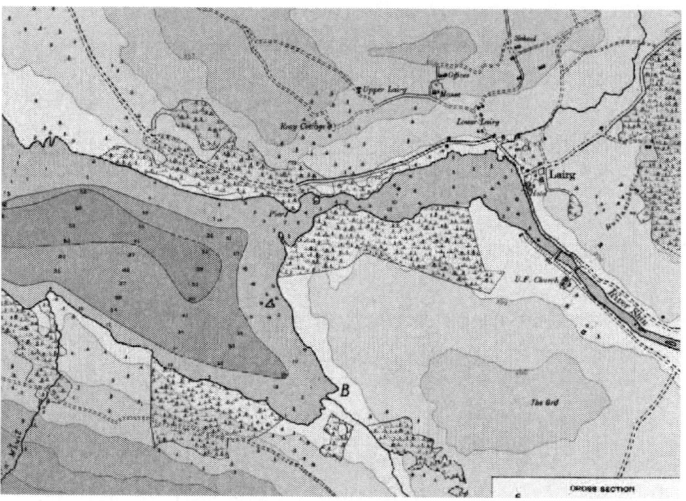

Figure 4. Bathymetric survey from 1897-1909

The survey pre-dates dam construction and some sedimentation may have occurred in the reservoir since construction. However due to the nature of the catchment this was judged to be minor.

Another problem faced with the data was correlating the survey datum, referred to in the survey just as "sea level", with Ordnance Datum (Newlyn). The offset between datums was not known so an attempt was made to correlate known features between the bathymetric survey and Ordnance Survey mapping to estimate a datum offset. Correlation of these features showed that the two datums matched within a few metres, though more accurate estimation of the datum offset was not possible. Instead the sensitivity of wave height to changes in the datum was tested by raising or lowering the bed level relative to water level.

The bathymetric survey was scanned and then digitised on-screen using the Arcview GIS package to develop a digital terrain model (DTM). The reservoir water edge as shown on OS mapping was digitised and overlain into the DTM both as additional water edge data and to confirm the correlation between the two maps. The DTM was developed by interpolating a 20 m x 20 m grid of bed levels from the bathymetric data. The DTM was then used as input to MIKE21.

Wind speeds and directions were developed along the reservoir for different wind and water level conditions. The wind directions adopted assumed that the topography would funnel the wind along the axis of the reservoir towards the dam. A number of different wind directions were analysed to find the worst case. Wind speeds were developed taking account of the increase in the speed of wind as it travels over-water.

The results showed the general features of wave generation that one would expect:

- Narrow reservoir widths limit wave generation

- Headlands and an uneven reservoir shore limit wave generation and can provide a considerable sheltering effect

- Wave heights generated are slightly greater than those calculated assuming a straight line fetch length confirming the anecdotal evidence that waves to a certain extent can be steered around the "banana shaped" reservoir

- Wave heights generated are significantly lower than those calculated assuming the total length of the reservoirs for wave generation confirming that judgement is required in applying the bent fetch length principle.

For large scale modelling it would be normal practice to calibrate the MIKE21 model with observed wind/wave data. In this case no calibration data was available. The cost of collecting the data would be high, of the order of £ 20,000. It was not considered necessary for this study to undertake data collection and calibration. Instead sensitivity tests were carried out and the model output was used with the understanding that there is a degree of uncertainty in the wave heights calculated and safety factors should be applied.

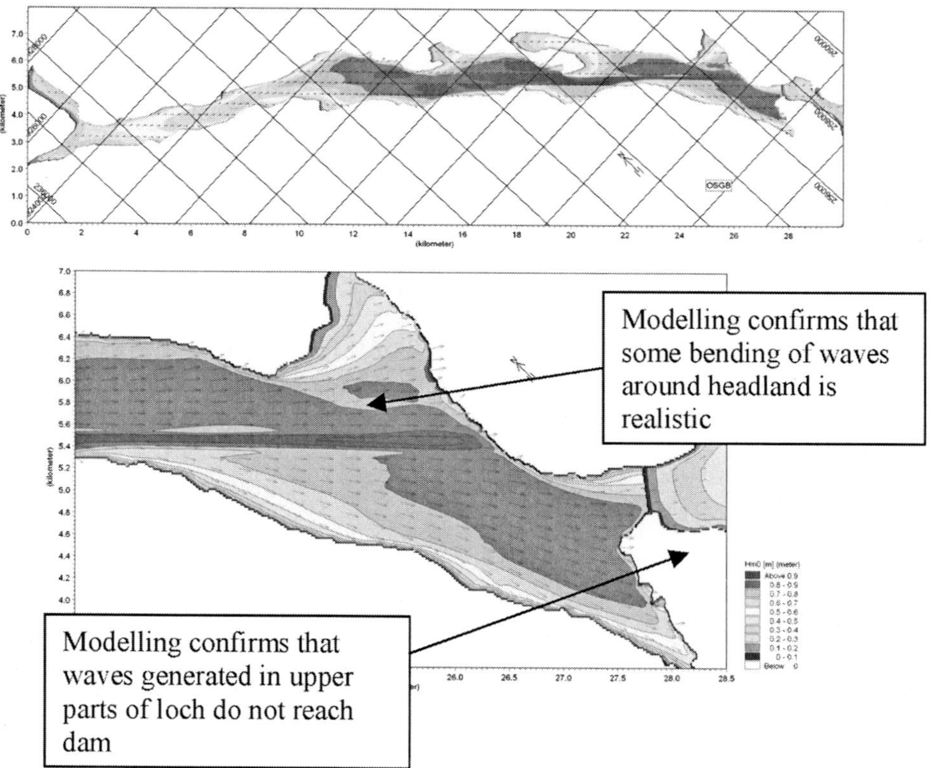

Figure 5. Wave modelling – Loch Shin

There would be opportunities to give additional confidence in the model outputs by using the wind and wave data collected by HR Wallingford for Glascarnoch and Megget as calibration of a model of Glascarnoch or Megget reservoirs but this has not been carried forward in this study.

CONCLUSIONS

The wave modelling reduced the freeboard deficit to a more realistic level. A range of remedial works to wave walls at these dams are now planned into SSE's works programme covering wave wall raising and strengthening.

The study confirmed the need to apply common sense and judgement when assessing wave surcharge for existing dams. For long narrow reservoirs, with fetch length to width ratios of about 10 or more, use of a standard straight line fetch along the full reservoir axis may be inappropriate for estimating wave surcharge. Numerical modelling techniques provide an alternative method of analysis, and are particularly cost effective where existing bathymetric data is available.

Improved methods for assessing overtopping have been developed since publication of the latest edition of Floods and Reservoir Safety, such as the Environment Agency's wave overtopping manual. Further research in this area is ongoing in the coastal sector such as the development of a European Overtopping Manual. Such knowledge can be transferred and applied to the dams and reservoirs sector when detailed wave surcharge assessments are required.

REFERENCES

Besley. 1999. Wave Overtopping of Seawalls Design and Assessment Manual. R&D Technical Report W178. Environment Agency.

British Research Establishment. 1989. BRE Digest 346 The assessment of wind loads.

CIRIA/CUR.1991. Manual on the use of rock in coastal and shoreline engineering. Special Publication 83. Construction Industry Research and Information Association.

CIRIA (1987) Design of Reinforced Grass Waterways. CIRIA Report 116

Franco, L, de Gerloni, M and van der Meer, J. W. 1994. Wave runup and overtopping on coastal structures. Proceedings of the Coastal Engineering Conference, Kobe, Japan, 1994.

Herbert, D. M. 1993. Wave overtopping of vertical walls. Report SR 316. HR Wallingford.

Herbert, D. M. et al. 1995. Performance of Blockwork and Slabbing Protection for Dam Faces. Report SR 345. HR Wallingford.

ICE. 1996. Floods and reservoirs safety. 3rd Edition.

Murray, J., 1897-1909. Bathymetrical Surveys of the Freshwater Lochs of Scotland

Owen, M. W. 1980. Design of seawalls allowing for wave overtopping. Report EX 924. HR Wallingford.

Owen M. W. 1987. Wave Prediction in Reservoirs: A Literature Review. Report Ex1527. Hydraulics Research.

Owen M. W. and Steele. A. A. J. 1988. wave prediction in Reservoirs – Comparison of Available Methods. Report EX 1809. Hydraulics Research.

Owen, M. W. and Steele A. A. J. 1993. Effectiveness of recurved wave return walls. Report SR 261. HR Wallingford.

Ruel, J-C., Pin, D. and Cooper, K 1998. Effect of topography on wind behaviour in a complex terrain. Forestry, Volume 73.

Saville, T. et al. 1962. Freeboard Allowance for Waves in Inland Waterways. Journal of the Waterways and Harbours Division, Proceedings American Society of Civil Engineers.

USACE, 1997. Engineering and Design - Hydrologic Engineering Requirements for Reservoirs EM 1110 2 1420

USBR. 1981. Freeboard Criteria and Guidelines for Computing Freeboard Allowances for Storage Dams. U.S. Department of the Interior, Bureau of Reclamation.

Yarde, A. J. Banyard, L. S. Allsop, N. W. H. 1996. Reservoir Dams: wave conditions, wave overtopping and slab protection. Report SR 459. HR Wallingford.

The Kielder Water Scheme: the last of its kind?

CS MCCULLOCH, University of Oxford, UK

SYNOPSIS.   The peculiar history of the Kielder Water Scheme provides insights into the operation of democracy, the politics of promotion of mega-projects and the problems of their subsequent assessment and accountability. Two public inquiries were held before the Scheme was approved but the industry it was planned to supply was already reducing its water requirements before construction started.   Opposition to the reservoir, particularly from those whose homes it would displace, was strong and divided Conservative political allegiances. The controversies, which led the Director of the Water Resources Board to claim that schemes like Kielder would never be repeated, continue even today. The history of the Scheme has been explored by examination of the records of the public Inquiry and by interviews with some of the principal actors involved in the drama.

INTRODUCTION

The Kielder Water Scheme was conceived in the mid 1960s at a time when the power and autonomy of water engineers in England and Wales had risen to levels never before, nor since, attained.   Unfortunately, engineering accomplishments were often marred by economic miscalculation.   The resulting mismatch between vastly increased water supply at a time of diminishing rise in demand, together with huge debts incurred at a time of rapid inflation and high interest rates, had lasting effects on the state's management of water resources.   Political intrigue and the overbearing ambition exhibited by the Northumbrian River Authority may have contributed to the replacement of river authorities by water authorities in preparation for privatisation of water supply in 1989.

A regional-scale scheme such as a water transfer network on the scale of Kielder, involving a large storage reservoir and inter-river transport of water through a tunnel, requires support politically and financially at more than the local scale.   Big schemes require big finance and state backing to proceed with schemes in the face of strong local opposition.   The political setting which encouraged engineers to think of major schemes such as the

Kielder Water Scheme will be described before its history and implications for water governance.

WATER RESOURCES BOARD

In England and Wales in the 1960s, post- War belief in national planning of resources was undiminished. As late as 1965, the idea of public ownership of all water supplies still appeared in the Labour Party's manifesto but successive governments failed to nationalise water because of fear of antagonising municipal and local authority water undertakings as well as private owners (Hassan, 1994). However, Members of Parliament (MPs) dealing with Private Bills for many reservoir schemes during the 1950s and early 1960s called for a national strategy for water resource development against which individual projects could be assessed. Reservoir construction in England was growing exponentially and many feared the associated damage to amenity and loss of farmland if this upward trend should continue unabated. Parliament wanted professional advice.

Following the 1963 Water Act, a national planning body dedicated to water resources was set up to strengthen the slow-moving Ministry of Housing and Local Government (MHLG) with its multiple responsibilities. This newly-formed body, the Water Resources Board (WRB), was an unique experiment in self governance of water engineers by water engineers and, remarkably, it survived for nearly a decade. Emerging from their customary position in the background, water engineers were allowed centre stage to proclaim their ideas of rational planning at regional and national scales. In typical British fashion, though, their power was limited to the giving of advice and was hampered by being confined to considerations of water quantity, crucially omitting quality.

The WRB's first Annual Report (WRB 1965) announced its role as "the master planner of the water resources of England and Wales", although implementation of its plans was not straightforward. In England and Wales, supplied by many rivers, the case for national planning was not obvious. River catchments appeared to be more suited for management purposes because of the interrelationship between water flowing from the tributaries into the principal rivers, on the way to the sea. Even Barry Rydz, Director of WRB planning, conceded that large areas of England and Wales were best served by local planning of water resources (Rydz 1971). Yet, many of the questions raised by reservoir construction impinged on national policies for industrial growth or for support for agriculture. The WRB was faced with a challenge to reconcile local issues with national policies, based on persuasion rather than authority.

The WRB built up to a staff of around a hundred and had a modest research budget but its influence conceptually was far greater than its size or budget might suggest. The Third Annual Report of the WRB stated:

> 'The Water Resources Act 1963 with its emphasis on collection of data provides the incentive to apply scientific and engineering principles to achieve logical development, so that water can be made available in the quantities required where and when it is needed (WRB, 1967,29).'

Supply "where and when it is needed" was the objective rather than adaptation of human developments to water availability.

With this objective, the WRB privileged quantifiable information. Their confidence in "logical development" allowed the rubbishing of opposing arguments, which were not proved by hard data. Adverse reaction "to the exploitation of the water resources of an area for the benefit of water consumers far away" was deemed "irrational" (WRB, 1967, para 49). The Director of the WRB, Norman Rowntree, believed "maintenance of our standards of life depends on expanding industrial, commercial and agricultural activity" and the "maximum development of natural resources". "The solution of water supply problems…will require the construction and operation of large works and highly-developed technical control". He believed that his opponents should not have a monopoly on emotion, "Enthusiasm and fervour" should be added to the water engineers' "cold calculations of safety, yield and cost" (Rowntree, 1962, 267).

Large-scale schemes such as interbasin transfers of water or even establishment of water grids on the model of the electricity grid certainly aroused enthusiasm and fervour and recognition for regional planning by the WRB. Without the WRB, and the financial arrangements endorsed by the 1963 Water Act, water resource development in the North East would have been very different.

NORTHUMBRIAN RIVER AUTHORITY

Another important player, this time with executive powers, was the Northumbrian River Authority (NRA), set up with 28 others, by the same Water Act 1963 which established the WRB. The ambitions of the NRA and the WRB reinforced each other. Both favoured large schemes to increase industrial water supply, mainly to Teesside. Like WRB, the NRA lasted only a decade and approval of their Kielder Water Scheme, achieved in 1973, was shortly followed by the taking over of their responsibilities and debts by the Northumbrian Water Authority (NWA), which became operational in 1974.

The Kielder Water Scheme involved construction of a large, remote storage reservoir and use of innovative tunnelling machines to link regulated rivers. Pipelines were not unusual but Kielder was revolutionary because it was "a very big tunnel and regulating" (Jackson Interview). In the words of the WRB, "The scheme is a bold and imaginative one: the largest single water conservation scheme yet undertaken in this country" (WRB, 1973, Appendix 2, 23).

In contrast, Pearce calls Kielder Water the "Cunningham reservoir" as a journalistic device to denigrate the whole Scheme as "an embarrassing and expensive monument to the follies of water planners in the 1970s" (Pearce, 1982, 8). Andrew Cunningham was jailed in 1974 for accepting bribes from John Poulson the notorious architect in return for a commission to design a grandiose headquarters for the NRA when he was its Chairman. Cunningham was influential in encouraging the ambition of the Kielder Scheme and dogged in its defence but it is wrong to attribute the vision to him. Credit should go to Urban Burston, Chief Engineer of the NRA[1], whose former colleague and successor, Nigel Ruffle, developed the plans soon after the formation of the NWA in 1974 (Rennison, 1979).

HISTORY OF THE SCHEME

In September 1965, the WRB set up a Northern Working Party of water engineers from River Authorities and water undertakers. Andrew McLennan[2], formerly Director of the Sunderland and South Shields Water Company was the Chairman and Burston was a member. Its role was to consider the possibilities of regional-scale cooperation in the development of water resources. The enthusiasm of Burston for planning water resources on a regional scale influenced the Working Party and his ambition was welcomed by WRB officials, who needed new ideas reaching beyond local water undertakings to justify their national planning role.

At first, the reservoir planned on the North Tyne was called Otterstone, rather than Kielder, and it was endorsed in the Interim Report of the Northern Working Party in 1967. To estimate water demand up to 2000, the Working Party used population projections of a 25% rise from the Office of National Statistics (in fact population declined!), per capita water consumption figures from the USA (despite differences in lifestyles and climate) and assumptions that water demand from industry would continue to grow rapidly. In the Interim Report, the use of aqueduct(s) linking the

---

[1] Two others, in addition to Burston, were attributed responsibility at the opening of Kielder in 1982, Ted Wrangham, farmer, and Andrew McClennan, Vice Chairman of WRB.
[2] McClennan was later succeeded as Chairman by J.F.Glennie and B. Rydz

principal rivers (the Tyne, Wear and Tees) in the North East, offered satisfaction to immediate deficiencies on Teesside but further reflection suggested that such aqueducts could not be built before completion of the newly-approved Cow Green reservoir in 1971 (NRA, 1973). The idea of linking the rivers was shelved only temporarily. If demand continued to rise as rapidly as predicted, even Cow Green would not assuage industry.

In 1969, the Scottish consulting engineering firm of Babtie, Shaw and Morton was appointed to prepare feasibility reports. The team's experience in building hydroelectric dams and tunnels in Scotland was pertinent to the new task. After the WRB's Interim Report, the Babtie consulting engineers made a desk study on alternative sources to meet the predicted rise in demand. The challenge of transferring water between rivers proved very attractive to the engineers involved. David Coats, the consulting engineer who had overall charge of the project, admired Urban Burston's big ideas: "he was very good in thinking forward and he was very keen that something be done". Coats himself was enthusiastic both about tunnelling and about thinking big. His previous work on the large and ambitious Loch Katrine tunnel and reservoir scheme had initially provided an embarrassing surplus of water for Glasgow but, eventually, it had proved to be valuable. Coats believed that "to think small and to assume that things are not going to change is wrong" (Coats interview).

In February 1970, the WRB's Final Report on "Water Resources in the North" recommended that the Tyne-Tees Link should be completed as soon as possible, "to be in operation by 1975"; two sites should be investigated as potential storage reservoirs Otterstone (Kielder) and Irthing. The report favoured Kielder, which could provide a yield of about 200mgd (910,000 m^3/d) and could meet all the projected water demands in the area until after 2001 or, alternatively, could meet the needs of both Northumbria and Yorkshire "for about 20 years". One very large new reservoir in Kielder Forest would solve other problems. The Forestry Commission had underestimated the rapid decline in labour requirements which had followed introduction of machinery; they had housing surplus to requirements in this remote area and construction of a reservoir would offer some alternative employment. WRB stated that 'A reservoir here would cause a minimum of disturbance and could be attractive in appearance, offering opportunities for a tourist centre (WRB, 1970a, 29).'

The WRB presented two alternative strategies. Plans for six new reservoirs in the West-East strategy (see Fig.1) looked more challenging politically and less engineeringly-exciting than a very large one at Otterstone (Kielder), with the possible addition of Irthing, with a tunnel linking three rivers as shown in the North-South strategy (see Fig.2) . NRA's Water Resources

Committee considered the two strategies with a report on comparative financial costs in July 1970. A recommendation "that powers be sought to develop the Otterstone Reservoir site and to construct an aqueduct tunnel linking the Rivers Tyne, Wear and Tees" was confirmed by NRA in September 1970 (NRA, 1973).

WRB suggested that the water grid proposed for the NRA should extend beyond its boundaries into Yorkshire but the political barrier of establishing co-operation between two neighbouring River Authorities proved insuperable. Recent experience of building reservoirs in the North East in rapid succession: Selset (1960), Balderhead (1965) and Cow Green (1971) for TVCWB, and Derwent Reservoir (1966) for the South Shields Water Company, suggested that fewer, larger reservoirs would avoid several long and expensive battles to gain permission as well as, more doubtfully, economies of scale. The prospect of raising regional finance and external funding following the 1963 Water Resources Act gave hope that the undertakers would not have to await last minute decisions by the main industrial beneficiaries for capital provision.

Enthusiasm for the Scheme was not just espoused in the local NRA and in the WRB but also within the MHLG. Senior civil servants were convinced that such schemes were the way forward. The Under Secretary, Jack Beddoe, wrote a memo to a colleague:

> 'within the next few years the most economic organisation of water conservation will require substantial transfer of water between the areas of the present River Authorities, the switching of sources between different distribution networks at different times, changing water undertaking sources to river regulation and the building of large-scale transmission links to supplement the transfer of water in rivers (HLG, 1970).'

The concept of large-scale water planning had come of age but Beddoe foresaw "major financial and administrative problems" (ibid, 1970). The limit to the extension of grids of water supply would be political more than technological.

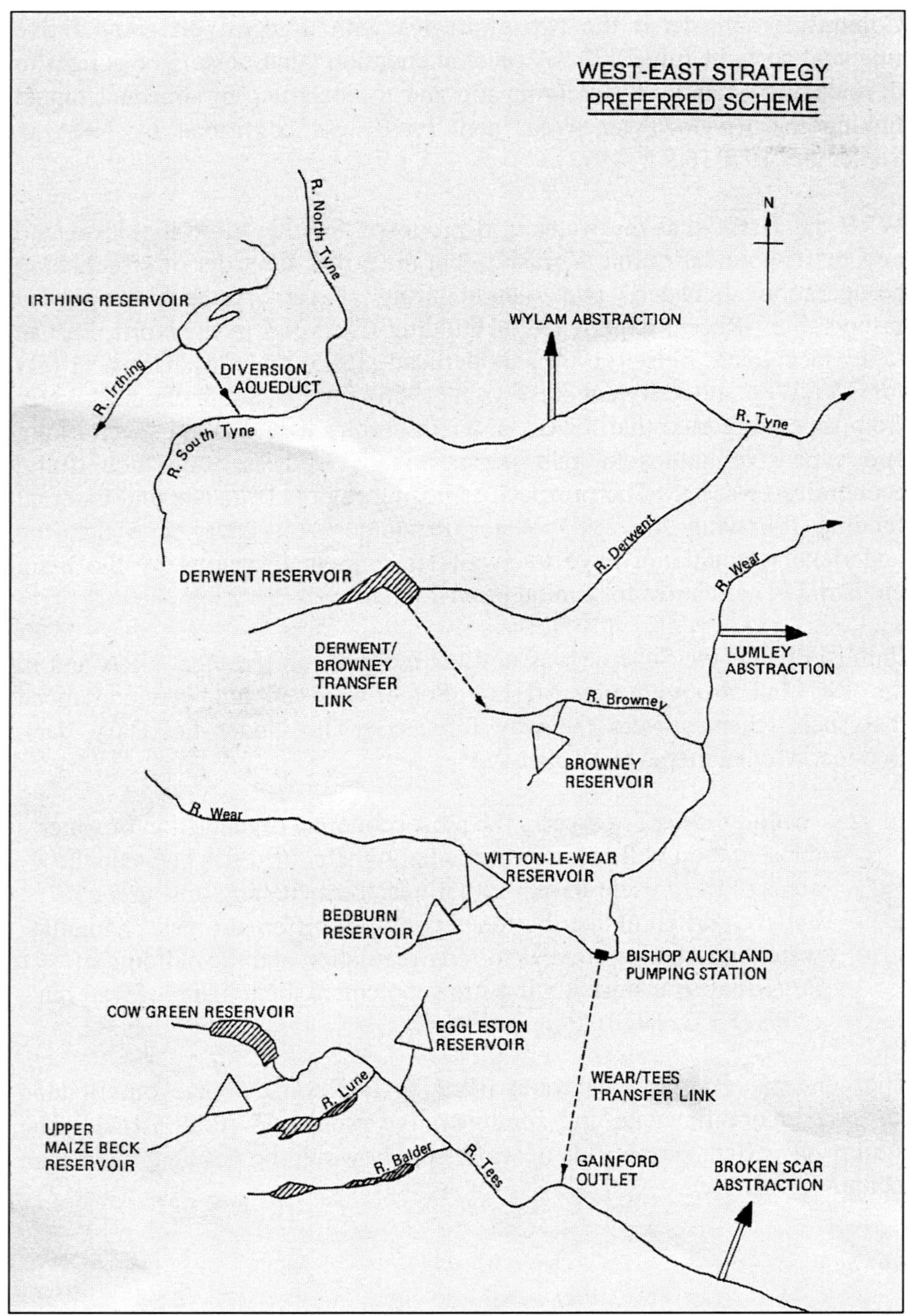

Source: WRB Water Resources in the North. Northern Technical Working Party Report 1970. Not to scale.

Figure 1:   WRB's West-East strategy. This strategy would have involved construction of 6 new reservoirs and 3 tunnels.

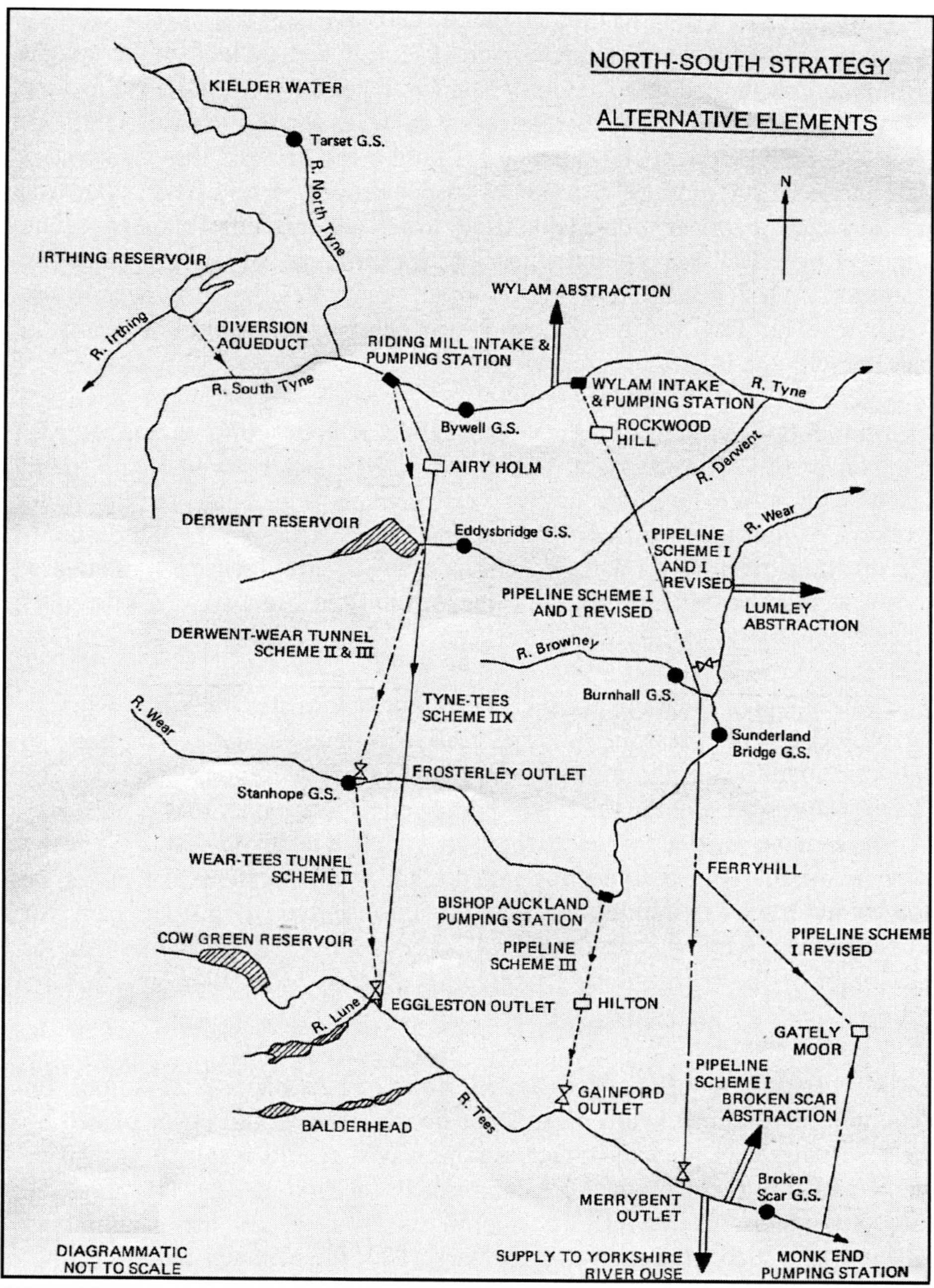

Source: WRB Water Resources in the North. Northern Technical Working Party Report 1970. Not to scale.

Figure 2 WRB's North-South Strategy: only 2 new reservoirs, Kielder and Irthing were needed to augment the Tees via the Tyne-Tees tunnel.

DECISIONS ON THE PROMOTION OF THE SCHEME
As soon as the Water Resources Act 1971 allowed powers to be sought without a Private Bill, the Kielder Water (draft) order was published in June 1971. In December 1971, the Secretary of State for the Environment, Rt. Hon. Geoffrey Rippon MP called for a public Inquiry and this was held 3 February to 15 March 1972 under an Inspector, Mr. A.R. Chaun, who was not a water engineer but a qualified town and country planner. The engineering case was strengthened by appointment of an "Engineering Assessor", Mr J. Keith Jackson, a former Superintending Engineer in the MHLG. The Engineering Assessor was allowed to submit a report in parallel with the Inspector's report.

The context of the public Inquiry was a time of great political tension with the Conservative Government, led by Prime Minister Edward Heath, being faced with strikes by the miners, railwaymen and other public sector workers, violence in Northern Ireland and rising inflation. In the drama of the opening meeting on a dark February day, Keith Jackson remembered trying to read documents by torch and candlelight because of a power cut (Jackson interview).

The political complications for the Environment Minister, Geoffrey Rippon, did not end with pressure to quell restive miners in the North East, to support manufacturing industry and to increase employment. The Conservative Party, at the time, was perceived as a defender of the rural way of life and the Hexham constituency, in which Kielder lay, was Geoffrey Rippon's seat. Yet, when he was called upon to adjudicate over Kielder, he confronted by his retired predecessor as Conservative MP for Hexham, Sir Rupert Spier, who was leading local opposition to Kielder reservoir as Chairman of the North Tyne Preservation Society. The NRA was led by strong Labour politicians.

Faced with such conflicts, retreat into compromise had its attractions for Rippon. The Inquiry heard pleas from people who would be displaced by the reservoir; fears that the Scheme's high cost would result in expensive water and that there might be an industrial depression. Despite this, the Inspector recommended the Scheme in its entirety and the Engineering Assessor was also enthusiastic. In January 1973, the Minister made a ruling: he agreed to the tunnel and the North-South strategy, but asked for a reconsideration of the Kielder reservoir site, which would drown 58 homes, displacing 130 people; he called for an investigation of the remote Irthing site, affecting 'two families at most', as an alternative. The Secretary of State's letter stated:

> '...the degree of hardship, particularly for those who would have to leave their homes and the damage to the environment which would result from flooding the site ought not to be accepted without first testing more fully the case for and against constructing a reservoir on the River Irthing...(NRO, 1973)'

His response was a shock for the engineers promoting this pioneering attempt to provide the first regional water grid: technological progress threatened to be impeded by social considerations for a small minority of affected locals. However, by being persuaded that the tunnel was a necessity, the Minister eventually lost any argument against the Kielder reservoir because only the huge quantity of water that Kielder could yield would justify the large tunnel. His verdict was only a temporary delay.

The NRA was not to be thwarted. Not only would Irthing produce less water than Kielder, it was in the region of the Cumberland River Authority and would not be fully under NRA control. The financial and administrative barriers foreseen by Beddoe were formidable. Jackson (Personal communication) reports that the NRA sent a sharp response on 2 February 1973 to the Secretary of State saying that they had no intention of considering the Irthing site for the main reservoir. The need for the water, they said, was too urgent for any delays. This reassertion of the power of the NRA proved effective in getting the Inquiry reopened. The power of the petitioners against the Scheme was diminishing.

Almost simultaneously in February 1973, a White Paper was published: "Steel-British Steel Corporation: 10-year Development strategy" (Cmd. 5226). This promised that Teesside would have one of the largest and most modern steel complexes in Europe. Large quantities of water would be needed. The recommendation of Spens' report (Spens, 1947, 7) that "the North East Development Area is not an area into which really large water using industries should be encouraged to develop" was ignored.

Faced with these pressures, in April 1973, the Minister ruled that the Inquiry be reopened. The second Inquiry, 19/06/73 to 09/07/73, had a different Inspector, this time a barrister, Sir Robert Scott, but the same Engineering Assessor. Scott tried to avoid going over old ground and at the end refused to make a recommendation. He wrote, "There was no new Application before the inquiry and therefore no occasion for formal recommendations." One of the intended main beneficiaries, ICI, did not bother to send a representative. The spokesman who forecast a huge increase in water for

British Steel Corporation " did not stay to be cross-questioned, perhaps because the plans on which he based his estimates had not been, and never were, approved " (Charlton, 1982, p.16). Scott stated in the final paragraph of his report.

> 'With Kielder the centre of interest, lukewarm support for alternatives except as lesser evils than Kielder, and the need for further site investigations, the reopened inquiry cannot be considered a satisfactory test of the case for an alternative scheme (Scott, 1973, para. 90).'

Despite this equivocal ending, Scott reflected that the North-South strategy already had the stated preference of the Minister based on the first Inquiry and that the weight of evidence was in favour of a tunnel in the light of WRB figures suggesting that the tunnel would cost £26m (£193m)[3] set against pipelines at £39m (£289m).

The Minister approved the Scheme in October 1973 and the Kielder Water Order was made in April 1974. The newly-formed Northumbrian Water Authority, with wider responsibilities, crucially including sewage and water quality, took over the Scheme. Perhaps in response to the behaviour of the River Authorities, the Water Authorities were set up with fewer local politicians and a majority of Ministerial appointees. The Kielder case threw question marks against the Ministerial Order procedure, which replaced the previous adjudication by Select Committees.

TUNNEL TEMPTATION
Throughout the discussions, the Tyne-Tees tunnel was key. Once the 32 km long, 2.9 metre diameter tunnel was approved, then a massive water discharge was needed to justify its huge size. The tunnel was tempting both technically and politically.

An attraction for ambitious engineers was the innovation of powerful tunnelling machines which performed "full face penetration" and which could be imported from Germany and the U.S. (Brown, 1975). The long and large tunnel would allow the three rivers to be used conjunctively.

Politically it was also attractive. Whereas a conventional pipeline would require way-leave permissions and construction disruption along the A68 main road, a tunnel would be bored underground and cause little visual upset on the surface, requiring few negotiations with landowners. In a depressed region, with declining coal mines and shipbuilding industries, infrastructure investment would provide some employment, and gave a little

---

[3] Figures in brackets indicate conversions to 2002 prices http://www.eh.net

hope that new industries might be attracted, or at least not deterred by lack of water.

Financially, the proposition was less appealing. The NRA was suspected of having a gung ho attitude towards costs, knowing that the Water Authority would soon supplant it. At the time of the Inquiry in 1973, the costs quoted were £26m (£193m) for the tunnel and £13m (£97m) for Kielder reservoir. By 1978, the cost estimates had risen to £70m (£391m) (Lambert, 1978, p.32). Later, the tunnel costs rose further because an unexpected outcrop of extremely hard dolerite of the Little Whin Sill was encountered (Coats, Berry and Banks, 1982) and a new £1m (£2.2m) boring machine had to be brought in, setting back completion by a year (Newcastle Journal, 1982). High capital costs were only part of the problem. Running costs would be high also because water abstracted from the Tyne 56km downstream of the Kielder dam at Riding Mill, Britain's largest pumping station, had to be pumped up 200 metres over a distance of 6.2km to the highest point of the aqueduct near Airy Holm whence the water could flow without further assistance as far as Teesside (NWL, 1993). Before the Scheme was built but only after the decision to go ahead had been made, doubts were expressed about the likely high operating costs, which would dwarf even the high capital costs (Ray discussion of Burston and Coats, 1975, p.149). When industrial water demands fell and the hoped for expansion of the steel industry in the North East did not materialise, local residents were faced with large increases in their domestic water bills.

IMPLICATIONS AND RECRIMINATIONS
The Scheme is described on a bronze plaque at the reservoir site as one of the biggest water projects ever undertaken in Europe and Kielder Water as the largest man-made lake in Europe. Today, the reservoir rests mostly idle. The water is rarely needed for supply and then mainly for transfer to the Wear rather than the Tees.   The hydroelectricity, generated as an afterthought to the original plans, is a small contribution to the National Grid and the reservoir's claims as a tourist attraction are hampered by its remoteness, rainy climate and monotonous coniferous plantations with associated populations of vicious midges.

Teesside continues to be supplied from the Teesdale reservoirs without supplementation from Kielder because the cost of pumping water from the Tyne to the highest point of the Tyne-Tees tunnel is greater than the cost of supply by gravity flow from Upper Teesdale dams. Only twice in its history has the Tyne-Tees transfer tunnel been used to transfer water to the Tees, first in 1983 and then in 1989, (FOE, 2003) although water has been transferred as far as the Wear to supplement the underperformance of the

Derwent reservoir (Soulsby *et al.,* 1999). David Archer, a former employee of Northumbrian Water and the Environment Agency, asserts that Kielder has "saved the North East from serious water restrictions during the droughts of 1989 and 1990s", although he concedes that a smaller reservoir would have sufficed (Archer 2003 155).

Such over-investment was enabled by separation of the industrial consumers from funding of water supply infrastructure. Rather than continued iteration with the industrial consumers to judge its effectiveness in promotion of economic development, the dedicated focus on water supply made it an end in itself and safeguards against overinvestment were weak. Uncritical extrapolation of water demands at the outset was not corrected at later stages when British Steel failed to expand on Teesside. Unlike the financial arrangements in Teesdale, reformed regional funding meant that those industries which demanded more water at the Kielder Inquiry made little or no contribution to the capital costs of the Scheme (McCulloch 2004 59-60). Brady concluded that major industries should have a direct financial stake in such resource developments and "pay the cost of the works whether or not their share of the increased demand is taken up, provided that the industries remain solvent" (Brady, 1985, p.140).

Instead, the Government in the 1980s decided that the "consumers of the NWA should meet the charges incurred by Kielder and that the costs should be borne regionally" other than "Temporary assistance given one year by way of a repayable grant"(Sir Peter Harrop's evidence to the House of Commons Committee of Public Accounts 1984-1985 para1166). The suffering of the regional domestic consumers was somewhat lessened by the writing off of some of the debts in preparation for the privatisation of water supply in 1989 and continuing subsidies from the Environment Agency (NSL Group 2003).

CONCLUSION
At the opening ceremony in 1982, banners decried the reservoir as a White Elephant but the Chairman of the NWA, Sir Ralph Carr-Ellison gave reassurance:

> 'Beyond any shadow of doubt, it was correct to go ahead with this scheme. Not only have we got a reservoir to serve the needs of the region for the 1980s but we have a reservoir that will serve its needs until 2050…The price we have paid will turn out to be cheap (Newcastle Journal 26/05/82).'

Yet, even in the engineering press, there was scepticism:

> 'It is possible that no water from Kielder will be required for consumption within the next decade with the scheme not being fully utilised until the second half of the next century (Hayward, 1982, 27).'

The controversies over the Kielder Water Scheme led the Director of the WRB, Norman Rowntree to doubt whether such ambitious water schemes would ever be repeated (Wolf interview). The cost of the Scheme, both its capital cost and the running cost, threw doubts on the sufficiency of checks and balances on infrastructure expenditure by public bodies, once the main industrial beneficiaries were not obliged to fund the construction. Privatisation post-1989 has been accompanied by re-regulation of the water supply industry. Now the plans of engineers are overseen by economists, accountants and others in the Office of Water Services (OFWAT) and by biologists and engineers within the Environment Agency. Political and financial barriers to the exercise of engineering technology remain strong today. Sir Norman Rowntree may well have been prescient in his belief that the Kielder Water Scheme was likely to be the last of its kind in England and Wales.

ACKNOWLEDGEMENTS

The author is most grateful to David Coats, Consulting Engineer; Keith Jackson, Engineering Assessor; David Newsome, Barry Rydz and another former senior member of the staff of the WRB; Professor Peter Wolf and a member of staff of Northumbrian Water for their kindness in providing information. Any errors are the author's own.

REFERENCES

Anon 1985 'The Kielder Water Scheme' *House of Commons Committee of Public Accounts*, HMSO, London:

Archer, D 2003 Kielder Water: white elephant or white knight? In Archer, D. (Ed) 2003 *Tyne and tide. A celebration of the River Tyne*, Daryan Press, Ovingham, UK, 138-156

Brady, J. A. 1985 Uncertainty in water demand forecasting, *Proc. Inst. Civ. Eng. Pt.1* **80**(Dec), 1383-1401.

Brown, W M 1975 Full face penetration, *Consulting Engineer* **29**, 31-33.

Burston, U and D J Coats, 1975 The Kielder Water Scheme, *Journal Institution of Water Engineers and Scientists* **29** (5), 226-243.

Charlton, W. 1982 Spending money like water, *Spectator* (22 May), 16-17.

Coats, D J, Berry, N S M and D J Banks, 1982 The Kielder transfer works, *Proc. Instn. Civ. Engrs. Part I* **72** (May), 177-208.

Cmd 5226, 1973 White Paper. Steel- British Steel Corporation: 10-year development strategy, HMSO, London.

FriendsoftheEarth 2003 Kielder transfer scheme
http://www.foe.co.uk/resource/briefings/Kielder_transfer_scheme.html
Hassan, J A 1994 *The water industry 1900-51: a failure of public policy*,
Manchester Metropolitan University, Manchester.
Hayward, D 1982 Kielder holds water. *New Civ. Engr* **492**(20 May), 26-28.
HLG 1970 Memo: J. Beddoe to Chilver 04/08/70 *Ministry of Housing and
Local Government HLG/127/1184*, PRO Kew.
Lambert, J 1978 Kielder 2000, *Water and waste treatment.* 32-33.
McCulloch, C S 2004 Political ecology of dams in Teesdale, in H. Hewlett
(Ed) *Long-term benefits and performance of dams*, Thomas Telford,
London, 49-66.
Newcastle Journal May 26 1982 *Kielder.* Supplement to mark the official
opening by H.M. The Queen. *Newcastle Journal*, Newcastle-upon-Tyne.
NRA 1973 Report on survey of water resources. Water Resources Act
1963. HMSO, London.
NRO 1973 Decision letter of the Secretary of State to the Parliamentary
Agents 3/10/1973 *Northumberland Record Office*, Newcastle-upon-Tyne.
NSL 2003 Northumbrian Services Ltd Group Annual Report and accounts
2003. Northumbrian Services Ltd Group.
NWL 1993 *Riding Mill Water Transfer Pumping Station*, Northumbrian
Water Ltd. Newcastle-upon-Tyne.
Pearce, F 1982 *Watershed. The water crisis in Britain*, Junction Books,
London.
Rennison, R W 1979 *Water to Tyneside. A history of the Newcastle and
Gateshead Water Company*, Northumberland Press, Gateshead.
Rowntree, NAF 1962 Presidential Address *J. Instn. Wat. Engrs* **16** 265-70.
Rydz, B 1971 Regional water resources analysis, *Proc. Instn. Civ. Engrs.* **49**
129-143.
Scott, R H 1973 Letter to the Right Honourable Geoffrey Rippon, M.P.,
Q.C. *Northumberland Record Office,* Newcastle-upon- Tyne.
Soulsby, C, Gibbons, C N and T Robinson, 1999 Inter-basin water transfers
and drought management in the Kielder/Derwent system. *J.Ch. Instn Wat.
Engrs. and Manrs* **13**(213-23).
Spens, C H 1947 Water Supply survey, N. E. Development Area.: HLG
113/49 & 50 PRO, Kew.
WRB 1965 *First Annual Report for the year ending September 1964,*:
HMSO, London.
WRB 1967 *Third Annual Report for the year ending 30 September 1966*,
HMSO, London.
WRB 1970 *Water Resources in the North*, HMSO, London.
WRB 1973 *Water Resources in England and Wales*, HMSO, London

Glendoe Hydroelectric Scheme, Optimisation and Dam Selection

MIKE SEATON, Scottish & Southern Energy, Perth, UK
JOHN SAWYER, Jacobs, Reading, UK

SYNOPSIS.   The Glendoe Hydroelectric Scheme will be the first major conventional hydro scheme to be built in the UK for over 40 years. The scheme will include a 1km long, 35m high rockfill dam, over 18km of hard rock tunnelling, and an underground powerhouse with an installed capacity of 100MW and an operating head of over 600m.

The project was tendered on a design-build basis in order to proceed to contract award in parallel with the planning process. This led to a phased development of the design through a two stage tender process leading onto the detailed design.

The dam is located in the Monadliath mountains at an elevation of 600m. The natural materials available are the bedrock and glacial fill, which is overlain by peat. A range of dam types were considered through the design process, with the final design developed by the successful construction tenderer. As with any dam a number of alternatives are viable and the final design reflects a judgement on the most efficient construction in the particular site, with due regard to the particular skills of the contractor.

This paper gives a brief overview of the background to the project, the approach to procurement, the scheme optimisation and the choice of dam type.

INTRODUCTION

As part of their commitment to generating electricity from renewable sources Scottish and Southern Energy (SSE) are constructing a new hydroelectric project in the Scottish Highlands. With an installed capacity of 100MW and an operating head of over 600m Glendoe will be the first major conventional hydro scheme to be built in the UK since the completion of the North of Scotland Hydro Electricity Board schemes in the early 1960's.

Improvements in reservoir construction, operation and maintenance, Thomas Telford, London, 2006, 211–223

The Glendoe hydro scheme is located in the Monadliath mountains to the south east of Fort Augustus, overlooking Loch Ness. The reservoir is at an elevation of 630m, with a direct catchment of 15km^2 supplemented by transfer aqueducts in pipeline and tunnel to give a total of 75km^2. The indirect catchments are all on tributaries which naturally discharge to Loch Ness in the vicinity of the scheme. From the reservoir the pressure tunnel leads to a power house and tailrace. The total tunnel length between the reservoir and Loch Ness is 8km.

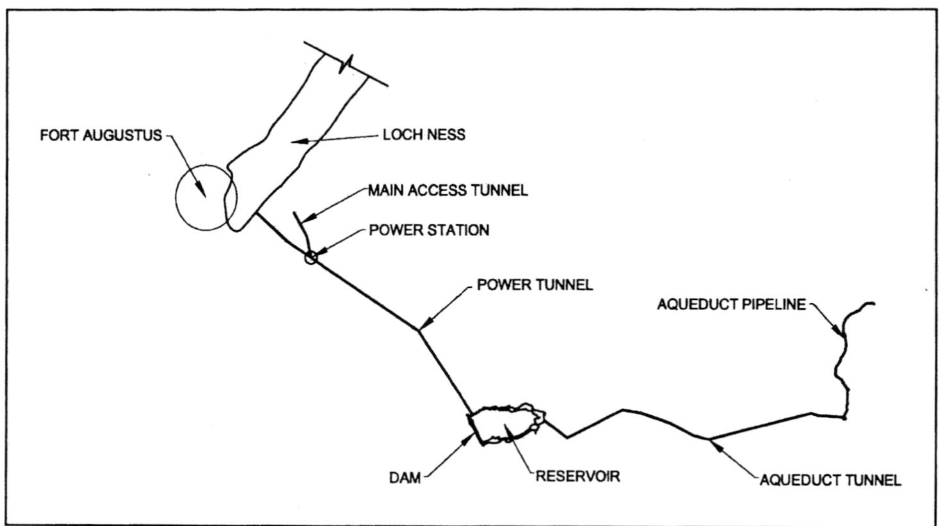

Figure 1. Glendoe Scheme Layout

SCHEME OBJECTIVES

As part of the Scottish Climate Change Programme, the Scottish Executive is committed to raising the overall proportion of electricity generated from renewable sources in Scotland to 18% by 2010. In 2001 the Renewables Obligation confirmed that new Hydro Electric schemes of any magnitude would qualify under this incentive, whereas this had previously only been expected to be applicable for new hydro schemes of less than 10MW. Under the Renewables Obligation, all licensed suppliers are required to source an increasing amount of their electricity from renewable sources or alternatively make a payment into the "buy-out" fund. Hence the size of the "buy out" fund is directly related to the level of suppliers compliance with the Renewables Oblgation (bearing in mind that the target levels of renewable generation increase year on year). A Renewable Obligation Certificate (ROC) is evidence that a supplier has sourced a megawatt hour (MWh) of its electricity from renewable sources. Once ROCs and buy-out fund payments have been produced Ofgen then re-distributes the buy-out fund collected to the holders of each ROC. In 2003/04 this gave a net benefit of around £54 per ROC , which is considerably greater than the energy value of the electricity.

SSE revisited the large schemes that had previously been identified, but not constructed, during the North of Scotland Hydro Electricity Board days. A review of these schemes was carried out taking into account current environmental, planning and construction practice. Of the areas previously identified for large hydro development, all bar Glendoe were shelved due to the potential environmental impact in areas highly valued for recreational and habitat uses.

Options considered for Glendoe, ranged from a sub -10MW run of river scheme to a large pumped storage scheme. The output from the smaller schemes could not justify the substantial capital costs of transporting the water from the plateau down to Loch Ness. The pumped storage option was discarded for two reasons:
- Because the electricity generated would not qualify for Renewable Obligation Certificates (ROCs) as pumping may use non-renewable sources of energy.
- The horizontal distance between the upper and lower reservoirs was excessive which would lead to hydrualic and construction issues which would be uneconomic to overcome.

The selected scheme at Glendoe involves a 100MW power station which will produce 180 Gigawatt hours (GWh) in a year of average rainfall. The net head of water at Glendoe will be greater than 600m – the highest head of any hydro scheme in the UK. In terms of water to power, the Glendoe hydro scheme will be the most efficient in the UK.

PLANNING PROCESS
In October 2001 the Glendoe area was identified as the preferred location for the development of a large scale hydro scheme and initial contact meetings were arranged with the statutory consultees and other interested parties. The preparation of the environmental statement and the refinement of the outline design for Glendoe were carried out from Spring 2002 until May 2003 when an application was made for planning consent under Section 36 of the Electricity Act 1989. An addendum to the Environmental Statement was issued in January 2004. The Highland Council recommended acceptance of the application (with conditions) on the 20th April 2004. A second addendum was submitted in January 2005. Final consents were granted in June 2005.

PROJECT STRATEGY

It has generally been expected that the level of compliance with the Renewables Obligation will improve with time, hence the buy-out fund is expected to reduce towards zero over a period of several years. In addition, there is uncertainty in future UK electricity prices. In order to minimise income risk there was a need for SSE to minimise the overall project programme. The greatest programme uncertainty for Glendoe was the time taken to obtain all necessary planning consents. At Glendoe even though there were very few letters of objection to the scheme, and the local community have generally been supportive, the Section 36 application still took over 2 years to be granted approval.

The strategy taken was to initially concentrate on submission of the planning application and finalising landowner deals. While the planning application was being considered the design and construct tendering process was to be taking place.

One reason why the design and construct route was selected for Glendoe was that until deals with the landowners had been finalised SSE were not permitted to carry out intrusive site investigation works on the land. Deals with the landowners were not concluded until early 2004. To minimise the overall project programme SSE could not afford to wait until then to start the design process.

This strategy meant that the planning application was based on SSE's preliminary design work. As there were no site investigation results available at this time, it was essential to keep adequate flexibility within the planning application to enable a detailed value engineering exercise to take place during the procurement phase. As a result, the initial planning application covered a relatively wide scenario of development including options for the power station to be underground or on the surface, the output from the station ranging between 50MW and 100MW and the likely dam construction technique being either roller compacted concrete or rockfill.

PROJECT DEVELOPMENT PROCESS

By electing to proceed with the design and procurement of the project in advance of planning approval, SSE were under a number of constraints relating to site access, budget availability and timing of the various activities. The project development therefore consisted of a series of increasing levels of expenditure and commitment as confidence in gaining approval increased.

The process adopted involved:

Activity	Resources	Programme
SSE concept development and planning submission.	SSE, planning & environmental consultants.	September 2001 - May 2003
Environmental surveys.	environmental consultants	2002/2003
Concept optimisation, scheme definition through workshops.	SSE and Jacobs	September 2003
Prequalification of Civil and Plant Contractors.	SSE	December 2003
Phase 1 Reference Design and Tender documents.	SSE and Jacobs	October 2003 – January 2004
Phase 1 Tender period to select 2 contractors for subsequent bid.		January – May 2004
Aerial geophysics survey.	Fugro	February 2004
Intrusive Site Investigation.	Fugro	May - July 2004
Phase 2 Tender Design and bid.	Skanska-Morgan Est jv. Hochtief	July – November 2004
Tender Assessment.	SSE and Jacobs	December 2004 - July 2005
Planning Approval and decision to proceed with project.	Scottish Executive / SSE	Final decision July 2005
Notification of preferred Contractor.	SSE	August 2005
Finalisation of Contract.	SSE / Jacobs / Hochtief	August – December 2005
Contract Award	SSE	December 2005
Detailed Design and Construction.	Hochtief	January 2006 onwards

The objective of the phased tender process was partly to respond to the increasing availability of information and client confidence in the project, but also to balance the SSE requirement for competitive tenders with the desire to minimise the tenderer's costs as far as practicable.

SSE adopted the NEC as their standard form of construction contract, and aim for a fixed price basis as far as practicable. Option A of the NEC (Priced contract with activity schedule) was therefore used. A variable payment mechanism which allows predetermined compensation for the ground conditions actually encountered in the tunnels is incorporated through 'Z clauses' {Seaton & Hobson 2005}.

CONCEPT OPTIMISATION

The planning application was made in such a manner to enable the optimisation of the scheme to be delayed so that SSE could get a better assessment of the likely capital costs of the scheme against the potential income that the scheme would generate on the renewable energy market.

For the Reference Design to be used for the first tender phase, the key issues were:

- Aqueduct System: intake type, aqueduct alignment, tunnel construction;
- Dam: dam type, spillway and outlet works layout;
- Power Tunnel System: alignment (vertical and horizontal), tunnel construction, surge system;
- Power Station and Generation Plant: location, turbine type and size, transformer location, operating regime

Jacobs were appointed in September 2003, and tasked with working with SSE to prepare the Phase 1 Tender by the end of the year. To achieve this, the basic scheme parameters had to be fixed within 4 weeks, which precluded detailed feasibility and optimisation studies. The approach taken was to run a series of workshops, interspersed with technical studies, in order to make best use of the diverse experience and skills available within the combined SSE/Jacobs team. The workshops were attended by representatives of all the engineering disciplines in the Jacobs team, with SSE staff representing planning, construction, operation, maintenance and energy trading disciplines. Achieving a balance between the physical constraints of the site and the objectives of the end users was particularly critical for the optimisation of the power system and the multidisciplinary workshop approach worked well in this area.

SSE 'end user' objectives

SSE gain significant value from the scheme by the ability to provide peaking power when required by the grid. The most important factors are:

- reliability – the certainty of being able to supply power at the time promised
- rapid response (how quickly the scheme can come on or go off load)
- efficiency – obtaining the maximum power for the water available
- flexibility – the ability to generate efficiently over a wide output range
- ease of operation and maintenance, minimum downtime

One of the benefits of hydro generation over other forms of power production is its ability to store power in a reservoir, ready to be used when required. This flexibility has significant benefits to the environment, transmission system and developer. 'Peaking power' attracts a significant premium value compared with base energy prices. Although the proposed volume of the reservoir is small in comparison to SSE's existing stock, due to the altitude above the turbine and the relatively small amounts of water required to generate large amounts of electricity, Glendoe does have the ability to contribute substantially. The reservoir has the capacity to store water to generate over 10 GWh of electricity ready for dispatch when the demand requires, thus complementing other forms of "must take" renewable energy.

Physical Constraints

The reservoir is located on a plateau approximately 8km from Loch Ness. After crossing a ridge adjacent to the River Tarff the ground profile above the pressure tunnel falls relatively regularly towards the loch. Hence there is no suitable site for a surge shaft, except close to the reservoir where it has limited hydraulic benefit. The rock cover is also relatively low over the downstream section of the tunnel. Where the reservoir head gives a water pressure greater than the confining rock pressure, rock joints will be opened by the water and leakage will result. Over time the water will penetrate throughout the rock leading to increased leakage and in areas of low cover could create slope instability. It thus becomes necessary to provide a steel lining wherever the water pressure could be greater than the confining rock pressure. This situation is exacerbated by transient pressures arising due to the operation of the system.

Turbine Characteristics

With 600m head the two most viable types of turbine are either Francis (where the runner is enclosed within a fully pressurized hydraulic system) or Pelton (where the runner is in air and turned by the impact of water jets). To achieve a station which could come on or off load within a matter of seconds would require a Francis turbine. However it is impossible to avoid the potential to trip the guide vanes on the turbine, giving substantial transient effects. This would require substantial engineering to ameliorate the transient pressure in the absence of a surge shaft (pressure release valves, air pressure chamber or steel linings to the tunnels were considered). A Francis turbine requires submergence and a long tailrace would also lead to the need for a downstream surge shaft.

A multiple jet Pelton turbine has a good efficiency over a broad range, but cannot be brought on load quite as quickly as a Francis. The Pelton turbine jets have deflectors which can isolate the runner, without instantaneously stopping the water flow, thus reducing the transient effects. SSE determined that the additional commercial benefit of a very fast response time did not justify the increased complexities required in adopting a Francis turbine.

The main civil works for the scheme are substantially the same for either a 100MW or a 50MW scheme, as the tunnels are close to the minimum diameter for efficient construction. There was therefore relatively little cost increase in adopting the 100MW scheme. There are significant operational benefits in the increased capacity (water management is greatly improved as SSE will have far more control the upper reservoir water level, thus being able to hit more periods of peak energy prices) , although the total annual energy production is constrained by the water available to around 180 GWhr/year.

<u>Selected Power System</u>
For the Glendoe site the physical constraints and operational priorities lead to the following system:
- an underground power house, located to minimise the length of steel lining
- a headrace tunnel inclined at the maximum gradient for constructability, to maximise rock cover
- a tailrace at near-horizontal gradient, to maximise head on the power station
- a 6-jet pelton turbine, which can minimise transient pressures and which has a broad efficiency range, while providing secondary response
- a 100MW capacity, with a load factor of 20%

Although the optimum power house location was considered to be at the point minimizing the steel lining, for the Phase 1 tender it was moved downstream in order to obtain additional cost information. Both Phase 2 bidders moved the power station back upstream in their later tenders.

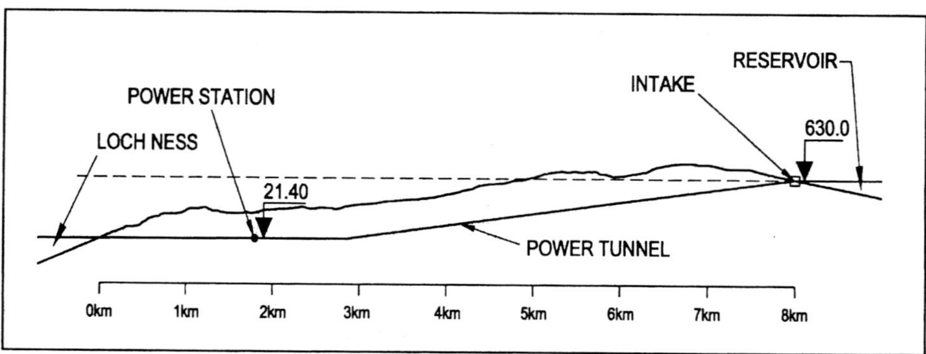

Figure 2. Section on Power Tunnel.

DAM SITE GEOLOGY AND MORPHOLOGY

The reservoir is on the river Tarff, which eventually passes through Fort Augustus to discharge into Loch Ness. The reservoir has a full supply level of 630m, a total storage volume of $11.5M,m^3$ and a live storage volume of $5.8M.m^3$. The area at full supply is around $1.5km^2$. The reservoir water level is restricted to a 6m operating range on environmental grounds.

The river is in a steep sided channel, leading to a rocky gorge. The surrounding area is a broad plateau of peat-hag, with further hills rising beyond the reservoir margin. Thus the dam is around 35m high for the 100m wide river channel section, and typically 10m to 15m high over most of the remaining 900m length.

The bedrock is schist with granite intrusions, overlain with glacial till and peat. The till is a variable granular material, largely sand and gravel with some boulders and little clay.

A number of major faults have been identified, including the Stronlairg fault which passes through the left abutment of the dam.

The construction materials which will be readily available at the site are:

- tunnel spoil : variable rock
- quarried rock
- glacial till : variable granular material

No reliable source of clay has been identified, and only limited quantities of natural river sand have been located. Aggregates and fill are therefore likely to be produced from processed rock, including tunnel spoil where practicable.

DAM AND SPILLWAY OPTIONS

The principle constraints on the dam selection are:
- profile : long, relatively low flank sections, with higher river section
- materials : no clay for core, tunnel spoil and rockfill readily available
- access : high transport costs for off-site materials
- climate : severe winter environment, restricted access for operations staff
- hazard rating : Fort Augustus lies on the river downstream of the dam
- limited site investigation : need for adaptable design
- environmental : maximum re-use of site produced material

For the Reference Design it was also appropriate to use generally available technology and avoid specialist operations which would limit competition between contractors.

Prior to developing the Reference Design an 'Options Study' was carried out for all the major elements of the scheme. The dam types which were evaluated included:
- Concrete: roller compacted concrete (rcc), gravity
- Rockfill Embankment:
- Upstream membrane: concrete faced rockfill (cfrd), asphaltic face, geomembrane,
- Central cutoff: asphaltic core, geomembrane, diaphragm wall

In addition a range of spillway layouts were considered:
- gravity section ogee weir in river channel
- rcc weir in river channel
- weir at right abutment, with channel to river
- drop shaft spillway at river channel
- side-channel weir at right abutment.

The PMF flood inflow is relatively modest at around 200 m^3/s, which is routed to around 160m^3/s depending on spillway design. The climate is extreme and access can be difficult in winter so no gated designs were considered. One tenderer offered a vented siphon spillway option, which had not been seriously considered by Jacobs.

DAM AND SPILLWAY ADOPTED DESIGNS

For the Reference Design, Jacobs considered issues of buildability, use of local materials, uncertainties in foundation conditions, conformity with the planning application, integration of the outlet structures and suitability for competitive tender. The rockfill embankment was the clear favourite for the long sections on the flanks. Asphaltic options would have restricted competition, so the CFRD was adopted. This was detailed to the typical details of ICOLD Bulletin 70 (1989). The low embankments are not ideal for slipforming the concrete face efficiently, but this is a well proven design, with good precedent within the SSE portfolio of dams. For the Stage 2 Design-Build tender, one contractor adopted the CFRD, with minor modifications, and the other proposed an asphaltic core rockfill embankment. This is a standard solution in Scandinavia under similar conditions, and the contractor had good access to this expertise.

The spillway adopted for the Reference Design was a gravity concrete ogee weir in the river valley. In order to set the wing-walls out of the river channel to reduce cost, the weir was 100m long. This had the added benefit of minimising the flood rise and hence freeboard requirements, albeit the weir concrete volume was substantial. This central spillway facilitated the integration of diversion and low level outlet works. The tenderers identified the spillway as an area for cost saving and almost all the layouts considered in the options study were proposed by various contractors. The two preferred solutions on the Stage 2 tender were either the drop-shaft spillway in the river channel, or a side channel weir at the right abutment. The significant potential cost of the channel from the abutment to the river has been mitigated by excavating the channel into rock and leaving it largely unlined. This is the solution adopted by the winning tenderer.

PHASE 2 TENDER AND CONTRACT AWARD

While the Phase 1 Tender assessment was in process an intrusive site investigation was carried out. This followed consultation with the contractors to ensure that any particular information they required was targeted. Access to the upper plateau was only available on foot and by helicopter, which severely constrained the amount of drilling work which could be carried out. Reasonable access was available on the power tunnel alignment, permitting a number of 400m deep boreholes, with hydrofracture water pressure testing of the insitu rock stresses at full depth. This indicated that the horizontal confining stress is at least as great as the vertical rock stresses, reducing possible requirements for steel lining to the tunnels. The intrusive investigation was supplemented by geophysical techniques on the dam alignment and a helicopter mounted resistivity survey of the entire project area. The aerial geophysics gives a broad indication of the depth of cover to sound rock and of major geological features across the entire site. The combination of techniques permitted increased confidence in the Phase 2 design, though detailed investigations will still be required at key structure locations during the construction period.

Figure 3. Dam Site Investigation

Two consortia were invited to submit detailed tenders during the second phase of the tender process. The tenderers were encouraged to optimise their design by being given tender assessment criteria identifying the monetary value (identified in terms of NPV over the project life) that SSE would place on their tender designs in respect of the following:

- water yield from catchment, achieved by:
  - additional intakes (limited by the planning application)
  - moving intakes downstream by reducing aqueduct gradients
- net head on turbine, improved by:
  - reduced headloss through larger diameter power tunnel or revised tunnel construction method
  - raising reservoir storage level by up to 6m by a limited relocation or construction of a larger dam
  - lowering the turbine closer to the Loch Ness flood level.
- efficiency of plant
  - Tenderers encouraged to offer high quality plant to increase reliability
- useful volume of reservoir storage (scheme flexibility), increased by:
  - raising dam height
  - relocating dam.

This is a balance between the various elements as changes to any part of the hydraulic system affect the others. Hence for example raising the reservoir storage level could improve head on the turbine and increase storage volume, but may reduce the available catchment. Taking the above criteria into account together with the tender price and programme enabled the project team to accurately establish the optimum design for the site for the assumed electricity market conditions.

The successful tenderer, Hochtief, has optimised the scheme, with modest adjustments to the Reference Design tunnel and dam alignments and various detailed design changes, in particular moving the power station upstream to minimise the length of steel lining, and relocating the aqueduct intake structures to give the maximum practical yield.

A letter of intent was placed with Hochtief in November 2005 and design work started immediately. The construction contract was awarded in December 2005. Tree felling commenced in December as soon as site possession was available. Mobilisation and enabling works started in January 2006 with a start to work on site roads, assessment and upgrading of the River Tarff Road Bridge to accommodate all predicted construction traffic, placing a contract with Herrenknecht for provision of a 5.0m diameter refurbished TBM and submitting a planning application for the work camp. Construction started in the Spring allowing the full labour force to be mobilised by early summer. Contract completion is currently planned for 28 February 2009.

REFERENCES

Seaton, M, Hobson DA (2005). *Glendoe Hydroelectric Scheme – Planning and Procurement.* Underground Construction 2005, UK

Assiut Barrage, to rehabilitate or to rebuild

T.J.F. HILL, Mott MacDonald Ltd, Cambridge, UK

SYNOPSIS. Assiut barrage, 400 km upstream of Cairo, is the last barrage downstream of the High Aswan Dam (HAD) before the Nile reaches Cairo. It was built between 1898 and 1902 in order to divert Nile river flows to the Ibrahimia canal. The barrage was remodelled extensively between 1934 and 1938, increasing the annual discharge to the Ibrahimia canal which, in its present form, has a length of about 350 km and irrigates an area of 690,000 ha.

The barrage was designed as an arched viaduct founded on a mass concrete floor, with a 16 m wide lock positioned on the extreme left bank. The overall length of the structure is 820 m with a water-way capable of discharging 14,000 m^3/s provided by 110 individual openings of 5 m width. Each opening contains a double leaf vertical lift roller gate designed for a maximum head difference of 4.2 m.

This paper describes the results and conclusions of a feasibility study carried out by Mott MacDonald in association with CES Salzgitter, Fichtner and Inros Lackner, all of Germany, and Hamza Associates of Egypt, to investigate the present structural and operational conditions at the barrage and to outline options for the future. The principal conclusion of this feasibility phase, completed in December 2005, was a recommendation to construct a new barrage downstream of the existing one rather than rehabilitating the existing barrage.

A HISTORY OF THE BARRAGE

Original Construction, 1898-1902

The Ibrahimia canal was excavated in 1873 to serve the cultivated area on the left bank of the Nile as far as Giza to the north, and Fayum to the West. When the river was high, during the annual Nile flood in August and September, the canal easily supplied enough water to satisfy the requirements of its command area. However, during the earlier summer months it was often difficult to pass all of the water which could be allocated to the canal, due to low river levels at Assiut.

Towards the end of the 19th century proposals were put forward for a large storage reservoir at Aswan. The additional water supplies that would be made available to the Ibrahimia canal during the summer months as a result of the reservoir's construction made the need for a barrage at Assiut more of a necessity.

The original plans for the dam at Aswan and the barrage at Assiut were the responsibility of the Director-General of Reservoirs, Mr William Willcocks (later Sir William Willcocks). The Egyptian Government then appointed Sir Benjamin Baker (a past president of the Institution of Civil Engineers in London) as their Consulting Engineer on the project and he made considerable modifications to the original Willcocks' designs. In February 1898 Messrs John Aird & Co. were appointed by the Ministry of Public Works as the Contractor for both the dam at Aswan and the barrage works (which included a head regulator on the Ibrahimia canal) at Assiut.

The location of the barrage appears to have been determined on the basis of three primary requirements:

- The need to be downstream of the Ibrahimia canal offtake

- The provision of suitable foundation conditions

- A position which simplified construction operations

The second criteria proved to be non-critical since no changes were identified in possible barrage foundation conditions throughout the river reach under investigation. However, both the first and third requirements were met at a position some 400 m downstream of the Ibrahimia offtake. Here a seasonally revealed sand island covered the western (left) third of the river channel (Leliavsky, 1934) affording improved river diversion and cofferdam construction possibilities. It may be noted that the sedimentation now prevalent along the left side of the river immediately upstream of the existing barrage may be a result of the river's natural inclinations towards silt and sand deposition at this location.

At Assiut construction began in June 1898, and was completed in March 1902, one year in advance of the contract period. The total cost of the works at Assiut, including the head regulator, was 921,772 Pounds Sterling (Stephens, 1904) (one-quarter of the expenditure on the earlier Delta Barrage).

The barrage was designed as an arched viaduct founded on a concrete floor, with a 16 m wide lock positioned on the extreme left bank. The barrage floor was 26.5 m in width and 3 m thick, the lower 0.9 m of which was cement concrete, the rest being rubble masonry set in 4:1 cement mortar. Cast iron sheet piles, extending 4 m below the underside of the barrage floor, were installed along the upstream and downstream edges of the foundation slab. The riverbed was protected for 20 m beyond both lines of sheet piles, giving an overall structure width of 66.5 m. The overall length of the structure was 820.2 m between abutment faces with a water-way provided by 111 openings of 5 m each. Each water-way, or vent, had a maximum height of 10.7 m to the springing of the arches. The vents were divided into groups of nine by 12 abutment piers each 4 m thick. The intermediate piers were 2 m thick. The upstream faces of all the piers were vertical, with the downstream faces inclined at approximately 6.6V:1H. The barrage roadway was 4.5 m wide, between parapets.

The Ibrahimia head regulator structure was of similar design to the barrage except having only nine 5 m wide sluices, and a 9 m wide lock.

<u>Interim Remedial Measures</u>
The barrage's primary purpose was to ensure adequate irrigation supplies to Middle Egypt during the early summer, as such it was not intended to be used during the annual Nile flood. The original 1902 design assumed that all gates would be fully raised to allow the flood to pass unheeded.

It is reported that in 1902, the first year after completion of the barrage, the annual Nile flood was very low and exceptionally late. All through August the river levels remained stubbornly low, leading to the potentially catastrophic loss of a significant portion of Egypt's crops going unirrigated. On August 15th Sir A. L. Webb, Director General of Reservoirs (he had succeeded Mr W J Wilson in the post upon the latter's death in August 1900, who in turn had succeeded Mr Willcocks in 1898), travelled to Assiut, and on arrival decided to utilise the barrage gates to raise the Nile level upstream of the barrage by some 1.50 m. This was a risky decision but it was crowned with success. The resulting irrigation of the crops was estimated at the time to have saved Egypt 600,000 Pounds Sterling. Leliavsky (1934a) records that this is an instance unique in the field of irrigation engineering, where a structure repays two-thirds of its capital costs within a few months of completion.

While a 1.50 m head drop across the structure was significantly less than the design head, the potential for downstream erosion as a result of the decision to partially close the gates was significant. This is because the required energy dissipation (proportional to head and discharge) at the barrage was much greater than envisaged in the design. To his credit Sir A. L. Webb was well aware of the risks he was taking. He arranged for soundings to be taken on a weekly basis to gauge the extent of downstream erosion and about 4,000 m^3 of rubble was eventually dumped to replace areas of damaged pitching.

The need for partial gate closure during the rising or falling flood period became a regular occurrence. Design curves for allowable head against river discharge were developed and modified throughout the 1920's, however scour remained a serious problem. In 1912 rubble was used to fill "an exceptionally deep scour hole" on the eastern side of the barrage and between 1920 and 1925 an average of 2,600 m^3 of stone was placed annually downstream of the barrage to control the erosion. During the 1926 flood 6,650 m^3 of rubble was placed.

In 1927 it was decided that a more permanent solution was required and so concrete blocks (1.5m x 1.0m x 0.7m) were prepared and placed downstream of the barrage by divers. 1,772 blocks were placed in 1927, 849 in 1928, 1,523 in 1929, and 2,041 in 1930 making a total of 5,685.

Remodelling, 1934-1938
The need for a more permanent solution to the problem of downstream erosion was evident and work began on finding an answer in the early 1930's.

Consulting Engineers Coode, Wilson, Mitchell and Vaughan-Lee were appointed to design remedial works to the barrage to enable the structure to be operated in the manner that was needed. The opportunity was also taken to further increase the level of the river upstream of the barrage to facilitate required increases in Ibrahimia canal flows.

On the 9th October 1934 Messrs John Cochrane & Sons Ltd were appointed by the Egyptian Government to carry out the remodelling works (Bostok 1940). These were completed on the 14th July 1938, 3 months ahead of the contract date, at a cost of 1,115,979 Egyptian Pounds[1]

The principal aspects of the remodelling works were stated to be as follows:

- New sluice gates and operating machines installed
- New lift bridges provided over the lock

---

[1] One Egyptian Pound was equivalent to approximately £1 0s 6d

- Extensions to both sides of the existing barrage floor (14.5m u/s, 19m d/s)

- Cement grout injected under the barrage floor extensions

- Flexible concrete block aprons added to both sides of the new barrage floor

- Three rows of sheet piling added parallel to the barrage: 3.5m deep steel sheet piles along the u/s edge of new floor. 6.5m deep interlocking r/c piles under the new upstream floor. 2.5m deep steel sheet piles along the d/s edge of the new floor

- New raised granite weirs, concrete sills and floors added on top of the existing floor

- Barrage piers and arches lengthened on the d/s side

- Roadway widened to 8m and resurfaced

- East wall of lock widened and strengthened (resulting in filling in of vent No. 1)

- Barrage lock gates overhauled and the bottom bearings renewed

Figure 1 shows a general view of the barrage as it stands today.

Figure 1: Assiut Barrage, View from Upstream

Figure 2 shows the design cross section for the remodelled barrage. Except for grouting works in the mid 1980's and replacement of the lock gates in the 1970's, no further major works have been carried out at the barrage since 1938.

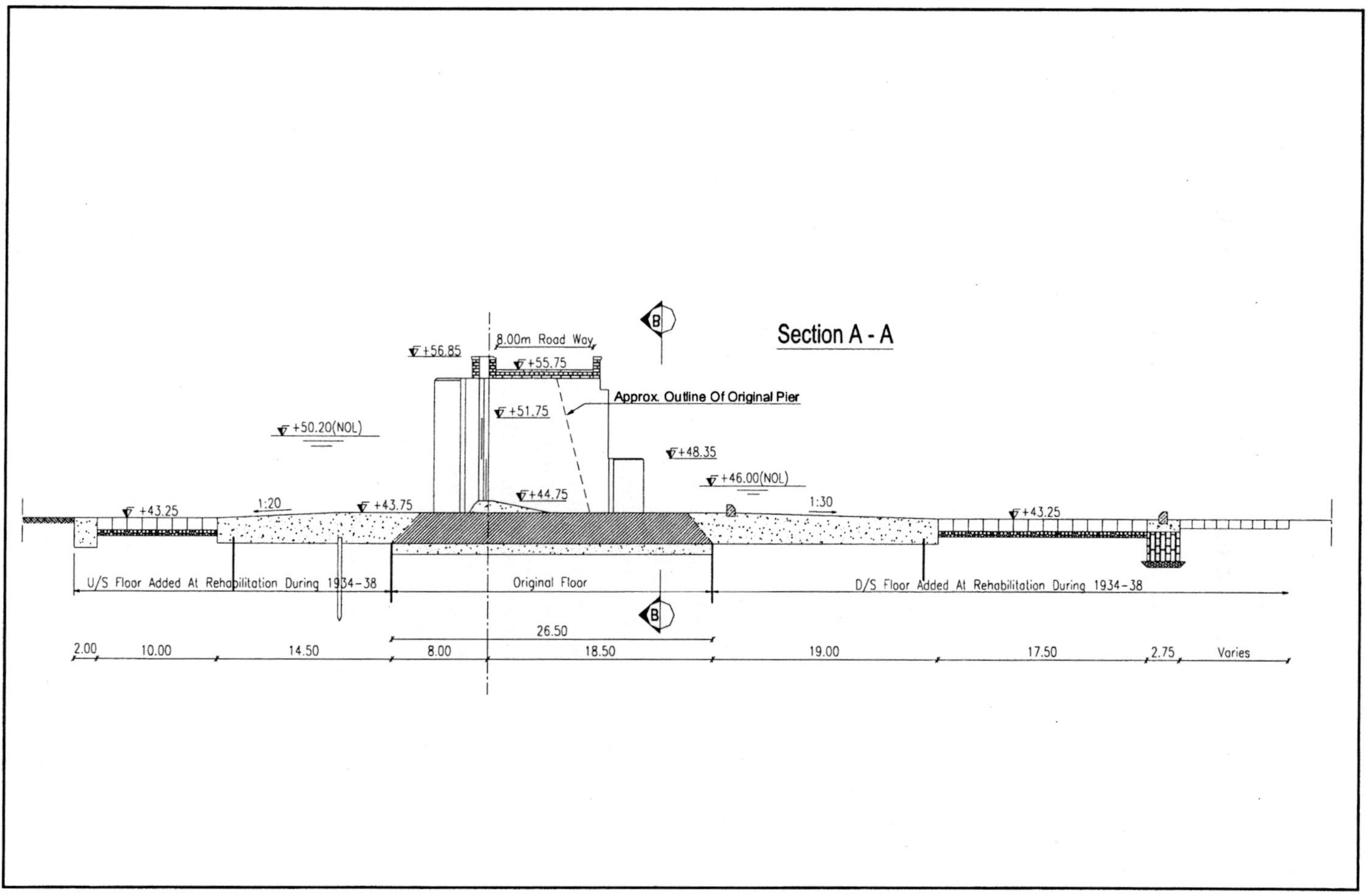

Figure 2: Assiut Barrage Design Cross Section, 1938

CONDITION SURVEYS

The Barrage Superstructure

The feasibility study commenced with a detailed condition survey of the barrage. In all, four different survey methods were adopted:

1. Walk-over visual inspections of the barrage and head regulator superstructures;

2. De-watering of four barrage vents behind newly procured stoplogs and detailed inspection in the dry;

3. Underwater diver surveys conducted within the barrage vents and both upstream and downstream of the barrage;

4. On-site investigations and laboratory analyses of samples. These included drilling works, lugeon testing and measuring, sampling and testing of steelwork.

The key conclusions of these extensive studies were that:

- Apart from some abrasion damage from shipping and trash clearing barges both the barrage and head regulator civil structures are in a generally good condition. There are no signs of structural distress, excessive cracking or settlement.

- Some high Lugeon values were observed, especially in the head regulator structure. This indicated that an extensive grouting exercise is required to ensure future structural integrity.

- Apart from some of the lock equipment, which has been relatively recently replaced, all hydro-mechanical equipment at the site is considered to have reached the end of its useful service life. Rehabilitation of the existing equipment is not economically feasible and full replacement is recommended.

Scour Erosion

Annual bathymetric surveys at the site revealed that an unusually deep (8m in 2004 reduced to 6m in 2005) scour hole had developed in the centre of the river between 50m and 100m downstream of the barrage structure. In addition, overall scour appeared to be developing at a faster rate than was recorded between 1991 and 1997. This is most likely a result of increased river flow in the period from 1997.

There is no direct evidence that the barrage is in imminent danger from excessive downstream scour erosion, however, the client was advised to consider the need for a stone infilling exercise on the downstream side of the barrage in the near future. Such an operation should secure scour protection at the barrage until the planned rehabilitation or new barrage works commence.

<u>Conclusions from the Condition Survey</u>
The overall condition of the barrage and head regulator structures is considered to be sufficient for a rehabilitated barrage solution to be a technically viable alternative.

FUTURE BARRAGE OPERATING STRATEGY
The client defined a complex future operating strategy for the new or rehabilitated barrage. The key changes from the existing strategy being an increase in the maximum allowable head across the barrage, from 4.2 m to 7 m, and a rise in the maximum upstream water level from around 50.5 m asl to 51.6 m asl.

These changes allowed the client significantly more operational flexibility at the barrage for irrigation purposes while also providing enhanced head for a proposed low-head hydropower plant at the site. On the negative side, increased upstream water levels will result in a commensurate rise in the surrounding groundwater table and the increased head will result in more severe structural and energy dissipation conditions at the barrage.

Although the hydropower plant, groundwater modelling and assessment of the resulting environmental and sociological mitigation needs did form a key element of the feasibility study, they are not reported in this short paper.

THE REHABILITATED BARRAGE SCHEME
The design work for the rehabilitated barrage scheme was undertaken with the aim to provide a robust design, suitable for the future barrage operating strategy envisaged by the client and with a design life commensurate with that of a new barrage.

This latter requirement became the driving force behind key decisions such as whether to patch and mend the less damaged aspects of the hydro-mechanical equipment, or to go for wholesale replacement. In the majority of cases the replacement option won through.

The following elements apply to the rehabilitation alternative:

Barrage Rehabilitation. The barrage superstructure is considered to be in a good condition. Minor rehabilitation works are proposed, primarily to the upstream pier nosings and vent soffits. The hydro-mechanical equipment is

70 years old and has reached the end of its useful service life. A complete replacement of these items is required.

New Barrage Navigation Lock. The construction is needed of a new 120m x 17m navigation lock on the left side of the river, immediately to the East (riverward side) of the existing lock. This location will require the demolition of some 9 barrage vents. Constructing the new lock on the landward side would have resulted in difficult land acquisition issues and the demolition of a number of houses and hotels.

Rehabilitation of the Existing Barrage Navigation Lock. The existing 80m x 16m lock will be fully rehabilitated, though its overall dimensions will remain unchanged. In the future the lock could be used to allow passage of vessels with a shallow draft throughout the year and for large cruise ships during the summer (deep water) months. It would also offer the possibility of continued navigation during maintenance of the new lock.

Downstream Weir. A concrete weir is to be constructed immediately downstream of the barrage apron slab. The purpose of the structure is to expressly limit future head differentials across the barrage to those already experienced (4.2m).

In addition to the above, there is also a need for consideration of a hydropower plant. It is proposed to position the HPP on the right side of the river, approximately 150m downstream of the existing barrage. Optimisation studies showed a clear preference for a 32 MW plant with four, 8MW bulb turbine sets.

THE NEW BARRAGE SCHEME
The design of the new barrage at Assiut commenced with an extensive qualitative assessment of thirteen potential sites, ranging from a location some 3.5 km upstream of the existing barrage, to a position some 2.5 km downstream. The limiting factors on the location were the need to construct a separate link canal once the new barrage position extended upstream of the Ibrahimia Canal, and the extensive remedial works to lower groundwater levels in a dense urban environment once the barrage location is downstream of the existing structure. The finally accepted position was between 200 m and 300 m downstream of the existing barrage.

Although the global position of the new barrage has been determined, key decisions regarding the location of the new lock are yet to be finalised by the client. As a result there remain four alternative arrangements of the key scheme components: spillway, HPP and new lock. These are described in Table 1 below.

Table 1. Description of New Barrage Scheme Layouts

Scheme	New barrage	HPP	Sluiceway	Barrage Lock	Road crossing
Alternative 1	200m d/s	Yes Right	Yes Right	Left inline	2-lanes on new and 2-lanes on existing barrage
Alternative 2	200m d/s	Yes Centre	Yes Centre	Left inline	2-lanes on new and 2-lanes on existing barrage
Alternative 3a	300m d/s	Yes Right	Yes Right	Right d/s	4-lanes on new barrage
Alternative 3b	200m d/s	Yes Right	Yes Right	Right d/s	4-lanes on new barrage

Alternatives 1 and 2 have the barrage new lock structure located on the left side of the river, adjacent and immediately East of the existing lock. The layout of the locks is essentially identical to that proposed for the rehabilitated barrage scheme. For alternatives 3a and 3b the lock is positioned on the right side, downstream of the existing barrage. Upon conclusion of the works, therefore, all schemes offer a new (120m x 17m) navigation lock in addition to a refurbished (80m x 16m) existing navigation lock.

The advantages of moving the new lock to the right side of the river lie primarily with ease of construction - all three major scheme components: HPP, sluiceway and lock, can be constructed within a single construction pit. The main disadvantage is the need to use the right channel around the downstream Bani Murr island for navigation purposes. This channel is narrower and deeper than the current navigation channel on the left, and it is likely to exhibit high surface velocities in the future because of the proximity of the new HPP immediately upstream.

In all cases the sluiceway and hydropower structures are located adjacent to each other. Separating them is considered to cause unnecessary complication both during construction and for future operation. Two locations have been examined for these structures: adjacent to the right bank and in the centre of the river. The former alternative offers ease of access to the structures both during construction and also during operation. On the negative side are the less than ideal hydraulic approach conditions to the HPP and the potential for increased erosion downstream of the barrage along the right channel of Bani Murr island. A more central location for the HPP/sluiceway improves both the approach conditions and the downstream flow split around the island. However, construction is more complex (there is no direct access to land) and there is limited room available around the structures for maintenance and operational needs.

Consideration has also been given to future road traffic requirements and for each new barrage alternative a possible scheme for providing a second 2-lane road crossing has been developed.

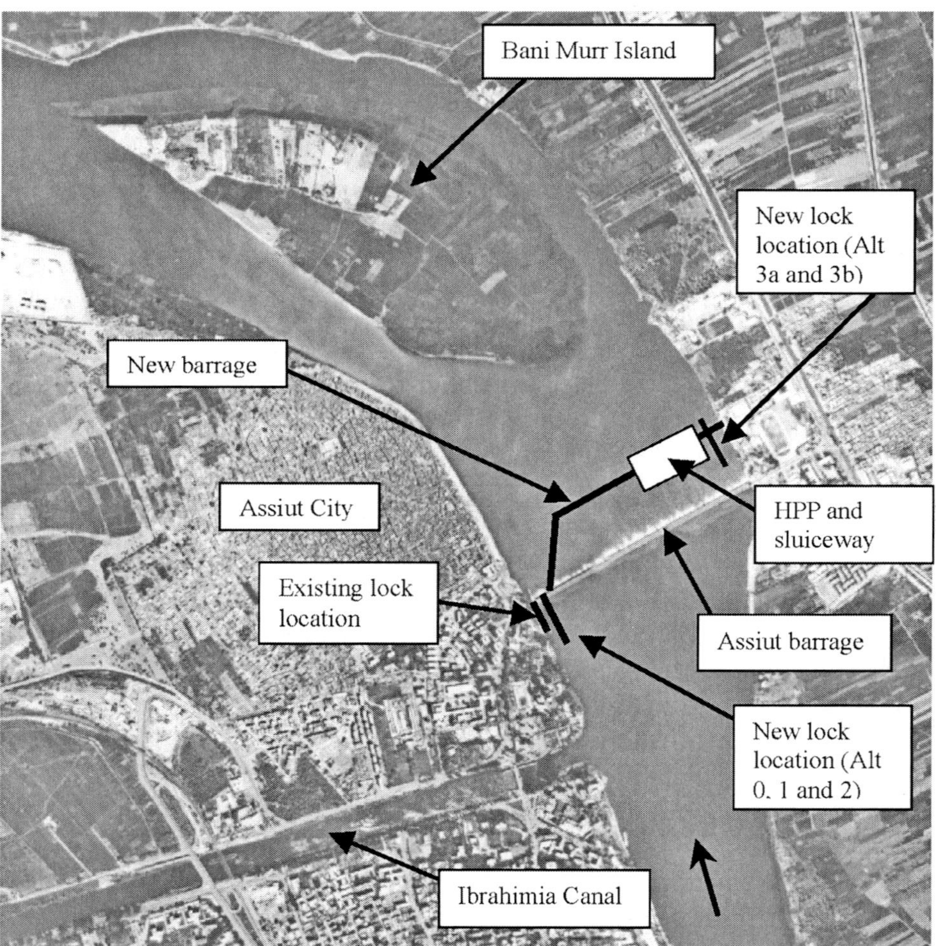

Figure 3. Assiut Barrage from the air

INVESTMENT COSTS

Detailed cost estimates for the rehabilitated and new barrage alternatives have been prepared. These are presented in Table 2 below.

Table 2. Summary of Scheme Investment Costs

Scheme	Total Investment Cost (Million Euro)		Cost of HPP component (Million Euro)
	With HPP	Without HPP	
Alternative 0	255.5	141.3	114.2
Alternative 1	284.6	175.5	109.1
Alternative 2	283.1	171.6	111.5
Alternative 3a	279.5	173.2	106.3
Alternative 3b	277.9	172.8	105.1

CONCLUSIONS

The cost of the rehabilitation scheme (Alternative 0) is some 8-12% cheaper than the new barrage alternatives. However, the cost of the hydropower element of the scheme is some 2-9% more expensive.

Of the new barrage options, those with the new lock on the right side of the river (Alternatives 3a and 3b) are found to be around 2% cheaper than the two left bank alternatives. This is primarily due to the reduced temporary works costs associated with combining the three key scheme components: HPP, sluiceway and new lock; into a single construction pit. There remain concerns regarding the suitability of the downstream right channel around Bani Murr island for lock traffic. Until this issue is resolved in the forthcoming physical hydraulic modelling, the concept of a lock on the right side of the river cannot be confirmed.

If the future of Assiut Barrage is to solely provide continued irrigation supplies for the Ibrahimia Canal then the rehabilitated barrage alternative without HPP is seen as the most appropriate way forward. However, the Client's vision is to see the barrage as a multi-purpose structure with a combined irrigation and power generating function. With this in mind the recommended future direction for Assiut is as a new structure, to be positioned between 200 and 300m downstream of the existing barrage.

REFERENCES

Leliavsky, S (1934). The Assiut Barrage Remodelling Problem, Figure 5.

Stephens, G H (1904). The Barrage across the Nile at Asyut, Proceedings of the ICE

Leliavsky, S (1934a). The Assiut Barrage Remodelling Problem, Lecture given at the Civil Engineering Society Royal School of Engineering, 24th to 31st January 1934, Cairo Al-Ettemad Press

Bostock, JE (1940). Remodelling of the Assiut Barrage, Egypt, Proceedings of the ICE

5. Rick assessment and dam break analysis

Preliminary feedback on the Interim Guide to Quantitative Risk Assessment for UK reservoirs, 2004

AJ BROWN AND JD GOSDEN, Jacobs Babtie

SYNOPSIS The Interim Guide to Quantitative Risk Assessment for UK Reservoirs (the Interim Guide) was launched at the last BDS Conference, in June 2004 at Canterbury, for a five year period of extended trialling. After a year of use of the Interim Guide feedback was sought on the use of the Guide and its application through a series of face to face feedback sessions with a sample of All Reservoirs Panel Engineers, supplemented by a questionnaire. This paper summarises this feedback and then discusses both the role for QRA in dam safety management and how the Guide may be finalised to produce the definitive Guide to Quantitative Risk Assessment for UK reservoirs.

INTRODUCTION

The Interim Guide to Quantitative Risk Assessment (Brown & Gosden, 2004a) was launched in June 2004 at the last BDS Conference at Canterbury. The programme for review and updating of the Interim Guide was given on page 11 of the Interim Guide, being anticipated as the five years to 2008. In a letter to Panel Engineers in July 2004 Defra stated that *"The Interim Guide is a tool for the management of reservoir safety enabling a screening level assessment to be made to inform decision-making by dam professionals on the annual probability of occurrence of reservoir failure, the consequences and the tolerability of that risk."*

The Water Act 2003 amends the Reservoirs Act 1975 to give the power to the Secretary of State to require dam owners to prepare flood plans, with the requirements for such flood plans being currently under development (Brown & Gosden, 2006). It is proposed that the overall consequence class embodied in Section 11.2 of the Interim Guide is adopted as the basis to evaluate whether a reservoir will be required under the Water Act 2003 to have a flood plan.

Improvements in reservoir construction, operation and maintenance, Thomas Telford, London, 2006, 239–250

In parallel the authors are producing guidance on the early detection of internal erosion, which includes relating the surveillance regime to the risk posed by the dam.

It was therefore decided it would be timely to seek feedback on the first year of use of the Guide, and at the same time seek views on what criteria should be used to determine which reservoirs would have flood plans and the surveillance regime adopted at any dam.

STRATEGY FOR OBTAINING FEEDBACK
The issues on which it was wished to obtain feedback comprise:
- the principles of the application of quantitative risk assessment (QRA) to dam safety management,
- the detailed features in the Interim Guide,
- use of overall consequence class to determine whether a flood plan is required
- use of some measure of risk to determine the level of surveillance (Brown & Gosden, 2004b)

Questionnaires were used in early 2003 to elicit the opinions from 120 dam professionals on approaches to incident reporting (Gosden and Brown, 2004) and the possibilities for the early detection of internal erosion (Brown & Gosden, 2004b), achieving a response rate of 43%. However, questionnaires on a proposed draft strategy for early detection of internal erosion given out at a BDS meeting and accompanying material on the BDS website only achieved two responses.

It was therefore decided that a proactive approach would be adopted to obtain feedback, with the first step being a number of face to face feedback sessions with selected All Reservoirs Panel Engineers. At these sessions a short (two page) questionnaire was handed out at the end of the meeting for later completion, both to obtain a written summary but also to allow feedback to be sent directly to Defra and thus to remain anonymous to the authors of the Guide. This questionnaire was also sent to all other Inspecting Engineers and the Reservoir Safety Managers of major dam owners.

FEEDBACK OBTAINED
Eight face to face meetings were held, two with groups of independents and six with all the Panel AR Engineers at a particular consulting engineering company. Each meeting typically lasted about three hours and comprised about one third the presentation of the author's experience with the application of the Interim Guide and two thirds, structured discussion on the experiences of the other attendees with the Guide. It is noted that these

were carried out under a Defra research contract which was novated to Jacobs Babtie in December 2005, along with the TUPE transfer of consultancy staff. As well as feedback on the use of the Interim Guide, the sessions provided an opportunity to resolve queries on the Interim Guide and discuss the level of accuracy of particular elements of the calculations.

Questionnaires were sent to all Inspecting Engineers, and the Reservoir Safety Managers of the companies who own the greatest number of dams, with responses as shown in Table 1.

Table 1: Summary of numbers of feedback questionnaires

Group	Number		% response
	In group	Returning questionnaire	
Panel AR attending face to face meeting	18*	8	44%
Panel Engineer who did not attend meeting (AR, NI, SR), for whom email address available	36	8	22%
Reservoir safety managers for major dam owners	13	3	23%
Total	67	19	28%

* 30 invited, some could not attend.

SUMMARY OF FEEDBACK FROM MEETINGS AND QUESTIONNAIRE

Use of QRA for reservoirs

All respondents considered that the use of a rapid (screening level) method of quantitative risk assessment for reservoirs should be encouraged. Half had used the Interim Guide for Inspections and a further 26% in some other context. Only three respondents had used the ALARP approach to determine upgrading works. In terms of promoting use of the Interim Guide as part of a Section 10 Inspection 56% had a strong or slight preference for this, 11% were neutral and the remaining 33% would not promote its use as part of a Section 10 Inspection. Comments generally supported the principle of QRA, but included a "note on client resistance", and the wish for "a simple non computer based system".

Excel workbook

The great majority found the workbook complicated initially, often preventing them using it, or requiring several determined attempts. However, in general once they had managed to put aside some uninterrupted time to work through it they found it useful, with 70% of those using the workbook having unprotected the sheet and modified it for their own use. It was generally felt that it would take about 2 days to complete the assessment for a dam, once familiar with the workbook. Feedback on the format of the spreadsheets is given in Table 2; the majority of those who had used the workbook considered that only minor improvements were required.

Table 2: Summary of feedback on format of Excel spreadsheets

Sheets relating to	Feedback score			
	0	1	2	Blank
Probability of failure (Sheets 2 - 7)	1	6	3	9
Consequences of failure (Sheets 8-10)	1	7	2	9
The tolerability of risk (Sheets 11-12)	3	5	2	9

Note: Score : 0- means no immediate improvement required to Interim Guide; 1- minor changes; 2- major changes;

In terms of ease of use the majority asking for the spreadsheets to be made simpler, although all but one of the respondents considered that some form of workbook was worthwhile. The approach of colour coding in yellow the essential data input cells was considered very helpful, with some respondents just completing these as an initial pass, to gain an overview of the workbook. Suggestions of how to make it more user friendly included

- adding a "black box" interface from which prompts asked for specific data,
- adding more comments within workbook cells,
- providing the facility to skip sheets where not applicable e.g. no upstream reservoir, spillway designed to pass PMF with full wave freeboard and no risk of blockage where probability of failure would be less than 10^{-6}/ annum
- adding more explanation within the sheets.

<u>Technical content</u>
In terms of the improvements required to the technical content the response is summarized in Table 3. It can be seen that of those who completed this section the majority felt no, or only minor, changes were required. The responses to Q10, asking for comments on features that need improvement and how this could be done were generally limited to criticism of particular features, rather than providing any suggestions of how improvement could be made. This outcome is probably a reflection that the Interim Guide has only been available for one year, most respondents are starting to explore its use and have not yet had time to formulate opinions of the reliability and uncertainty of estimates made by the workbook.

<u>Extension of the Interim Guide to Concrete Dams and Service Reservoirs</u>
The questionnaire also asked what priority should be put on extending the Interim Guide to concrete and masonry dams, and service reservoirs, with the responses as shown in Table 4. Twelve and eight respondents respectively considered this should be by collecting data on historical incidents and developing event trains and associated guidance; the remainder leaving this question blank. There is some ambiguity in the question, but the majority indicated this should be in the medium term, implying there was a desire that a supplement covering these should be issued in advance of the review and updating of the Interim Guide.

Table 3: Summary of feedback on technical content in Interim Guide

Element of calculation in workbook (Section number; ref Figure A.1 in Interim Guide)	Feedback score			Blank	Research priority (R)
	0	1	2		
Event trains (Sections 2.2, 2.7.1, sheets 2.2, 3.2, 4.2, 5.2, 6.2, 6.3)	4	6	3	6	1
Annual probability of failure due to extreme rainfall (S 2)	3	7	2	7	1
Annual probability of failure due to upstream reservoir (S 3)	5	6	1	7	1
Annual probability of failure due to internal threats (S 4, 5)	1	11	1	6	4
Inclusion of other threats (S 6)	3	7	3	6	0
Rapid inundation analysis (S 8)	1	6	4	8	3
Consequence assessment – likely loss of life (S 9)	2	6	3	8	3
Consequence assessment – third part damage £M (S 10)	2	6	3	8	1
Assessment of "tolerable risk" (S 11)	5	4	3	7	1

Key: Feedback Score 0- means no immediate improvement required to Interim Guide; 1- minor changes; 2- major changes;
R - the priority order for future research (the number of responses putting this element in the top three, noting that only 8 respondents completed any of this question; with one noting it was too early to say)

Table 4: Summary of feedback on priority for extending to concrete/ masonry dams and service reservoirs

	Not required	Medium term	Prior to issue of definitive Guide	Left blank
Concrete/ masonry dams	1	9	5	4
Service reservoirs	7	7	1	4

<u>Risk management of UK reservoirs</u>
The other key area for feedback was the extent to which quantified estimates of risk should be used to determine dam safety management issues, such as which reservoirs should have flood plans and the level of surveillance for internal erosion (Brown & Gosden, 2004b). The feedback for these is summarised in Table 5.

It can be seen that there are approximately equal numbers supporting Consequence Class and risk, as alternative criteria to determine which reservoirs should have flood plans. In relation to the basis for determining surveillance there is a wide disparity of views, with 50% giving "other criteria" and the remainder spread over the three options in the questionnaire. The suggestions for "other" for the level of surveillance included
- "amplification of condition" to include vulnerability to erosion,
- judgement of the panel engineer
- while risk is the theoretically best criterion, I do not have sufficient confidence in the assessment of risk to advocate using it in this way.

Table 5: Summary of feedback on criteria to determine level of measure to control risk from a dam

Preferred dam classification system	Determine which reservoirs have flood plans?	Determine the level of surveillance?
Overall Consequence Class (A1 to D)	9	3
Conseq. Class x Condition	Not app	1
Risk (probability of failure x conseq.)	7	4
Other	0	7
Left blank	3	4

OTHER SOURCES OF FEEDBACK

<u>Experience in application of QRA</u>
Early experience comprised a trial of the prototype system on ten dams, written up in a research report on the Defra website (KBR, 2002) and for Dam 4 in Brown & Gosden (2005). Since the launch of the Interim Guide several major dam owners are using the system in a variety of ways, including as part of Section 10 Inspections (Gosden & Dutton, 2006), to prioritise surveillance and to produce a portfolio risk assessment. The authors also now routinely use the consequences element as part of Section 10 Inspections, to provide the dam consequence category which is then used to inform the decision as to which recommendations made in an inspection are "in the interests of safety". This has identified a number of areas for improvement, the main points being:

Section	
2	A methodology for quantifying the risk of flows down the spillway chute causing failure by scour along the sides of the structure (Row 24 in sheet 2.3)
4, 5	More detailed guidance is required on scoring current condition, to deal with both indicators smaller than the guide value and where the magnitude of the indicator is not known
8	In some cases the attenuation length, la, over which flow attenuates to 37% of its initial value is excessively long (in excess of 100km)
9	More detailed guidance would be helpful in estimating the number of houses, area of non-residential property and population at risk

<u>Review of technical aspects</u>
Eddleston & Carter (2006) present a comparison of three methods of estimating the annual probability of failure due to internal threats, one being that given in the Interim Guide. It is noted that the UK experience of a large number of serious incidents leading to emergency drawdown with few failures leads to a modest annual probability of failure for dams in current condition score 8. This relies on a high standard of surveillance and prompt intervention, as some of these incidents would have developed into failure if there had been no intervention.

Ackers et al (2006) present an updated summary of detailed inundation analysis, including valuable case history data on the attenuation length estimated from detailed dambreak.

<u>Guide to Emergency planning</u>
In developing the forthcoming Engineering Guide to Emergency Planning (Brown & Gosden, 2006) the authors have applied the consequence elements of the QRA workbook to the example to be included with the Guide to Emergency Planning. This has provided useful feedback on technical aspects of the consequences estimation, which it is anticipated will be published as a supplement to the Interim Guide.

DISCUSSION
The following text sets out the authors' views on the possible ways in which the Interim Guide could be improved. These are to promote debate, and hopefully promote detailed feedback to Defra by others, using the sheet on page xiii of the Interim Guide.

<u>Benefits of QRA for dam safety management</u>
Some of the strategic drivers for QRA were given in Brown and Gosden (2002); including the benefits of a documented safety case for dam operation, and transparency in the level of risk which is considered tolerable. At an operational level QRA provides value in

- promoting critical consideration of potential modes of failure
- determining which recommendations in a Section 10 Inspection under the Reservoirs Act should be in "the interests of safety"?
- determining which upgrading measures are proportionate in terms of cost relative to the reduction in risk achieved (ALARP analysis)?
- targeting the surveillance regime, and other dam safety management measures
- for commercial companies, to rank risk from their dams with the other infrastructure that they are responsible for

The feedback supported this view of the potential benefits of QRA, whilst requiring further experience of the use of QRA to decide whether the Interim Guide in its current state was the vehicle to deliver them.

<u>What is a proportionate level of technical detail in analysis?</u>
QRA can be carried out at many different levels of detail and sophistication, with a categorisation of risk analysis levels given by McCann (1998) reproduced in Table 6. The authors' suggest that the Interim Guide is intended to be a screening level of quantitative analysis, appropriate for use as part of Section 10 Inspection and/ or portfolio risk assessment on most UK dams and not taking more than say two days to complete. As such it would be equivalent to Levels 2 to 3 in Table 6. On this basis it might be argued that CIRIA Report C542 (2000) represents Level 1 in Table 6.

Table 6 : Summary of risk analysis levels as McCann, 1998

	Level	Scope/ Application
1	Scoping	Qualitative assessment of failure modes
2	Ranking	Quantitative analysis of all elements of a risk analysis
3	Detailed	Results can be used to justify dam safety modifications
4	Comprehensive	A higher degree of defensibility than level 3
5	Full scope	Where the highest degree of defensibility is required due to the level of consequences ($billions), the technical complexity

Some of the feedback suggests that the respondents consider the analysis in the Interim Guide is too complex; although the comments often do not differentiate between the technical content and the format of the spreadsheet. It is therefore unclear whether the feedback suggests QRA at this level should not be used for inspections, or whether the use of Excel as a medium for calculations is too complex.

Various tests to assess whether the technical complexity of the analysis incorporated in the spreadsheet is appropriate for use in a Section 10 Inspection are discussed below:

1. One comparison is with the rapid method for floods, which comprises six sheets in Floods and Reservoir Safety (ICE, 1996). The Interim Guide may be viewed as comprising nine separate calculations, five relating to probability of failure from different threats, three relating to consequences and one of tolerability of risk. The 57 sheets in Appendix C of the Interim Guide therefore represent an average of 6.3 sheets per calculation, comparable with the rapid method for floods.

2. Another test is to consider what would be a reasonable cost and complexity for a periodic (10 yearly) safety review of a hazardous installation which if it failed could lead to loss of life. An additional two days to add QRA does not seem excessive. The total time for an Inspection is likely to be less than the time spent on comparable safety reviews in the chemical industry. It is marginally less than required for a principal bridge inspection (Highways Agency, 1994, 1995), which has a six year inspection cycle and is applied to all the many hundred of thousands of bridges in UK. It is acknowledged that the cost of a QRA assessment is probably disproportionate for a Category C or D dam, but the author's experience is that a significant proportion of such dams when the consequences are assessed quantitatively often move up to be Category B.

3. A third test is the level of detail adopted overseas. As well as the Interim Guide there have been three other recent publications on estimation of risk posed by dams, namely the ANCOLD Guidelines (2003), Risk and Uncertainty in dam safety (Hartford, 2004) and ICOLD Bulletin 130 (2005). Of these only the Interim Guide provides a spreadsheet methodology, with the others limited to principles only; generally implying a more complex analysis.

On balance the authors consider that the level of analysis is appropriate for most UK reservoirs as part of a Section 10 Inspection, or portfolio risk analysis. Where a dam retaining a reservoir is Category C or D, and there is a documented auditable case for this, then it may be reasonable to either rely on judgment alone, or perhaps carry out a qualitative analysis, using the event trains only.

<u>Future research and development</u>
The main item identified from the feedback is the need to extend the Interim Guide to cover concrete and masonry dams, preferably by the issue of a supplement prior to issue of the definitive Guide.

<u>Excel workbook with the Interim Guide to QRA</u>
Most of the comment relates to the format and ease of use of the Excel workbook included with the Interim Guide. Table 7 sets out three options, at a strategic level, for improving the ease of use of the spreadsheet. The feedback favoured Option 2. The choice then reduces to the style to be adopted which the majority would favour. Design of the structure of the existing workbook took the view that

a) the input data was all input in Section 1, except that
b) input data which has a large effect on the analysis was better located in the sheet where the calculation took place (to facilitate the role of judgement, by seeing the effect of the input assumption on the output).

It is suggested that as part of the review of the Interim Guide a number of options for the structure and format of the Excel workbook be identified, and subjected to a consultation process to identify the option which the majority preferred.

Table 7 : Strategic options for improving ease of use of Excel workbook with the Interim Guide

	Option	Comment
1	Delete spreadsheet, rely on description of principle only	The feedback was overwhelming against this
2	Improve spreadsheet format and ease of use	The advantage of the spreadsheet approach is that it can be reviewed and edited by all those involved in the safety analysis, and does not require "specialist operators". Moreover it can readily be edited to suit an individual dam. It does, however, require familiarity and comfort in working in Excel.
3	Turn workbook into a black box, with simplified input requirements (similar to commercial software for slope stability and retaining wall analysis)	This would make it more difficult for the engineer to exercise his judgment, as the analysis could not be easily amended to suit the individual dam. This is therefore not favoured

SUMMARY AND CONCLUSIONS

This paper has summarised both preliminary feedback on the use of the Interim Guide, based on a series of face to face meetings and a questionnaire, and ongoing use of the Interim Guide which is providing additional feedback. This shows significant support for use of QRA for UK reservoirs. In relation to the Interim Guide 76% have used it in some context and 55% of respondents would promote its use as part of a Section 10 Inspection. Persistence is needed to use the Excel workbook included with the Interim Guide; all those who used the workbook considering it worthwhile with 70% of these having unprotected the workbook and modified it for their own use. It was generally considered that once familiar with the system it takes about 2 days to complete the assessment for a dam.

The paper then discussed feedback on ways in which the Interim Guide can be improved, the majority considering that only minor changes are required before the Interim Guide is reviewed and finalised. It is noted that the feedback suggests that a supplement should be issued prior to review of the Interim Guide, to cover concrete and masonry dams.

Finally there is a spread of views on the extent to which QRA should be used to determine the type and magnitude of dam safety measures, with approximately equal numbers supporting risk and consequence class for flood plans, and no clear outcome to determine the level of surveillance.

ACKNOWLEDGEMENTS

The work described in this paper was carried out as a research contract for Defra, who have given permission to publish this paper. However, the opinions expressed are solely those of the authors and do not necessarily reflect those of Defra.

The advice and assistance of the Steering Group (given in the preface to the Interim Guide) who oversaw production of the Interim Guide on behalf of Defra is gratefully acknowledged. Thanks are also extended to all those who participated in the feedback session, particularly those who took the time to complete the questionnaire

REFERENCES

Ackers J.C., Pether R.V., Tarrant F.R., 2006, Reservoir hazard analysis and flood mapping for contingency planning. In *Improvements in reservoirs*, Thomas Telford.

ANCOLD, 2005, Guidelines on risk assessment. 156pp excl appendices

Brown AJ & Gosden JD, 2002, A review of systems used to assess dam safety. Proc BDS Conf. pp 602-619

Brown, A.J. and Gosden, J.D., 2004a, Interim Guide to quantitative risk assessment for UK reservoirs. Thomas Telford.

Brown, A.J. and Gosden, J.D., 2004b, The early Detection of Internal Erosion ", *Dams and Reservoirs,* Vol.14, No.1.

Brown, A.J. and Gosden, J.D, 2005, An example of application of the Interim guide to QRA. *Dams and Reservoirs.* 15(2) pp11-13

Brown, A.J. and Gosden, J.D., 2006, Development of the requirements for Flood Plans under the Reservoirs Act 1975 (as amended). In *Improvements in reservoirs,* Thomas Telford

CIRIA, 2000, Risk Management for UK Reservoirs *Report No.C542* p.213.

Eddleston & Carter, 2006, Comparison of methods used to determine the probability of failure due to internal erosion in embankment dams. In *Improvements in reservoirs,* Thomas Telford

Gosden, J.D, and Brown, A.J., 2004, An incident reporting and investigation system for UK dams. *Dams and Reservoirs,* Vol.14, No.1.

Gosden, J.D. and Dutton D, 2006, Quantitative risk assessment in practice. In *Improvements in reservoirs,* Thomas Telford

Hartford DND, Baecher GB, 2004, Risk and uncertainty in dam safety. CEA technologies Dam safety Interest group. Thomas Telford. 391pp

Highways Agency, 1994, Highway Structures: Inspection and maintenance. BD 63/94, and BA 63/94. DMRB Volume 3 Section 1 Parts 4 and 5

ICE, 1996, Floods and reservoir safety, 2nd Edition. Thomas Telford

ICOLD, 2005, Risk assessment in dam safety management. Bulletin 130

KBR, 2002, *Floods and reservoir safety integration.* Defra research contract. Available at www.defra.gov.uk/environment/water/rs/index.htm

McCann MW, 1998, A framework for applying and conducting risk-based analysis for dams. USCOLD lecture series (conf). Pp 115-132

Quantitative risk assessment in practice

J D GOSDEN, Jacobs Babtie (previously KBR)
D DUTTON, British Waterways

SYNOPSIS.   In 2004 the Interim Guide to Quantitative Risk Assessment for UK reservoirs was published. This document gives a methodology for evaluating the risk posed by the principal threats to dam safety within a common framework using a series of Excel worksheets. The probability of failure of the dam is estimated and compared with the likely loss of life to evaluate the risk posed by the dam, and whether this is tolerable.

This paper describes one of the first uses of this methodology in practice. The system has been applied to 6 reservoirs owned by British Waterways in the United Kingdom which feed the Leeds and Liverpool Canal as part of the regular 10 yearly review of reservoir safety under Section 10 of the Reservoirs Act 1975.   The reservoirs are impounded by earthfill embankment dams constructed in the early 19th century and are all around 10 metres high.

The paper presents the results of the quantitative risk assessment and the criteria used to determine whether any works are required to improve dam safety. The benefits obtained from using quantitative risk assessment are evaluated from the perspective of both the dam inspecting engineer and the reservoir owner. The use of the quantitative risk assessment in reviewing the existing surveillance procedures for the reservoirs is also described.

The paper concludes with a review of the quantitative risk assessment methodology and identifies where there are opportunities for future improvement.

INTRODUCTION

Regulation of a high hazard civil engineering industry was first implemented through the Reservoirs (Safety Provisions) Act 1930. Although the Reservoirs Act 1975 added further measures to improve the management of reservoir safety, the system of reservoir inspection is largely unchanged since 1930 and has served the public well in this period, with no reservoir failures occurring which have resulted in loss of life. The inspection system places full reliance on the judgement, experience and knowledge of individual inspecting engineers. A consistent approach by different inspecting engineers has been promoted through a Government funded research programme of guidance documents. Prescriptive guidance is provided for two threats, floods and earthquake, with the remainder being open to wider interpretation.

In recent years four other high hazard industries have been regulated by the Health and Safety Executive; nuclear sites, onshore chemical plants, offshore and railways (HSE, 2000). The approach to regulation in these industries is broadly similar comprising four underlying principles which represent current best practice.

The most important principle is that the organisation which creates the hazard has a legal duty to manage the risk through the preparation of a safety case which describes how the risk is managed. The safety case involves the following steps:
- Identify the hazards
- Assess the risks
- Develop effective control measures in a coherent whole (i.e. an integrated approach)
- Keep a current documentary record.

The general approach to regulation is that a goal setting framework is preferable to defining prescriptive standards as it makes duty holders think for themselves. This flexibility leads to methods of risk control being tailored to particular circumstances.

Risk is the product of the probability of an event and its consequences. Quantitative risk assessment (QRA) allows risk to be quantified by assigning numerical values to both the probability and consequences to arrive at a risk value of £/annum and likely loss of life/ annum. QRA, as a tool for the safety management of high hazard industries, was pioneered in the nuclear industry but is now more widely used and facilitates preparation of a safety case.

The Interim Guide to Quantitative Risk Assessment for UK Reservoirs (Brown and Gosden, 2004) sets out a methodology for using QRA as part of the safety case for continued operation of a reservoir. This paper describes the application of that methodology

DESCRIPTION OF RESERVOIRS

Quantitative risk assessment (QRA) has been used as part of the inspections under Section 10 of the Reservoirs Act for 6 reservoirs supplying water to the Leeds and Liverpool Canal to inform the findings and recommendations.

The principal characteristics of the reservoirs are summarised in Table 1 below. Both Whitemoor and Rishton reservoirs are formed by continuous embankments which cross the catchment watershed. Depending on the location of any breach, failure could take place into either one of two separate valleys. This was not recognized in the last Inspection Report where only a single flood hazard category was determined (presumably the more severe) but two Consequence Class assessments have been made as part of the QRA, which apply to particular lengths of the embankment.

Reservoir	Height m	Reservoir capacity m^3	Catchment area km^2	Dam	Flood hazard category	Consequence Class
Upper Foulridge	12	430,000	3.5		B	A2
Lower Foulridge	9	1,490,000	4.8		B	A2
Whitemoor	10	640,000	1.7	East	B	A2
				South	B	B
Slipper Hill	7	165,000	0.3		C	B
Barrowford	9	450,000	Non-impounding		n/a	B
Rishton	10	615,000	0.7	West	C	A2
				East	C	B

Table 1: Principal characteristics of Leeds-Liverpool canal reservoirs

In carrying out the Section 10 inspections the recommendations were deliberately not formulated until the QRA had been completed.

QUANTITATIVE RISK ASSESSMENT

Data

The data collected on the condition of the dam was little different from that obtained in a normal inspection, with the differences identified below. The condition of the dam was described in the Section 10 report and used to develop the annual probability (AP) of failure for the internal threats.

In order to evaluate the AP of failure due to extreme rainfall, a level survey of the embankment crest is required to assess the extent of overtopping in order to derive the imminent failure flood. This has been carried out as standard practice in recent years in any case to check on crest settlement.

Greater attention was paid to visiting the downstream valley to assess flow routes, identify infrastructure which could reduce or increase the flood peak (in the event of failure) and potential properties at risk. In this case this was carried out during a second visit to the reservoirs to inspect them under high reservoir level when seepage was more likely to be evident. However what a number of recent inspections has shown is that a failure to properly consider the downstream valley has considerably underestimated the population at risk. We have found that carrying out a rapid dam break assessment forces this proper consideration to take place.

<u>Results of QRA</u>
The results of the QRA are shown on the Consequence Class diagram and FN chart in Figures 1 and 2.

<u>Improvements to dam safety</u>
For each of the dams the impact of the measures recommended in the interests of safety and the surveillance improvements on the annual probability of failure are shown in Table 2 below and also illustrated on Figure 2. The current AP of failure is shown in normal type and the AP of failure following the proposed works in italics. The principal measures recommended in the interests of safety are indicated in the final column.

Table 2: Impact of proposed works on the annual probability of failure

Reservoir	Annual probability x E-05					Principal works recommended in the interests of safety
	Extreme rainfall	Upstream reservoir	Internal stability embankment	Internal stability appurtenant works	Total	
Upper Foulridge	0.1	n/a	4.0	30	34	Repair draw-off upstream sluice gate
	0.1	*n/a*	*0.6*	*0.3*	*1.0*	CCTV survey of the draw-off pipe
						Fill crest depression; close gaps in crest kerb
						Seal leakage paths through spillway crest
Lower Foulridge	1.0	34	2.0	4.0	41	Repair bottom draw-off upstream sluice gate
	0.1	*1.0*	*0.6*	*0.1*	*1.8*	CCTV survey of upper draw-off culvert
						Clear strip along embankment d/s toe
						Reconstruct upper part of spillway chute
						Protect d/s toe from high spillway chute flow
Whitemoor	1.5	n/a	4.0	3.0	8.5	East embankment minimum freeboard of 1.5m
	0.1	*n/a*	*0.3*	*0.8*	*1.2*	Increase spillway chute capacity
						Rebuild spillway chute floor
Slipper Hill	0.01	n/a	0.2	10	10.2	Construct cut-off below spillway crest
	0.01	*n/a*	*0.2*	*0.85*	*1.0*	
Barrowford	Wind 5		10	10	25	Reduce current overflow level
	Wind 0.1		*2.0*	*0.2*	*2.3*	CCTV survey and repair draw-off pipework
						Review surveillance frequency
Rishton	0.05	n/a	9.0	0.8	9.9	Investigate toe drains and measure flow
	0.01	*n/a*	*0.3*	*0.8*	*1.2*	Embankment minimum freeboard of 1.1m

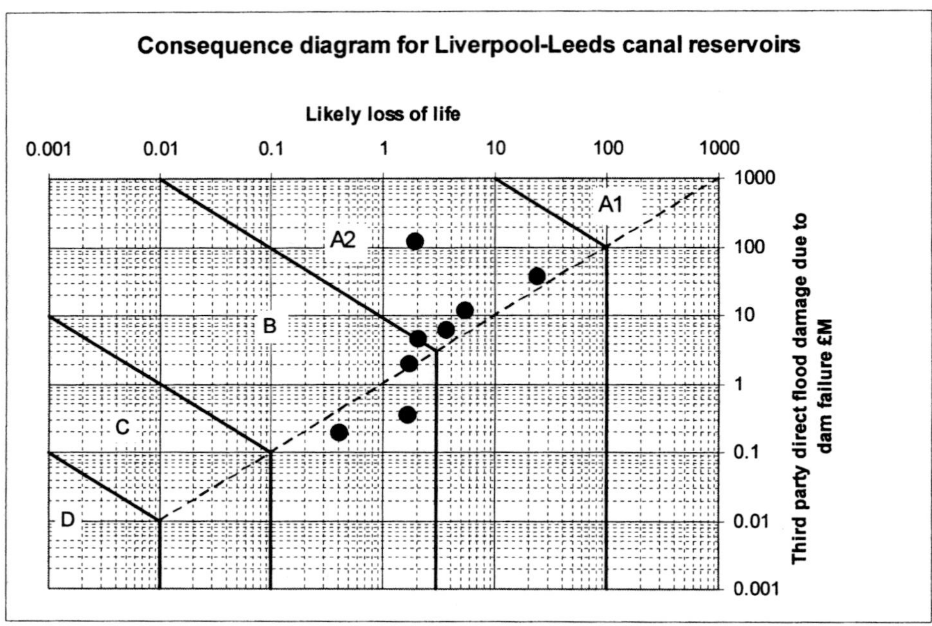

Figure 1: Consequence Class diagram for Leeds-Liverpool canal reservoirs

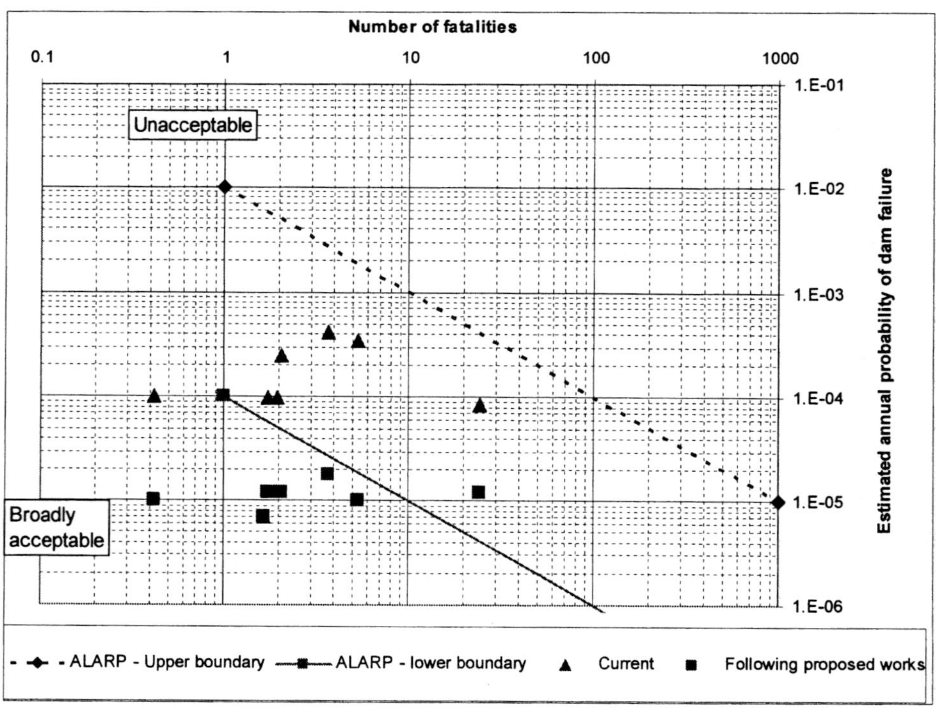

Figure 2: F-N chart for Leeds-Liverpool reservoirs

BENEFITS TO THE INSPECTING ENGINEER

I found the following benefits through carrying out the QRA as part of the Section 10 inspection:

1) Carrying out a rapid dam break assessment ensured that potential flow paths were assessed and a quantitative estimate of the depth of flow assisted the visual identification in the valley of potential properties at risk. This process highlighted for two of the reservoirs that were close to the watershed that breach flows would take very different paths depending on the breach location resulting in different Consequence Classes for particular sections of the embankment, as shown in Table 1. This resulted in different minimum freeboard requirements for the east and south embankments at Whitemoor reservoir and works only being required to raise the east embankment

2) Consideration of the event trains ensured systematic assessment of potential failure modes, including the addition of some not included as standard and helped to identify the most critical ones. This ensured attention was focused on where improvement was required. In particular this lead to the identification of a failure mode at Lower Foulridge reservoir of erosion of the embankment downstream toe from the limited capacity of the spillway chute. Previous upgrading works had increased the spillway crest capacity to in excess of the 10,000 year flood while the chute capacity remained around the 100 year flood. It was estimated that the annual probability of embankment failure from this mode was 1×10^{-5}.

3) Repairs to some of the bottom outlets had been postponed for many years. The improvement in AP of failure resulting from a reliable bottom outlet demonstrated clearly the need to include these as measures in the interests of safety

4) The rapid dam break analysis demonstrated that the population at risk had been optimistically assessed in previous inspections. Using the guidance given in Floods and Reservoir Safety (ICE, 1996) significant improvements in the spillway capacity to provide the recommended wave freeboard would have been required at several reservoirs. I did not believe that this was the best use of funds rather than expenditure on other items where there is no prescriptive guidance. The QRA provided a rational, defendable basis on which to recommend that no further upgrading in spillway capacity was required. This approach was adopted at Lower Foulridge and Whitemoor reservoirs.

5) Recommendations for improving surveillance were justified on the basis of reducing the AP of failure. This was more easily accepted by the reservoir undertaker. Examples of this included the

recommendation at Barrowford reservoir, which had suffered repeated embankment surface instability, to review the frequency of surveillance and nature of monitoring of the embankment to take into account the recommendations of the Early Detection of Internal Erosion research project, which are due to be published in 2006

6) The QRA provided a rationale for prioritising the required measures. The time suggested for completion of the measures in the interests of safety was judged from the absolute AP of failure and the extent of improvement required to achieve an acceptable condition. This resulted in the works at Upper Foulridge reservoir being prioritised ahead of the works at Lower Foulridge reservoir, despite Lower Foulridge reservoir having a higher AP of failure. Prior to carrying out any works the major contributor to the AP of failure of Lower Foulridge reservoir was the threat posed by failure of the upstream reservoir.

Overall the QRA provides an excellent audit trail establishing the basis on which recommendations have been made. This will have significant benefit to Inspecting Engineers in the future in our increasingly litigious society.

BENEFITS TO THE RESERVOIR OWNER AND SUPERVISING ENGINEER

Possibly the main benefits to the reservoir owner are that the QRA provides a consistent approach to the Section 10 Periodical Inspection, and makes transparent the reasoning behind the inclusion or exclusion of certain safety measures. Thus, for example, the owner is no longer faced with the dilemma of whether or not a particular inspecting engineer, relying on his own perception of risk, will insist upon the prescriptive use of Table 1 in the ICE Guide 'Floods and Reservoir Safety', or the provision of an upstream valve on a draw-off; the effect of the perceived shortcoming can be analyzed and arguments developed to justify particular recommendations which are proportionate in relation to the risk posed by the individual dam.

The QRA process provides the reservoir owner with a clearer understanding of where the risks to his reservoir lie and the potential hazard that the reservoir represents, both to his own undertaking and to others; in a business where there are competing claims on money, this proper appreciation of risk enables more realistic and responsible spending plans to be developed. The background knowledge that the QRA generates also allows the owner to comment authoritatively on others' proposals, for example planning applications, where responses often have to be made within a limited timescale. The undertaking of the dam breach analysis within the QRA of the Leeds and Liverpool Canal reservoirs has provided a salutary lesson on the likely level and extent of damage that would be caused in the event of a reservoir failure; the change in dam category that has been necessary at a number of these reservoirs suggests a certain lack of appreciation in the past.

Undertaking a QRA as part of the Section 10 Periodical Inspection provides a much more detailed record of the reservoir's condition than is normally given in inspection reports. This 'benchmark' information will be of benefit to both the supervising engineer and future inspecting engineers, as it allows the rate of change of any worrying or unusual feature to be readily assessed. Also, it can be used by the reservoir owner to set up a tailor-made surveillance regime, directing resources and attention in particular to those features which have a direct bearing on the safety of the dam.

The total cost of the six inspections including the QRA was £22,000. This is probably around 50% higher than the inspections alone would have cost. However this is an order of magnitude lower than the cost of potential spillway upgrading which might have been required without the QRA to support a different approach.

CONCLUSIONS

Quantitative risk assessment following the Interim Guide to Quantitative Risk Assessment for UK Reservoirs was successfully used during the Section 10 inspections of 6 reservoirs for British Waterways. It yielded a number of benefits for both the Inspecting Engineer and the reservoir owner, which have been outlined above, including a substantial cost saving by avoiding the need for further spillway upgrading works.

These assessments have demonstrated the robustness of the approach adopted by the Guide to QRA and have identified a small number of improvements which should be made when preparing the definitive guide.

The principal improvements proposed are:
a) Revision to the rapid dam break routing, which in some situations underestimates the amount of flood attenuation occurring down the valley. At present this is implemented by selecting values of the attenuation parameter La which are below the recommended range
b) Additional guidance on the scoring of current condition for the internal stability evaluation, in particular where indicators are either longstanding or have not been evident during surveillance visits since the previous Section 10 inspection

REFERENCES

Brown, A.J. and Gosden, J.D., 2004, Interim Guide to quantitative risk assessment for UK reservoirs. Thomas Telford.

HSE 2000, Regulating higher hazards: exploring the issues. Published as a draft for discussion 28 Sept 2000.(DDE15) 61pp. Summary of response 6 pages (Annex to HSC/02/51)

Institution of Civil Engineer, 1996, Floods and Reservoir Safety, Thomas Telford

FLOODsite (Integrated Flood Risk Analysis and Management Methodologies) - Research Relevant to the Dams Industry?

MW MORRIS, HR Wallingford Ltd, Wallingford. UK.
PG SAMUELS, HR Wallingford Ltd, Wallingford. UK.

SYNOPSIS.   The FLOODsite Project is an Integrated Project under the EC 6th Framework Programme (www.floodsite.net). The project value is ~€14M and is being implemented by 36 partner organisations, drawn from 13 different participating countries. To achieve the goal of integrated flood risk management, FLOODsite brings together managers, researchers and practitioners from a range of government, commercial and research organisations, all devoted to various, but complementary, aspects of flood risk management.

The FLOODsite project covers the physical, environmental, ecological and socio-economic aspects of floods from rivers, estuaries and the sea. Work is divided into over 30 project tasks with pilot applications in Belgium, the Czech Republic, France, Germany, Hungary, Italy, the Netherlands, Spain, and the UK. Many of the topics of research and the tools being developed are directly or indirectly relevant to the dams industry, since they address common issues such as hydraulic loading, structure performance, flood risk management, dealing with uncertainty and training / uptake of knowledge. Specific examples of research included within these areas are analysis of extreme loading conditions, breach initiation and formation, social, environmental and economic impacts, pre- event and event management, analysis of flood defence systems and dealing with modelling and decision uncertainty.

This paper provides a brief introduction to the FLOODsite Project and highlights key areas where the research is of relevance to the UK dams industry.

Improvements in reservoir construction, operation and maintenance, Thomas Telford, London, 2006, 261–270

WHAT IS FLOODSITE?

FLOODsite is the largest ever EC research project on flood risk management, with an EC "grant to the budget" of nearly €10 Million, with another €4Million from the research community involved. The project, which started in 2004, is scheduled to take 5 years to complete, and involves approximately 150 researchers from 13 countries. HR Wallingford leads the project consortium of 36 partners which includes many of Europe's leading institutes and universities and the project involves managers, researchers and practitioners from a range of government, commercial and research organisations, specialising in aspects of flood risk management.

FLOODsite is interdisciplinary, integrating expertise from across the physical, environmental and social sciences, as well as spatial planning and management. There are 35 project tasks divided into 7 research themes, also including pilot applications in Belgium, the Czech Republic, France, Germany, Hungary, Italy, the Netherlands, Spain, and the Thames Estuary in the UK (See Figure 1). FLOODsite covers the physical, environmental, ecological and socio-economic aspects of floods from rivers, estuaries and the sea.

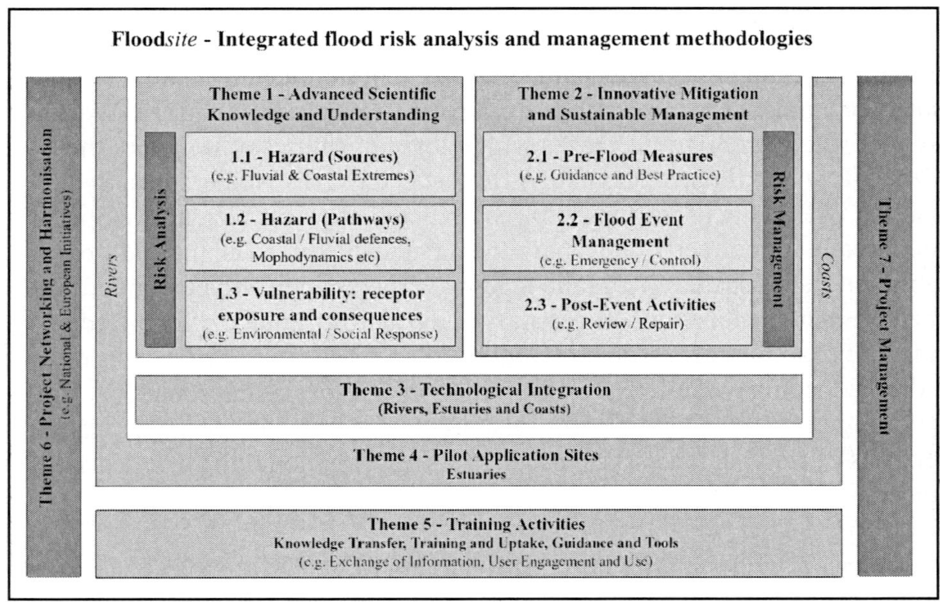

Figure 1 Research programme structure for the FLOODsite Project

The FLOODsite project is arranged into seven "Themes" covering the project science, integration, training activities, networking and management. Within these themes there are over 30 project tasks. The objectives of the Themes are as follows (see also Figure 1):

Theme 1 – Risk analysis: Scientific knowledge and understanding

1.1   To improve understanding of the primary drivers of flood risk (waves, surges, river flow etc.) through research targeted at key issues and processes that contribute most to current uncertainty in flood risk management decisions.

1.2   To Improve understanding, models and techniques for the analysis of the performance of the whole flood defence system and its diverse components, including natural and man-made defences (e.g. seawalls, embankments, dunes) and the extent of inundation.

1.3   To understand the vulnerability and sensitivity of the receptors of risk and to improve and harmonise the methods to evaluate societal consequences and to estimate flood event damages

Theme 2 – Innovative mitigation and sustainable flood risk management

2.1   To evaluate flood risk management measures and instruments after the event (ex-post) and to develop sustainable flood risk management strategies and evaluate these prior to implementation (ex-ante) under consideration of a wide range of different physical and societal conditions.

2.2   To improve flood risk mitigation measures that are applied during the flood event, through improved technology for flood warning in small flash-flood catchments and through measures for emergency evacuation.

Theme 3 – Frameworks for technological integration

3.1   To integrate the scientific, technological and procedural advances to support long term flood risk management decisions.

3.2   To integrate the scientific, technological and procedural advances to support flood event management decisions.

3.3   To develop a framework for the identification and quantification of the influence of uncertainty in the process of flood risk management.

Theme 4 – Pilot application sites

4.1   To provide real sites with real and specific problems upon which tools, techniques and decision support systems may be developed and tested.

4.2   To provide feedback into the research and development process from flood risk managers and river, estuary and coastal stakeholders.

4.3   To ensure the FLOODsite deliverables are of real value, practicable and usable.

<u>Theme 5 –Training activities (Knowledge transfer, training and uptake, Guidance and tools)</u>

5.1   To provide a series of Best Practice Guidance based upon the research outcomes

5.2   To disseminate, and support transfer of knowledge to the stakeholder communities

5.3   To provide public educational tools (web-based)

<u>Theme 6 – Project networking, harmonisation and monitoring</u>

6.1   Link with external research and policy development activities

6.2   Provide internal coherence within the FLOODsite consortium (e.g. through the development of a common language of risk for flood management)

6.3   Integrate review and assessment into the project activities

<u>Theme 7 – Project co-ordination</u>

7.1   To ensure effective and efficient overall management of the project, including administrative and financial aspects, communication with the commission, exploitation of results etc.

The project will deliver:

- An integrated, European, methodology for flood risk analysis and management
- Consistency of approach to the causes, impacts and control of flooding from rivers, estuaries and the sea
- Techniques and knowledge to support integrated flood risk management in practice
- Dissemination of this knowledge including the development of training media
- Networking and integration with other EC national and international research.

The FLOODsite Management Team has set out a vision for the project as:

- The results of our research will make a difference to the way flood risks are managed within member states of the EU and more broadly
- The partners will work together to support implementation of the research results in practice
- Implementation of the research results will ultimately benefit the citizens of Europe, through a reduction in flood risk and an improvement in resilience in the face of flooding
- The team will produce peer-reviewed scientific publications from their research.

The early outcomes of the project include a common definition of the concepts in flood risk management, improved flood forecasting in small catchments, guidelines on the estimation of flood damages and improved understanding of the failure processes of embankments. By the end of the second project year over 130 papers have been presented or prepared for publication on the project science. Access to this material is available via the project website at www.FLOODsite.net

RELEVANCE TO THE UK DAMS INDUSTRY?

Since the primary focus of the FLOODsite Project is in improving Flood Risk Management practice, there are many aspects of the research work that are of relevance to the UK dams industry.

Research does not address explicitly specific issues on the performance of dam structures but will offer tools and methodologies that will be relevant when considering flood risk issues associated with reservoir operation. For example, the research includes hydraulic loading, wave overtopping, breach formation, flood inundation modelling, flood mapping, estimating flood impacts (socio-economic) etc.

The following sections provide a brief introduction to some of the more relevant tasks within FLOODsite. More detailed information can be accessed through the project website at www.floodsite.net

Task 1 Identification of Flash flood hazards

The key objective of Task 1 is to advance the understanding of the major atmospheric and hydrologic factors leading to extreme flood events, especially those affecting small to medium ungauged basins (those with area less than 500 km^2). This topic reflects the prominence of flash flooding in European flood mortality statistics.

Task 2 Estimation of extremes

Task 2 is focusing on the improved understanding of extreme events (river, estuarine and coastal extremes, incl. marginal extremes, joint probabilities, temporal and spatial variability).

Task 4 Understanding and predicting failure modes

Research under Task 4 will gather the substantial body of existing information of defence failure mechanisms and extend knowledge in a number of critical areas including geotechnical instability, crest and rear face erosion by wave overtopping and block removal by wave impacts

Task 5 Predicting morphological changes in rivers, estuaries and coasts

The research in this task aims to improve our ability to predict morphological change in the short and longer term through detailed analysis of coastal and riverine morphological processes and changes and how these relate to flood defence structure performance.

Task 6 Modelling breach initiation and growth

Research here aims to extend existing capabilities in breach modelling with a particular focus upon breach initiation as well as growth.

Task 7 Reliability analysis of flood defence structures and systems

The potential complexity of the relationship between the condition of individual elements of a flood defence and its overall performance is poorly understood and difficult to predict routinely (i.e. the combination of failure modes and their interaction and change in time and space). This task will focus on developing reliability analysis techniques that incorporate present process knowledge on individual failure modes and interactions between failure modes (collated through Task 4 above), interactions between failure modes and length effects of flood defences.

Task 8 Flood inundation modelling methodologies

Over recent years many flood inundation models and have been developed using a range of modelling approaches (1-D, 2-D, 3-D and hybrid approaches). This topic aims to develop a consistent hierarchy of flood inundation methodologies exhibiting a range of complexities and capabilities linked with a range of data requirements.). It will then seek to develop these capabilities through the innovative use of remote sensed data to facilitate rapid model construction.

Task 9 Guidelines for socio economic flood damage evaluation

The aim of this task to is to develop harmonised guidelines on the socio-economic evaluation of the most important types of flood damage based upon systematic collation of existing experience and research results.

Task 10 Socio economic evaluation and modelling methodologies

The overall objective of this task is to focus research efforts on innovative methods to understand, model and evaluate flood damage.

Task 11 Risk perception, community behaviour and social resilience

The purpose of this task is to understand better the impacts of floods on communities, and their ability to respond and recover from the impacts that they bring. Major objectives of the task are (a) to characterise types of communities with regard to their preparedness, their vulnerability and the resilience related to flood events by means of indicator sets; (b) to understand the driving forces of human behaviour before, during, and after floods; (c) to summarise the fieldwork results in terms of acceptable flood risks; and (d) to learn lessons from case studies in Germany, Italy and the U.K.

<u>Task 14 Design and ex-ante evaluation of innovative strategies for flood risk management</u>
To assess the sustainability of different comprehensive strategic alternatives for flood risk management in different situations.

<u>Task 16 Real time guidance for flash flood risk management</u>
This task aims to identify the best method for allowing the evaluation of flash-flood risk at a regional level. The two existing approaches for this are: (1) using 'classical' detailed hydrologic models, and (2) using the so-called Flash Flood Guidance FFG concept (developed in the US) that directly links rain or discharge thresholds to levels of risk.

<u>Task 17 Emergency flood management – evacuation planning</u>
The objective is to develop a methodology and tools for identifying appropriate emergency evacuation and rescue plans. These plans will consider two cases, rapid-onset flash flooding with little warning and extensive inundation of defended lowland areas (e.g. the Dutch polders).

<u>Task 18 Development of a framework for long term planning</u>
To provide a framework for the long-term planning of flood risk management using the skills base developed in Themes 1 and 2. The framework will include:
- An ability to integrate information on hazard processes, vulnerability and mitigation measures to differentiate between alternative long-term flood risk management strategies.
- An outline computer-based Decision Support System (DSS) that is relevant to all those involved in developing long-term flood risk management strategies including policy makers, their technical advisors and the general public.

<u>Task 19 Development of a framework for flood event management planning</u>
Task 19 will link knowledge and models from Themes 1 and 2 in a Decision Support System in support of emergency management planning and practice.

<u>Task 20 Development of a framework for the influence and impact of uncertainty</u>
To develop an approach and prototype software for propagating integral (total) uncertainty through integrated flood models.

<u>Tasks 21-27 Pilot Studies</u>
Tasks 21-27 each relate to a different Pilot Site within the project. Pilot sites are spread across Europe and include rivers, estuaries and coastal sites. The pilot sites are used interactively with the various research tasks to help ensure that the research work directly reflects and meets industry needs. The River Thames provides one of these pilot sites (Task 24). Research work here is closely linked with ongoing Environment Agency / Defra flood defence research and the Thames Estuary 2100 project.

<u>Tasks 28-31 Training, Dissemination and Raising Public Awareness</u>
Tasks 28-31 form Theme 5 and are all related to communication and dissemination of the project work. An overall Communication and Dissemination (C&D) plan has been produced for the project under Task 28. Tasks 29, 30 and 31 deal with C&D via text based, web based and face to face (training) methods respectively.

Full details of the entire work programme may be found at: http://www.floodsite.net/html/project_overview.htm

OPPORTUNITIES FOR THE DAMS INDUSTRY?
HR Wallingford is acting both as overall project coordinator and also as technical researcher in many of the specific project tasks. In developing the proposal for this project, HR Wallingford also helped to ensure that the research goals were closely aligned with the recommendations of the DTI Foresight project on flooding, the themes of the Environment Agency / Defra joint research programme for flood and coastal erosion risk management and the Flood Risk Management Research Consortium (FRMRC) programme of work. Consequently, Defra and the Environment Agency are "affiliates" to the project consortium and also contribute to the overall project budget. An opportunity exists here to ensure that EA/Defra approaches to dam safety also build upon the extensive knowledge base that this work draws together along with the ongoing research work.

From the review of the project tasks above, particular topics in FLOODsite which would appear to be relevant to the dams industry would appear to be:
- Information from Tasks 1 and 3 for assess the probability of high inflows to the reservoir and joint probability with wind-generated waves
- The failure modes and probability for embankments (Tasks 4, 6 and 7)
- Methods for flood inundation modelling (Task 8)
- Emergency and evacuation planning (Tasks 17 and 19)
- Concepts of performance-based asset management (Task 24)

- Socio-economic appraisal including damage potential and public attitudes (Tasks 9 to 11)
- Link between spatial planning and flood risk (Task 14)

The challenge will be to see what sort of regulation, if any, is needed and how information from this project is used within dam safety assessment and reservoir management work.

More Information?

The project approach to dissemination of research is to make maximum use of the project website and to post research findings online at the earliest possible opportunity. An extensive website has been developed – and continues to expand – through which research reports, papers, conclusions etc may be found. In addition, users may also register to receive periodic newsletters which will advise upon research progress and key events.

WWW.FLOODSITE.NET

CONCLUSIONS

FLOODsite is an ambitious project which will maintain the world-leading position of Europe in knowledge and practice for flood risk management. The pilot studies draw together the development and testing of the project knowledge and will provide feedback from flood risk managers and river, estuary and coastal stakeholders. The use of the pilot sites and collaboration with executive agencies in several countries will ensure that FLOODsite deliverables are of real value, practicable and usable.

With the focus of research on Flood Risk Management there are many aspects of the research that should be of direct interest to the UK dams community. Research work is now well underway, with a research programme running from March 2004 until February 2009. Reports and associated research papers may be accessed online at www.floodsite.net

ACKNOWLEDGEMENT AND DISCLAIMER

The work described in this publication was supported by the European Community's Sixth Framework Programme through the grant to the budget of the Integrated Project FLOODsite, Contract GOCE-CT-2004-505420 and also the Environment Agency Contract SC0500052. This paper reflects the authors' views and not those of the European Community, Defra or the Environment Agency. Neither the European Community nor any member of the FLOODsite Consortium is liable for any use of the information in this paper.

REFERENCES
FLOODsite Consortium (2006). FLOODsite: Integrated Flood Risk Analysis and Management Methodologies. Annex 1 – Description of Work. 12th April 2006.

Numerical tools for dam break risk assessment: validation and application to a large complex of dams

B. J. DEWALS, FNRS & University of Liege, Belgium
S. ERPICUM, University of Liege, Belgium
P. ARCHAMBEAU, University of Liege, Belgium
S. DETREMBLEUR, University of Liege, Belgium
M. PIROTTON, University of Liege, Liege, Belgium

SYNOPSIS. The present paper first describes briefly the hydrodynamic model WOLF 2D. Secondly, the simulation of the Malpasset dam break enables to highlight the effectiveness of WOLF 2D. Lastly, the model is applied to simulate the flood generated by the hypothetic collapse of a large concrete dam, which would induce three dam breaks in cascade.

INTRODUCTION

For more than ten years, the HACH (Applied Hydrodynamics and Hydraulic Constructions) research unit from the University of Liege has been developing numerical tools for simulating a wide range of free surface flows and transport phenomena. Those various computational models are interconnected and integrated within one single software package named WOLF. After a series of validation tests (benchmarking) and numerous comparisons with other internationally available models, the Belgian Ministry for Facility and Transport (MET) has selected WOLF for performing flood risk analysis on the main rivers in the South of Belgium and for conducting dam safety risk analysis in the country. The present paper focuses on the practical application of the model WOLF 2D for predicting the flow induced by dam breaks and for performing risk assessment.

The main features of WOLF 2D are first depicted. Then, the simulation of the Malpasset dam break is exploited to demonstrate the efficiency and accuracy of the model. Finally, the paper details the application of WOLF 2D to the simulation of the flood wave generated by the hypothetic instantaneous collapse of a large concrete dam. As this dam failure would occur upstream of a complex of five dams, it would induce three other dam breaks in cascade, including the gradual breaching of a 20-meter high rockfill dam. Those dam breaks in cascade are considered in the computation.

Improvements in reservoir construction, operation and maintenance, Thomas Telford, London, 2006, 271–282

THE HYDRODYNAMIC MODEL WOLF 2D

The software package WOLF includes an integrated set of numerical models for simulating free surface flows (see Figure 1), including process-oriented hydrology, 1D and 2D hydrodynamics, sediment transport, air entrainment, … as well as an optimisation tool (based on Genetic Algorithms).

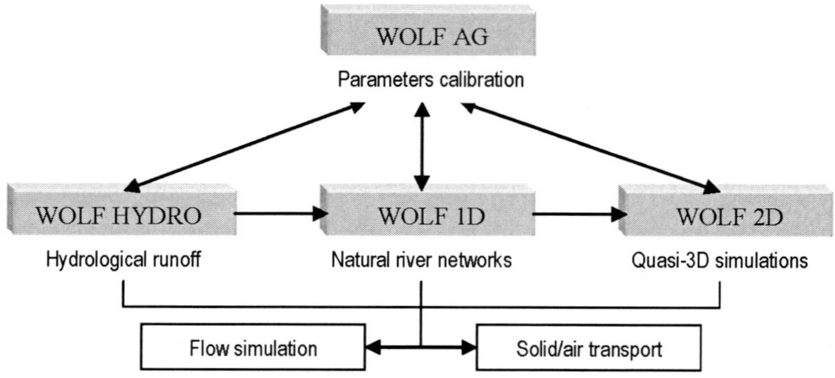

Figure 1: General layout of WOLF computation units.

The computation unit WOLF 2D described and applied in the present paper is based on the shallow water equations (SWE) and on extended variants of the SWE model, solved with an efficient finite volume technique. It achieves fast computation performances while keeping a sufficiently broad generality with regard to flow regimes prevailing in natural rivers, including highly unsteady flows.

Physical system and conceptual model

In the shallow-water approach the only assumption states that velocities normal to a main flow direction are smaller than those in the main flow direction. As a consequence the pressure field is found to be almost hydrostatic everywhere. The large majority of flows occurring in rivers, even highly transient flows such as those induced by dam breaks, can reasonably be seen as shallow everywhere, except in the vicinity of some singularities (e.g. wave front). The divergence form of the shallow-water equations includes the mass balance:

$$\frac{\partial h}{\partial t} + \frac{\partial q_i}{\partial x_i} = 0 \tag{1}$$

and the momentum balance:

$$\left[\frac{\partial q_i}{\partial t} + \frac{\partial}{\partial x_i}\left(\frac{q_i q_j}{h} \right) \right] + gh\left(S_{fi} + \frac{\partial H}{\partial x_i} \right) = 0; \quad j = 1,2 \tag{2}$$

where Einstein's convention of summation over repeated subscripts has been used. H represents the free surface elevation, h is the water height, q_i designates the specific discharge in direction i and S_{fi} is the friction slope.

Algorithmic implementation

The space discretization of the 2D conservative shallow-water equations is performed by a finite volume method. This ensures a proper mass and momentum conservation, which is a prerequisite for handling reliably discontinuous solutions such as moving hydraulic jumps. As a consequence no assumption is required as regards the smoothness of the unknowns.

Designing a stable flux computation has always been a challenging and tough issue in computing fluid dynamics, especially if discontinuous solutions are expected. Flux treatment is here based on an original flux-vector splitting technique developed for WOLF. The hydrodynamic fluxes are split and evaluated according to the requirements of a Von Neumann stability analysis. Much care has been taken to handle properly the source terms representing topography gradients.

Since the model is applied to transient flows and flood waves, the time integration is performed by means of a second order accurate and hardly dissipative explicit Runge-Kutta method.

Friction modelling

River and floodplain flows are mainly driven by topography gradients and by friction effects. The total friction includes three components: bottom friction (drag and roughness), wall friction and internal friction.

The bottom friction is classically modelled thanks to an empirical law, such as the Manning formula. The model enables the definition of a spatially distributed roughness coefficient. This parameter can thus easily be locally adjusted as a function of local soil properties, vegetation or sub-grid bed forms. An original evaluation of the real shear surfaces is realized and the friction along vertical boundaries, such as bank walls, is reproduced through a process-oriented model developed by the authors.

Multiblock grid and automatic grid adaptation

WOLF 2D deals with multiblock structured grids. This feature enables a mesh refinement close to interesting areas without leading to prohibitive CPU times. A grid adaptation technique restricts the simulation domain to the wet cells, thus achieving potentially drastic reductions in CPU times.

Besides, the model incorporates an original method to handle covered and uncovered (wet and dry) cells. Thanks to an efficient iterative resolution of the continuity equation at each time step, based on a correction of the discharge fluxes prior to any evaluation of momentum balances, a correct mass conservation is ensured in the whole domain.

User interface

A user-friendly interface makes the pre- and post-processing operations very convenient and straightforward to control, including a wide range of graphic capabilities such as 2D and 3D views as well as animations.

VALIDATION: MALPASSET DAM BREAK

Located in a narrow gorge of the Reyran River in the Department of Var in France, the double curved 66.5 metres high Malpasset arch dam has been built for irrigation purpose and for drinking water storage. Its crest was 223 metres long and the maximum reservoir capacity was 55 millions m³.

Description of the accident and collected data

At 21:14 on 2 December 1959, during the first filling of the reservoir, the dam broke almost instantaneously, inducing the catastrophic sudden release of 48 millions cubic metres of water in the 12-kilometer long river valley down to the town of Frejus and the Mediterranean Sea. A total of 423 casualties were reported [4, 6]. Further investigations showed that key factors in the dam failure were ground water pressure and left bank rock nature. As this dramatic real dam break case has been well documented, it has also been extensively used as a benchmark for validation of numerical models. Indeed, estimations of propagation times are available, since the shutdown time of three electric transformers destroyed by the wave are known. Similarly, a survey conducted by the police enabled to determine the flood level marks on both riverbanks. Finally, a non-distorted 1/400-scale model was built in 1964 by EDF and was calibrated against field measurements. This model has thus provided the modellers with additional data of water level evolution and propagation times [4, 7].

Characteristics of the simulation

By means of a multiblock regular grid, the mesh is refined close to strategic areas (dam location, sharp river bends), without leading to prohibitive CPU times. 180,000 cells, with three unknowns, are used to simulate the reservoir emptying and the wave propagation down to the sea (Figure 2).

Because of the important changes in the topography after the accident, an old map (1/20,000 IGN map of Saint-Tropez n°3, dated 1931) has been used by EDF to generate the initial bottom elevation of the valley. 13,541 points have been digitised to cover the area of interest. These points have been interpolated to generate a finer Digital Elevation Model (DEM) on the 180,000 cells finite volume multiblock grid exploited for the numerical simulation with WOLF 2D (Figure 2). The elevation of the domain ranges from minus 20 metres ABS at the sea bottom to plus 100 metres ABS, which corresponds to the estimated initial free surface level in the reservoir (with an uncertainty of about 50 cm).

From physical scale model tests, the Strickler roughness coefficient K has been estimated to be in the range of 30 to 40 $m^{1/3}$/s. Four simulations have been carried out with K values of 20, 30, 35 and 40 $m^{1/3}$/s.

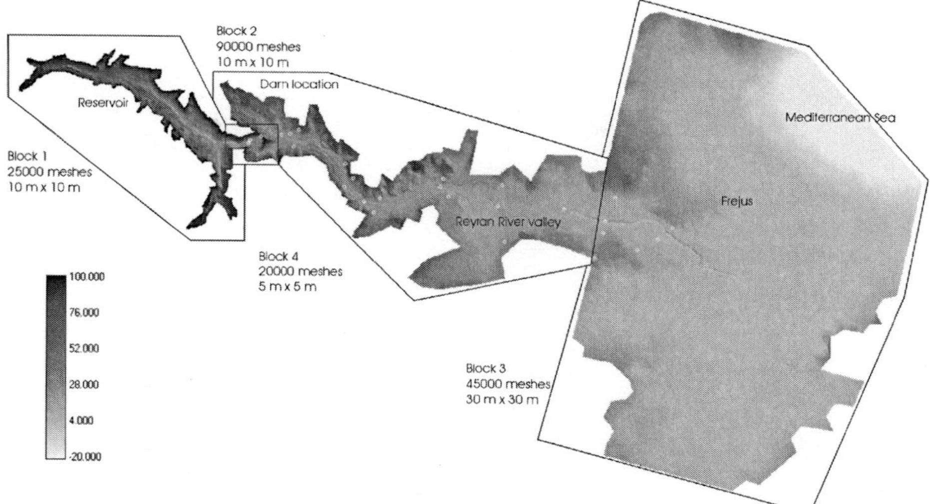

Figure 2. Shaded representation of the interpolated topography and definition of the blocks characteristics on the multiblock computation grid. Position of the field and laboratory measurement points.

Initially, the reservoir is at the level 100 metres ABS and, except in the reservoir as well as in the sea, the computation domain is totally dry, though in the reality a negligible but unknown base flow escaped from the dam. The simulations were run with a CFL number of 0.1 to maximize the accuracy.

Results

Figure 3 and Figure 4 enable to compare the computed results with the field and laboratory measurements, both in terms of wave arrival time and of maximum water levels. Indexes A to C refer the three electric transformers. Since transformer A is located in the bottom of the valley, its shutdown time corresponds indeed to the wave arrival time. For the two others, the shutdown time is assumed to be between wave arrival time and time of maximum water level. Therefore it is more relevant to compare the time interval between the two shutdowns (Figure 3). The computed arrival times are found to be in satisfactory agreement with the measurements. The value of $K = 35 \text{ m}^{1/3}/\text{s}$ appears to fit best reference data. This fact also corroborates the corresponding results from the physical tests performed by EDF.

Given the topography accuracy, the maximum water levels are found to be in good agreement with the measurement points, as shown by Figure 4. It must be outlined that the sensitivity of the results to the value of the roughness coefficient is particularly weak.

The user-interface of WOLF enables to easily produce and edit several risk maps, such as wave arrival time and highest water levels. These maps corroborate again the reliability of the simulation results, as shown by Figure 5 in terms of flood extension.

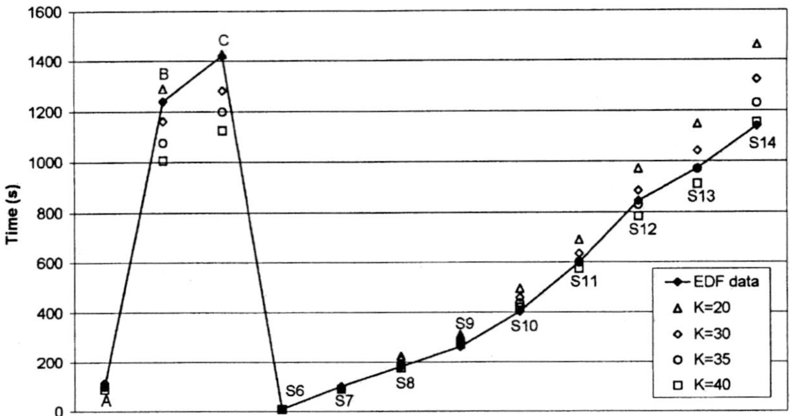

Figure 3. Wave arrival time to electric transformers (A to C) and to the gauges (S6 to S14) of the scale model.

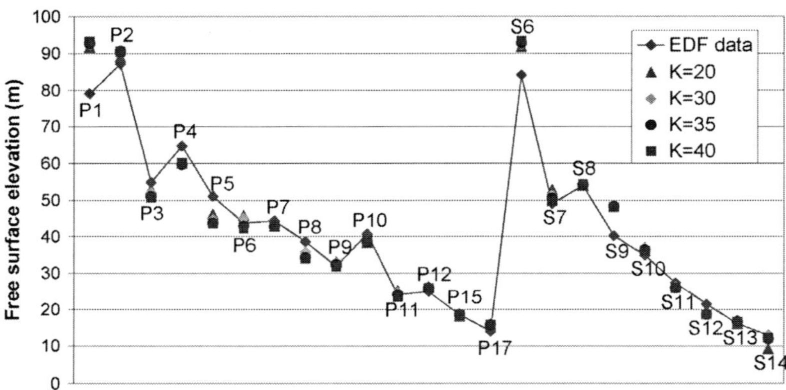

Figure 4. Maximum water level at police survey points (P1 to P17) and at the gauges (S6 to S14) of the scale model.

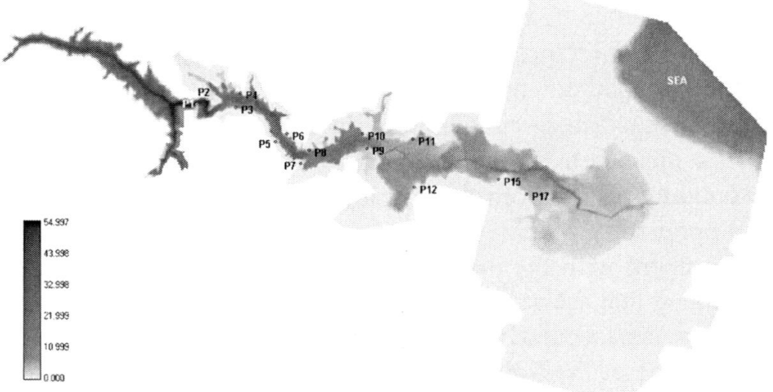

Figure 5. Map of highest water levels. Comparison of the flood extension with the police survey points.

APPLICATION: DAM BREAK ON A COMPLEX OF FIVE DAMS

The hydrodynamic model WOLF 2D has been applied in the framework of a complete risk assessment study of an important complex of dams.

The main dam (dam n°1), a 50 m-high concrete dam, is located upstream of a complex of five dams, including a 20-meter high rockfill embankment one (dam n°2). Figure 6 illustrates the global configuration of the complex of dams.

As a consequence of this layout of several dams, the hypothetical failure of the upstream concrete dam (dam n°1) is likely to induce other dam breaks in cascade. Therefore the risk analysis of such a complex must be performed globally and not simply for each dam individually.

In the following paragraphs, the failure scenario of dam n°1 is first detailed. Secondly, the propagation of the dam break wave is studied on the hydraulic plant itself and the potential breaching in cascade of other dams is analyzed. In a third step, the simulation of the propagation of the flood wave is performed in the whole downstream valley, taking into consideration the sudden collapse or the gradual breaching of the dams located downstream of the main one.

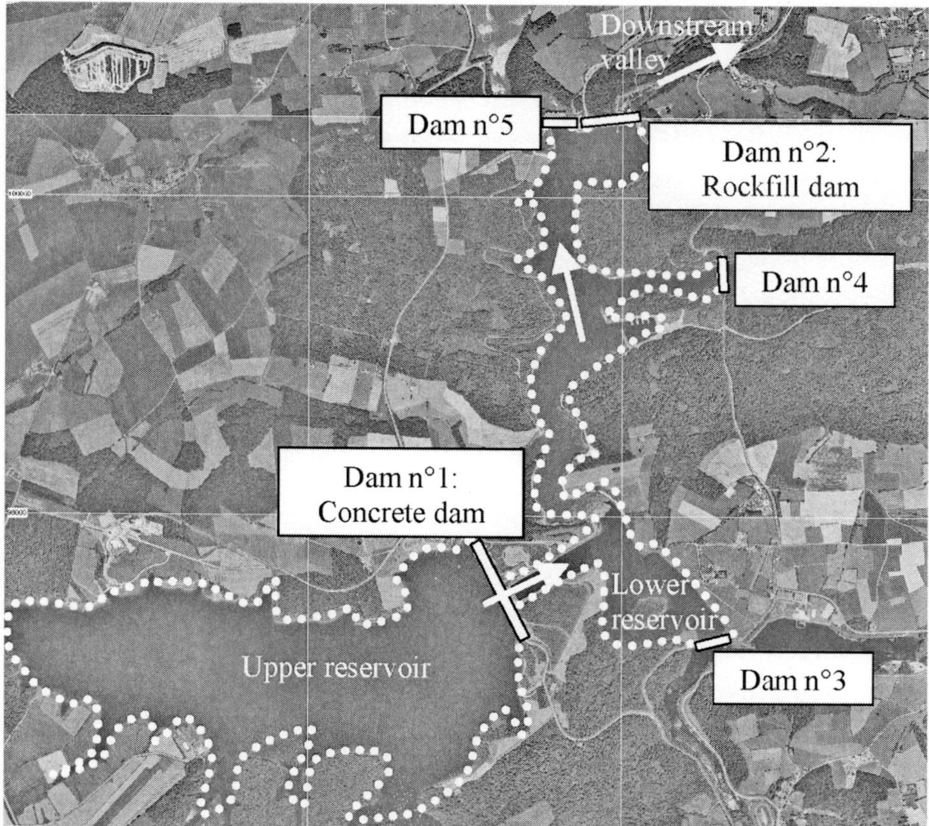

Figure 6: Global layout of the complex of five dams.

Dam break scenario

Defining the failure scenario requires to identify the proper breach formation time as well as the final breach size. According to ICOLD's recommendations, the failure of a concrete dam is supposed to be purely instantaneous. Regarding the final breach geometry in a concrete dam, the question is more complex. Indeed, the guidelines to select this parameter differ from one norm or legislation to another and there is no general rule. For this reason, a sensitivity analysis has been performed here by simulating the induced flow in the case of two different final breach widths: either a total collapse (width: 800 m) or a 200 m-wide breach (see Figure 7). The results of the computation reveal that the difference in the peak discharge value remains lower than 15 %. This can obviously be explained by the higher "efficiency" of the section of the valley in its centre (where the narrower breach is located). As a consequence the hypothesis of a total collapse of dam n°1 is selected, since it doesn't lead to an unrealistic overestimation of the flow.

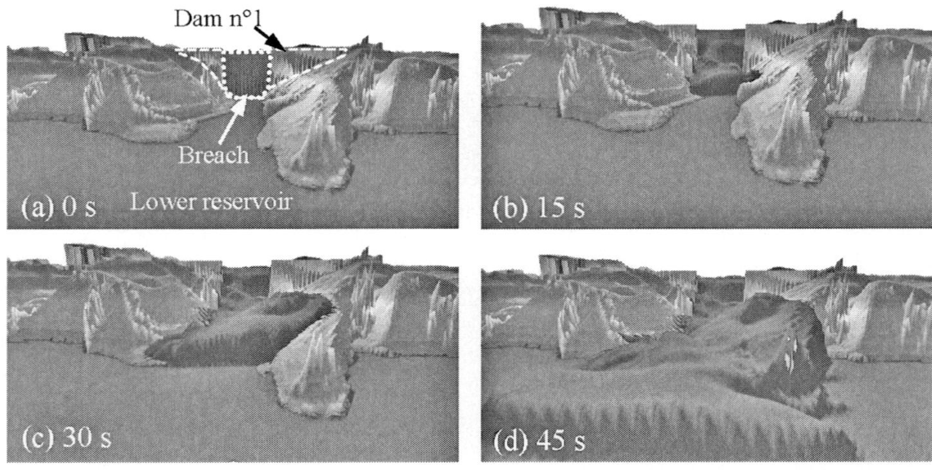

Figure 7: 3D view from downstream of the flow induced by a 200 m wide breach formed instantaneously in the concrete main dam (dam n°1).

Hydraulic and structural impact on the complex of dams

By means of a 2D numerical simulation with WOLF 2D, involving over 400,000 computation cells, the detailed propagation of the waves induced on the lower reservoir (see Figure 6) has been simulated.

Input data

Let's emphasize that the accuracy of the simulation results depends widely on the quality of available topographic data. In the present case, the simulation is based on a highly accurate Digital Elevation Model with a horizontal resolution of 1 point per square metre and a vertical precision of 15 cm. The

Belgian Ministry of Facilities and Transport (MET-SETHY) have acquired this high quality set of data on most floodplains in the South of Belgium, by means of an airborne laser measurement technique. Specific features of the flow in urban areas, such as the macro-roughness effect of buildings, are thus directly reproduced in the topography and there is no need any more for an artificial increase of the roughness coefficient in urbanized areas [1].

Simulation results

The computation performed provides an estimation of the overtopping wave, induced by the total collapse of the upstream dam, over each downstream dam. The overtopping height on the downstream rockfill dam (dam n°2) has been evaluated at 5 m, with a maximum discharge of about 10,000 m³/s. Similarly, the overtopping wave on the three other dams (n°3 to 5) has been characterized: the dams are respectively submerged by waves of 15 m (dam n°3), 8 m (dam n°4) and 5 m (dam n°5), with maximum discharges of 35,000 m³/s, 5,480 m³/s and 720 m³/s respectively. The question of the stability of those dams under such severe solicitations naturally arises.

Stability of the downstream dams

For the concrete dams (n°3 and n°4), a comparison has been carried out between the forces normally acting on the dams and those acting on the dams during their overtopping. This comparison has shown that the resultant force and the resultant moment differ so significantly that the dams would most likely not remain stable once the hydrodynamic waves reach them. As a consequence, in the final simulation, dams n°3 and n°4 will be artificially removed from the simulation topography as soon as they are overtopped.

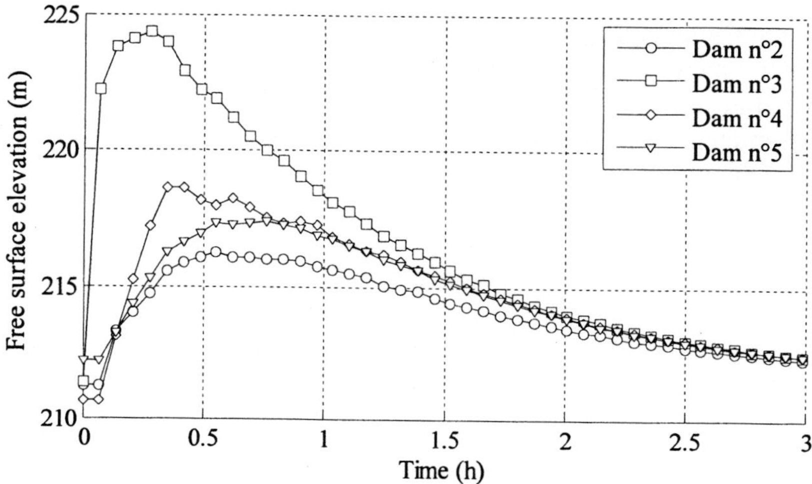

Figure 8: Free surface elevations in the reservoirs near the four downstream dams, after the instantaneous and total collapse of dam n°1.

On the contrary, the embankment dam n°5 is supposed not to be breached considering that its solicitation is significantly weaker than for dams n°3 and n°4. Moreover, the reservoir upstream of dam n°5 is smaller and would essentially act as a temporary storage area for the flood wave. Consequently, assuming that dam n°5 remains stable doesn't lead to an underestimation of the wave reaching the downstream valley.

Regarding the downstream rockfill dam (dam n°2), the challenge is more intricate because it will most probably be breached by the overtopping flow and the proper breach parameters have thus to be estimated. For this purpose, using a process-oriented computation approach was hardly possible because of the lack of reliable data concerning the embankment material. For this reason, the breach parameters have been evaluated on the basis of several empirical formulae [3]. Norms imposed in the regulation of several European and American countries or in national companies have also been considered to validate the selected breach parameters.

Modelling methodology

Afterwards, the breaching mechanism has been introduced in the DEM used for computation in the form of a transient topography. In other words, once the computation code detects that a dam is overtopped, it triggers a time evolution of the local bed elevation in the topography matrix, according to the breach parameters previously defined. This transient topography reproduces a very realistic breaching mechanism for embankment dams made of non-cohesive material (Figure 9 and Figure 10).

The progressive removal of the dam from the topography is defined by the instantaneous location of a virtual plan. This plan is initially tangent to the downstream face of the dam (plan α on Figure 9) and it progressively evolves towards upstream, rotating simultaneously until it coincides with the bottom natural topography at the bottom of the dam (plan β on Figure 9). For the two concrete secondary dams, which are assumed to break in cascade, the parameters correspond to a total collapse in two minutes, while the breach parameters selected for the rockfill dam (dam n°2) correspond to a total erosion in 30 minutes.

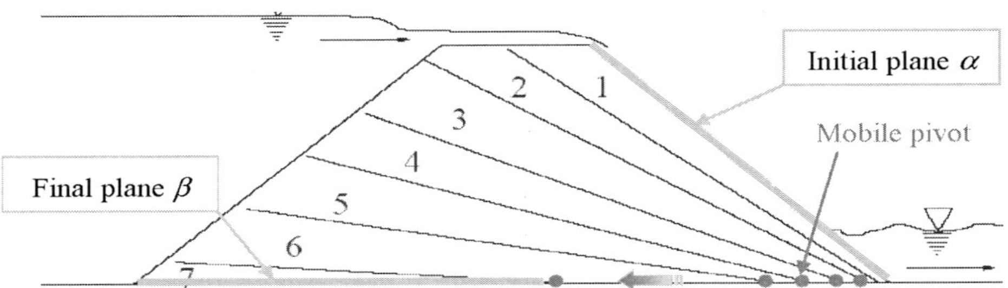

Figure 9: Schematic representation of the breach formation in a non cohesive embankment dam [5].

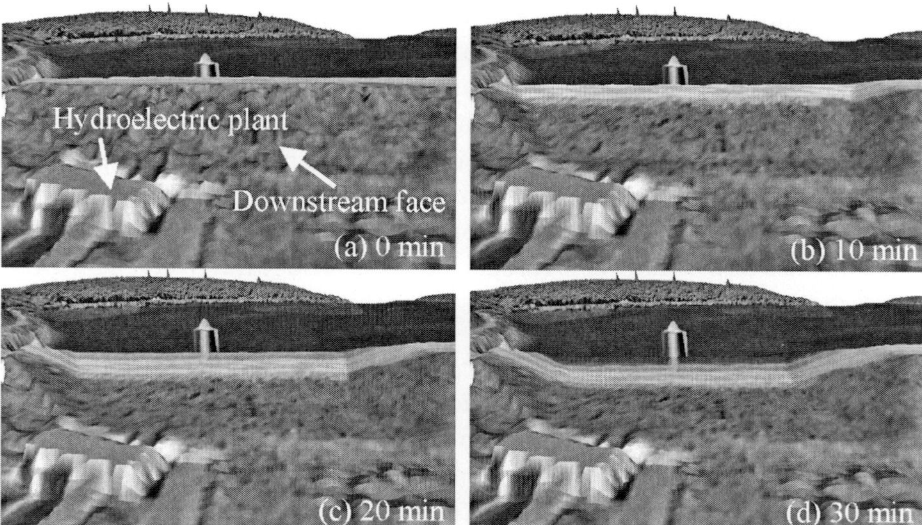

Figure 10: Breach formation process in the downstream rockfill dam (dam n°2), as reproduced in the DEM for hydrodynamic computation.

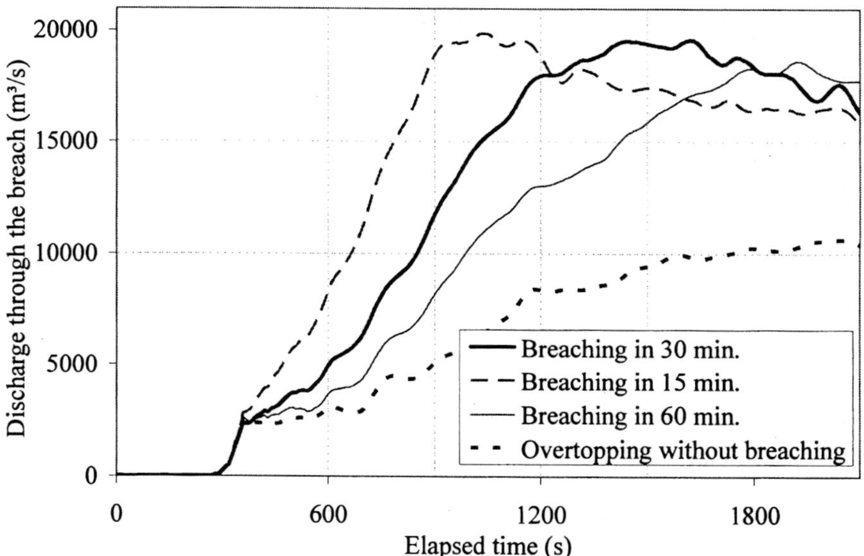

Figure 11: Hydrograph corresponding to the flow overtopping dam n°2 as a consequence of the total collapse of dam n°1.

Since uncertainties remain obviously regarding those parameters, an extensive sensitivity analysis has been carried out. In particular, Figure 11 represents the hydrographs obtained through the breach of dam n°2, depending on the formation time. For a breach formation time varying between 15 minutes and one hour, it appears that the peak discharge is only modified by a few percents. The sensitivity of the hydrodynamic results with regard to this breach parameter remains thus extremely weak.

<u>Hydraulic impact in the downstream valley</u>
Finally, a global 2D hydrodynamic computation has been carried out, coupling the flows in the reservoirs and in the whole downstream valley, and taking into consideration the transient topography (dam collapses in cascade). The simulation mesh is based on a grid of about 900,000 potential computation cells (8 m × 8 m). The obtained results enable to draw essential risk maps and to plot hydrographs or limnigraphs at numerous strategic points downstream of the dams (urbanized areas, bridges, ...).

CONCLUSION
The hydrodynamic model WOLF 2D, based on a multiblock mesh, has shown its ability to accurately compute extreme wave propagation induced by the instantaneous or gradual collapse of dams. Very satisfactory agreements have been found between computed results and validation data in the case of the Malpasset accident. WOLF 2D has also been successfully applied to the simulation of a major dam break on a complex of dams, considering three additional dam breaks in cascade.

Further developments are currently undertaken. In particular, new simulations taking into account sediment movements under dam break flows have been performed. The first results already obtained by the authors [2] demonstrate that these phenomena strongly affect both the wave propagation time and the maximal water levels.

REFERENCES
[1]   Dewals, B., *Une approche unifiée pour la modélisation d'écoulements à surface libre, de leur effet érosif sur une structure et de leur interaction avec divers constituants*. 2006, PhD thesis, University of Liege: 636 p (in French).

[2]   Dewals, B., P. Archambeau, S. Erpicum, T. Mouzelard et M. Pirotton, *Dam-break hazard mitigation with geomorphic flow computation, using WOLF 2D hydrodynamic software*, in *Risk Analysis III*, C.A. Brebbia (ed). 2002, WIT Press. p. 59-68.

[3]   Dewals, B.J., P. Archambeau, S. Erpicum, S. Detrembleur et M. Pirotton. *Comparative analysis of the predictive capacity of breaching models for an overtopped rockfill dam*. in *Proc. Int. Workshop "Stability and Breaching of Embankment Dams"*. 2004. Oslo, Norway.

[4]   Goutal, N. *The Malpasset dam failure. An overview and test case definition*. in *Proc. of CADAM Zaragoza meeting*. 1999. Zaragoza.

[5]   Hanson, G. *Soil parameter development for a breach headcut migration model based on experience from dam incidents and results from model tests*. in *Proc. Int. Workshop "Stability and Breaching of Embankment Dams"*. 2004. Oslo, Norway.

[6]   Marche, C., *Barrages : crues de rupture et protection civile*. 2004: Presses Internationales Polytechniques, 388 p (in French).

[7]   Soares Frazão, S., M. Morris et Y. Zech, *Concerted Action on Dam Break Modelling : Objectives, Project report, Test cases, Proceedings*. 2000, UCL, Belgium, CDRom.

Failure impact assessment of a mine site flood levee in Australia

RACHEL PETHER, Senior Engineer, Black & Veatch Ltd

SYNOPSIS.   Coal production from the 200 million tonne Curragh North coal resource in Central Queensland commenced in late 2005. The mine is located on the eastern floodplain of the Mackenzie River, and requires a 22 km long flood levee before operations can proceed. The earthfill embankment will generally be 1 m to 5 m in height; however, in eight locations the height will be between 11 m and 15 m. Although the levee's primary function is to protect the mine from flooding it will also form part of the site water management plan, retaining run-off across the 31 km2 site.

Under the Queensland Water Act 2000, a dam is referable and requires licensing if there is a population at risk (PAR) below the dam. Five potentially affected low-lying homesteads were identified over a 50 km downstream reach of the flood plain. An assessment of maximum breach characteristics was made using the mine layout and the State of Queensland Department of Natural Resources and Mines (DNRM) Guidelines. The unsteady hydraulic modelling program HEC RAS was used to model the breach impact at the homestead locations for both sunny-day and flood events. The assessment concluded that the levee is non-referable and a costly dam safety management program is therefore not required.

THE CURRAGH NORTH OPEN-CUT COAL MINE PROJECT

Australia has more than 74 billion tonnes of identified black coal reserves of which over 95% is located in either New South Wales or Queensland. Most black coal in Queensland comes from Newlands, Blair Athol or the Bowen Basin, extending south from Collinsville to Blackwater and Moura. Although the majority of the world's coal reserves are recoverable by underground mining, in Queensland approximately half of the 31,420 million tonnes of identified *in-situ* black coal resources is open cut coal. In 2004 the State of Queensland produced 169 million tonnes of saleable black coal.[1]

---

[1] Australian Coal Association website: www.australiancoal.com.au

Improvements in reservoir construction, operation and maintenance, Thomas Telford, London, 2006, 283–291

The Curragh mine is situated 200 km west of Rockhampton, 30 km north of the township of Blackwater within Queensland's Bowen Basin. It is owned and operated by Wesfarmers Curragh Pty Ltd. In 2004 the mine produced around 7 million tonnes of black coal. In 2004 Wesfarmers commenced development of the Curragh North coal site. This 200 million tonne resource will double the recoverable coal reserves currently available at Curragh and will extend the life of mining operations until at least 2025.[2]

A map showing the location of the Curragh North mine site is shown in Figure 1. The site is located on the eastern floodplain of the Mackenzie River, approximately 5 km downstream of the Bedford Weir.

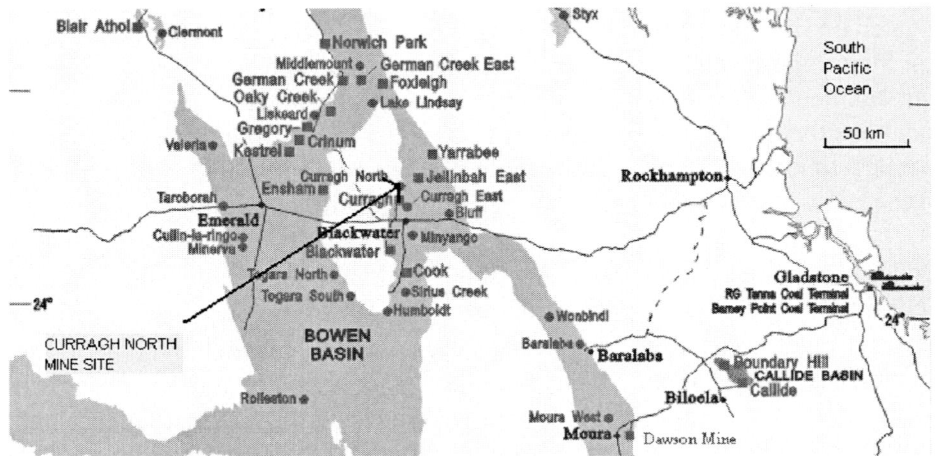

Figure 1: Location of the Curragh North coal mine

THE FLOOD LEVEE

The floodplain in the vicinity of the mine site is subject to flooding during events in excess of the 10 year average recurrence interval (ARI) event. It is therefore proposed to protect the mine site from river flooding by constructing a perimeter levee which will surround the mine and prevent regular river inundation. The flood levee is designed to provide protection to the mine up to the 200 year ARI event.

A location plan for the flood levee and a typical section are shown in Figures 2 and 3 respectively. The levee is an earthfill embankment. It has a length of 22 km, with 12.5 km running along the river. Although the levee will generally be 1 m to 5 m in height; in eight locations significant gullies drain into the river and the height will be between 11 m and 15 m.

---

[2] Wesfarmer's website: www.wesfarmers.com.au

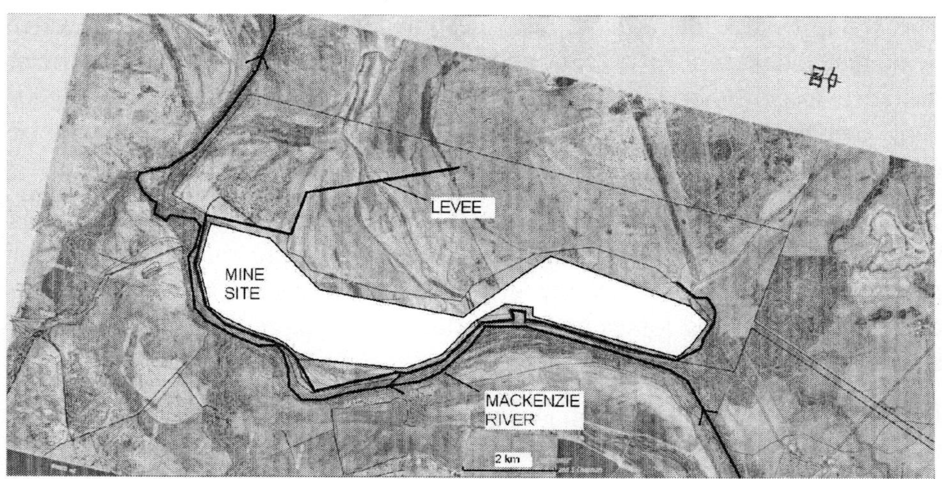

Figure 2: Curragh North mine site flood levee plan

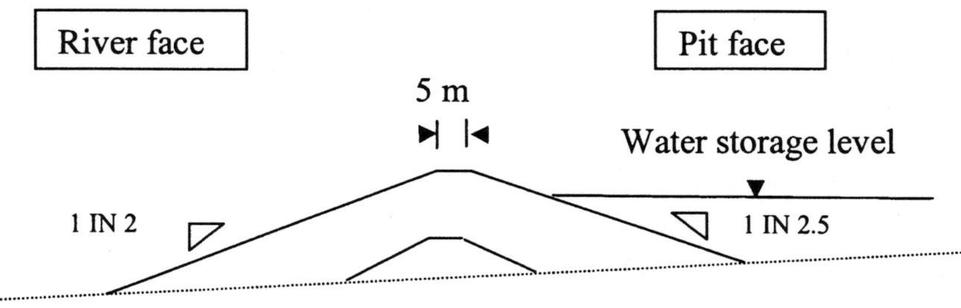

Figure 3: Curragh North mine site flood levee typical cross section

The levee will also form an important part of the mine water management system, as it will act to contain contaminated water from disturbed areas of the mine site, thereby minimising the frequency and volume of runoff to the receiving waters of the Mackenzie River. Water which is retained behind the levee will be harvested and used to satisfy mine site water demands.

WATER MANAGEMENT PHILOSOPHY

The philosophy behind the water management system is to retain as much runoff as possible within the levee system, throughout the 25-year life of the mine. Local runoff within the levee system is intended to be directed to a series of water storage dams via overland flow paths and drains.

The water storage dams have been designed firstly to satisfy mine water demands and secondly to control contaminated runoff. They have been designed to comprise a lower level sump storage component and a higher level overflow storage component. The sump storage component will be excavated below the general ground levels inside the levee.

The concept of providing sump storage and overflow storage for the dams was adopted to minimise nuisance flooding, by storing runoff from minor and moderate events below the general ground level. This concept also assists in minimizing losses to evaporation and seepage, due to the reduced surface area of the excavated storage.

During major events, the sump storage will fill and water will spill out onto the overflow storage component. The overflow storage comprises the natural ground beyond the extent of the sump storage. The areal extent of the overflow storage will therefore be constrained by the flood levees, spoil dumps, pit protection bunds and the access road.

The overflow storage will enable retention of the majority of site runoff within the levee system, by providing large attenuating storage volumes above the natural ground level. During large storm events, this will mean that large areas of the mine site will be inundated for prolonged periods (several months).

THE NEED FOR A DAM FAILURE IMPACT ASSESSMENT OF THE CURRAGH NORTH FLOOD LEVEE

Under Australia's federal system of governance responsibility for dam safety falls under state legislation. Dam safety in Queensland is regulated by the Water Act 2000. Owners of dams are required to assess the impacts of dam failure on the safety of people living downstream by way of a dam failure assessment to determine whether the dam is referable.

Under the Act, the chief executive of the Department of Natural Resources and Mines (DNRM) is responsible for the regulation of referable dams in Queensland.

A failure impact assessment is required when a dam is or will be:

- 8 m in height and with a storage capacity > 500 Ml

- 8 m in height and with a storage capacity > 250 Ml and with a catchment area > 3 times the surface area of the dam at full supply[3]

---

[3] Queensland Government Department of Natural Resources and Mines, Guidelines for Failure Impact Assessment of Water Dams, April 2002

The Curragh North flood levee will have a maximum height of 15.8m and could retain a maximum of 25,200 ML over its 2170 ha surface area. Therefore the Director of Dam Safety in the DNRM requested a failure impact assessment, assuming that the area inside the levee is filled by extreme rainfall to a level where additional runoff would discharge over the crest of the levee.

In accordance with the DNRM Guidelines, the assessment included the analysis of two cases:

1. clear or 'sunny-day' failure with no flow in Mackenzie River

2. failure coincident with the peak discharge during each of several floods of increasing magnitude in the river

If the levee embankments proved to be referable, Wesfarmers would be required to undertake the preparation of a dam safety management programme in accordance with Queensland Dam Safety Management Guidelines. This programme would involve the assembly of available design and construction documents, the preparation of standing operating procedures, operation and maintenance manuals, an emergency action plan and a schedule of surveillance and reporting. A referable dam in Queensland requires a safety review, generally at not more than 20 year intervals.

ANALYSES OF FLOOD INUNDATION

A hydrology and groundwater report for the Curragh North Mine Site was carried out by Parsons Brinckerhoff Australia (PB) in 2003. Estimated peak flow rates at Bedford Weir, 5 km upstream of the mine site, are shown in Table 1 below:

Table 1: Estimated peak flow rates

ARI (years)	Peak flow (m^3/s)
5	2600
10	4000
20	5600
50	7600
100	9200
200	11000
1000	15000

In order to determine the potential Population at Risk (PAR) as a result of a dambreak of the mine site levee, aerial photographs of the floodplain extending 110 km downstream of the area were inspected to identify homesteads. A vehicle inspection of the floodplain at various locations within this reach was then undertaken. Nine homesteads were identified that may be affected by a breach of the Curragh North levee.

Using the steady state hydraulic modelling program HEC RAS and a model developed as part of the levee design flood levels for the design floods in Table 1 above were calculated at each of the nine identified homesteads. Four of the homesteads were located above the 1000 year ARI flood peak and will be immune from dambreak during any flood. For the remaining five homesteads the critical flood ARIs are between 10 years and 100 years.

DAMBREAK ESTIMATES

Basis of estimate

In accordance with the DNRM guidelines the maximum breach discharge, Q_{breach} was determined using the following empirical formula for a typical homogeneous earthfill embankment:

$$Q_{breach} = 2.5FV^{0.76}H^{0.1} \text{ m}^3/\text{s}$$

Where:
F = 1.3 = Factor of safety to account for the simplified nature of the assessment
V = volume of water released (in megalitres)
H = maximum depth of water in the storage (in metres)

The maximum volume of water that could theoretically be stored within the levee system corresponds to the lowest crest level along the embankments. The maximum depth of water is taken as being based on the level of the most critical gully bed. For a sunny-day failure, this is the volume on which the breach discharge estimate is based.

For a concurrent river flood, the volume released from a breach will be the net volume stored above the river level at the breach location. The depth of water will be the difference between river level and crest level.

Sunny-day failure

For the sunny-day scenario a discharge hydrograph was developed for the peak discharge with an area equivalent to the corresponding PAR storage volume released. Table 2 below summarises the breach discharge.

Table 2: Breach parameters for sunny-day failure

Depth (m)	Volume (ML)	Qbreach (m3/s)	Breach development time (min)
15.7	41360	13810	100.0

The dambreak wave pulse along the river and floodplain was modelled using an unsteady HEC RAS model. The flood wave envelope remained within the river channel, well below the ground level at any of the five critical homesteads. There is therefore no PAR from a sunny dam failure of the levee.

Concurrent flooding failure

Under the DNRM Guidelines there is a PAR for a dam breach that occurs concurrently with the peak of a flood if the incremental rise in the water level caused by the concurrent dambreak exceeds 300mm. The most critical case for each dwelling involves the flood which alone would just reach the ground level at the dwelling. Dam breach scenarios were calculated for various stages during the development of the mine. Table 3 below shows the flood levels, depths and estimated storage volumes upon which concurrent flooding dambreak discharge estimates were based.

Table 3: Concurrent flooding dambreak discharge estimates

Concurrent Event (ARI-year)	Flood level RL (m)	Storage depth (m)	Storage volume (ML)	Q_{breach} (m^3/s)	Breach development time (min)	Concurrent flood flow (m^3/s)	Q_{total} (m^3/s)
10	124.4	3.8	30800	9565	108	4000	13565
15	124.8	3.4	30320	9355	108	5300	14655
20	125.0	3.2	30040	9235	109	5600	14835
50	126.0	2.2	27120	7530	110	7600	15100
100	126.4	1.9	24200	7420	109	9200	16620

The HEC RAS model was run with a constant flow rate equal to the peak flow during the critical floods for each of the homesteads. The dambreak hydrographs from Table 3 above were then routed through the model. Table 4 below shows the incremental flood levels at each of the homesteads during the critical flood.

Table 4: Incremental flood levels at critical ARIs for each homestead

Homestead	Concurrent event (ARI – year)	Homestead floor level RL (m)	Flood level (RL (m)	Dambreak level RL(m)	Incremental flood rise (mm)
Bingegang	100	117.91	117.90	118.06	160
Greenacres	15	112.54	112.82	113.05	230
Bundaleer East	20	109.92	109.50	109.60	100
Parker Creek	10	107.55	107.69	107.95	260
Honeycomb	50	109.06	109.21	109.36	150

The maximum incremental flood rise is 260 mm at Parker Creek homestead. The incremental rise in water level during a dambreak is therefore less than 300mm in all cases and under the DNRM Guidelines there is no PAR.

Sensitivity tests
The DNRM Guidelines require that sensitivity tests are carried out on the parameters used for the dam failure assessment. The effect of a variation in the following parameters was therefore investigated:

- Floodplain width
- Roughness values
- Breach development time

Variations in the floodplain width were investigated due to the limited accuracy of the DTM available. Cross sections within the HEC RAS model were widened arbitrarily by 2000m. This caused a decrease in the peak levels, for both the dambreak scenarios and the flood alone. This led to a slight increase in the incremental flood rise for the first two homesteads and a slight decrease for the homesteads further downstream. The results were not significant.

Sensitivity analysis was carried out for a 50% increase and a 50% decrease in Manning's n value. For the 50% increased roughness value there is no incremental flood rise. For the 50% decreased roughness value the flood rise is 0.6m. The roughness values used in the principal analyses were derived by calibration of the steady state model during the sizing of the flood levee. Values significantly less than these are therefore considered highly unlikely.

The dambreak hydrographs used in the principal analyses were assumed to be triangular. The DNRM Guidelines also describe a methodology for determining breach development time based on the volume of material eroded. Routing these alternative DNRM hydrographs through the HEC RAS model was found to have no significant effect on incremental flood levels.

The trends indicated by the sensitivity tests therefore confirmed that there is no PAR along the Mackenzie River due to a dam breach of the Curragh North mine levee. It will therefore not be necessary for Wesfarmers to undertake the preparation of a dam safety management program in accordance with Queensland Dam Safety Management Guidelines

CONCLUSIONS

The 22 km perimeter levee at the Curragh North mine site provides protection from inundation by the Mackenzie River during flooding events and contains contaminated runoff from the mine site for use during mine operations.

A dam failure assessment has been carried out on the levee in accordance with the Queensland Water Act 2000 using the DNRM guidelines. For the sunny-day scenario it was found that the flood wave envelope remained within the river channel and therefore there is no PAR. For a dam breach that occurs concurrently with the peak of a flood the maximum incremental rise at the homesteads was 260mm, which is within the allowable rise of 300 mm under the Guidelines. A robust sensitivity analysis was carried out on the floodplain width, the assumed channel roughness and the breach development time confirming that there is no PAR. Wesfarmers is therefore not required to undertake the preparation of a dam safety management program in accordance with Queensland Dam Safety Management Guidelines.

6. Refurbishment, construction and maintenance

Refurbishing and Upgrading Old Spillway Gate Installations

J. LEWIN, Independent Consultant, Richmond upon Thames, UK
G.M. BALLARD, Independent Consultant, Newbury, UK
P. TO, BC Hydro, Burnaby, Canada

SYNOPSIS   Many old spillway gate systems are approaching the limit of operating life in their present form.   The majority were designed and constructed robust, with little or no redundancy, and incorporate many single point and common cause failures.   Design deficiencies, age and degradation present a reliability risk to the dam system and the downstream population.   The paper sets out the requirements for safety and reliability, how these can be attained and to what extent, recognising the wide variations in installation design and condition.   The degree of reliability obtained as a result of upgrade and refurbishment must be consistent with that required by the dam consequence category.

INTRODUCTION

Many old spillway gate installations, some dating back to early last century, are still in operation. Installations in service dating from the 1930s to 1960s are mainly of the vertical lift type. Radial gates, called Tainter gates in the USA, became more widely used later on because of advantages such as absence of overhead structure, gate guide channels and reduced hoisting forces. Although J.B. Tainter obtained a patent for a segment gate in 1886, the first installations (by M.A.N. of Germany) were not constructed until the 1930s. Radial gates have become dominant for spillways and vertical lift gates have remained the first choice for low level outlets or gates in tunnels.

There have been failures of spillway gate installations, resulting in extreme cases in dam collapse, such as Machhu II (India, 1979) when gate malfunction during a catastrophic flood caused overtopping and washout. It was estimated that 2000 people were killed. At Tous (Spain, 1982), which failed by overtopping during an extreme flood event, 16 were killed. Foster et al (2000) suggests that about 13% of dam failures are associated with a spillway gate.

During the 1987 floods in southeastern Norway, 19% of dam owners experienced gate operating problems (Hinks & Charles, 2004). Other reported problems included communication (23%), damaged access roads (17%) and blockage of spillways (10%).

Some gate failure incidents outside Europe are documented in United States Society on Dams (2002).

In the last decades, dam safety has become more prominent and has been analytically examined. There has also been a greater realisation that spillway gate installations are part of the dam system, and need to be reliable so that the dam system can safely retain and effect controlled release of water. The analysis can become complex in the case of a river system comprising a number of gated dams, each one having different ownership and different reservoir storage capacity.

While old spillway gate installations may be robustly designed, they tend to reflect industrial practice which relied on timely repair of failed components, rather than on the availability of redundancy, diversity and independence. In a few instances it was recognised that some backup equipment was required, mainly in connection with electric motors.

Increased demands, greater complexity of operation, more stringent control systems and quantification of risks have shown that old spillway gate installations may not meet minimum reliability requirements. The operating mode of older facilities has been changed from local to remote automatic. Dam owners face increasing and more diverse demands on water use, both upstream and downstream of the dam. They can no longer assume sole proprietorship on the water, but have to exercise greater responsibility in properly controlling the release of water to satisfy society's expectations on safety and duty of care for the environment. This sometimes requires revision of the operating regime and relatively frequent updates to emergency response plans, e.g. for floods.

In recent years there have been considerable advances in seismic and flood design and data availability. In seismically active or flood prone areas, dam owners increasingly find that the initial design parameters may be too low to meet modern-day standards.

Additionally age, wear and design deficiencies may render it necessary to upgrade the installation as far as is practical. However, operational demands on spillway gate installations may be infrequent so the amount of effort spent on gate maintenance and testing may have diminished over the years, a situation often exacerbated by inadequate maintenance budgets, shortage of knowledgeable personnel due to staff turnover, and pressure from competing priorities.   This means that dam safety is not only threatened by an extreme load event (probable maximum flood or maximum design earthquake), but also by a relatively small normal load event (say 1 in 200) which could have been avoided by the dam owner exercising proper duty of care.

STANDARDS OF RELIABILITY

Ballard & Lewin (2004) proposed a set of reliability principles. The authors suggest that for systems intended to provide some type of standby function, where the reliability measure is probability of failure on demand, a well designed and operated system should be able to achieve a reliability of approximately 10^{-3}. A high integrity system intended for a safety function should aim to achieve a standard of approximately 10^{-4}.

It requires careful design to achieve reliability of 10^{-4} for the mechanical and electrical aspects of spillway gate installations. Upgrading an old installation to that standard could be difficult or uneconomical. Nevertheless, a reliability assessment can identify the significant contributions to failure. Figures derived for reliability should be of an order of quantity since wear and degradation are difficult to assess.

ISSUES REQUIRING INVESTIGATION AND RESOLUTION

Provision of adequately reliable standby electrical power arrangements

Mains supply is vulnerable during a flood or storm. This is particularly the case where hurricanes are an annual event, but also applies wherever a tendency towards more extreme weather has been predicted as a consequence of global climate change. At sites where there is an earthquake risk, overhead mains supply is often the first casualty.

Standby generators are a requirement at spillway gate installations. Manual winding is not, in the majority of cases, a realistic alternative. Diesel engine driven standby supply has a high probability of failure on demand. For good reliability two units are required, or two portable diesel engine drive units which can be coupled to the hoist. Diesel engines which are not regularly tested on load have a very poor reliability record.

Installation of robust electrical power distribution with redundancy for gate operating equipment

Older installations – and even some recent ones – were designed on industrial lines with no redundancy and are subject to single point failure. The exception is equipment that can be repaired in the time between receipt of information of a flood event and when gates have to be raised, provided the gates are tested at the appropriate time to identify any failures.

Adequate segregation and protection of standby and operating equipment
Standby generators are frequently located in the same chamber as electrical switchgear and distribution. In the event of a fire, the whole installation becomes non-operational. Segregation should also be provided for electrical feeds to the gates. Cable runs to several gates in the same duct or cable tray can result in a common cause failure.

Improved protection against hazards such as lightning and fire
Lightning protection is not provided at most spillway gate installations, even at locations subject to electrical storms. Grounding grids may not meet contemporary standards. For vertical lift gates with steel overhead structures and high level hoist equipment, it should not be assumed that a conducting path is provided even if earthed. Stray high currents occur during a lightning strike. They can cause welding together of moving parts, cable insulation failure or control cubicle breakdown. Portable fire extinguishers should be provided at all spillway gate installations, particularly near generator enclosures.

Comprehensive assessment of seismic resistance
Where there is a seismic risk, the possible effects on gates and overhead structures are usually investigated. The assessment is not always extended to include the security of bolting of standby generators, or the possibility of overturning of transformers and control cabinets. Electrical switchgear fixed to the wall using wood screws has been observed, as well as fuel tanks supported by an inadequate base. Batteries for starting diesel alternators are frequently positioned so that they can slip off their support bank.

Maintenance of gate roller bushes at vertical lift gates
Lubrication facilities were provided for roller bushes at the guide wheels of lift gates. Self-lubricating bushes were fitted in North America from the 1940s onwards. However, in some cases the bearing pins were steel and were subject to corrosion. Even if bearings were originally sealed to prevent water ingress, seals could have degraded and may not have been replaced. Seizure of bearings is frequent. Replacement by self-lubricating bushes and fitting new roller axles machined from austenitic stainless steel is required at many vertical lift gates.

Maintenance of seals
At some old vertical lift gates where staunching bars seal the sides, flats have developed. Elastomeric seals may last 20 to 25 years, longer at bottom outlets where they are not subjected to sunlight. Delaying seal replacement

can cause local erosion where persistent leakage occurs, and at higher head gates vibration can occur. The potential to cause gate vibration can be recognised when side seal leakage becomes periodic. Discharge at sill seals is a cause of gate vibration (Lewin, 2001). Appreciable side seal flow discharge at staunching bars has also been observed.

<u>Reliable reservoir water level instrumentation</u>
Upstream water level is a key operation indicator. Two instruments are a minimum, arranged to compare output. Three instruments, controlled using PLC (programmable logic control) on a voting basis should ensure high reliability. Location should also be reviewed to ensure that instrument outputs are not affected by velocity head.

<u>Formal staff training</u>
Periodic tests give operating staff an understanding of and familiarity with the normal functioning of the flood discharge facilities and any alternative modes of operation. They also provide opportunities for fault diagnosis. All staff who may be required to operate gates should be rostered to take part in regular tests. A related issue which must be taken into account is the possibility of human error in gate operation (Hinks & Charles, 2004).

<u>Planning of major maintenance and replacement operations</u>
Elastomeric gate and machinery seals are subject to degradation and replacement should be pre-planned before serial failures are likely to occur. Couplings, brakes and limit switches without backup are sources of single point failures. Monitoring should provide data for planned replacement. Electric heaters to prevent freezing of seals, to maintain lubricating and hydraulic oil temperature and to avoid ice formation which can impede gate operation are subject to low reliability. Pre-planned replacement can reduce operational failures.

<u>Analysis of operational and maintenance records</u>
Operating and maintenance staff should be instructed to provide reasonably detailed reports of the wear of components, operational deficiencies and failures. The analysis of such information is a valuable tool for planning frequency of maintenance and forward replacement.

<u>Spares availability</u>
Spares for old installations are rarely available. In some cases this can necessitate early replacement of major components because sub-assemblies or components are no longer available. A review of spares holding can assist forward planning.

Design problems

At old vertical lift spillway gates which were designed to overflow under some discharge conditions, flow breakers were not correctly designed to break up the nappe. Replacement based on published experience is required in these cases.

Another example of a design problem which needs to be rectified is where structural skin plate stiffener beams located close to the bottom of the gate cause flow re-attachment of the discharge under the gate lip. This can be responsible for serious gate vibration problems (Lewin, 2001; Naudascher & Rockwell, 1994).

PRIORITIES FOR IMPROVEMENT

A reliability assessment will identify a range of deficiencies in a spillway gate installation which have a detrimental effect on the reliability of flood discharge. Not all issues will have the same significance. The cost of improvements will vary appreciably and can be a limiting factor. The most important deficiency may be the one that has the largest single effect, however this may also be the most difficult and expensive to rectify. It may be preferable to address a group of individually smaller, but collectively significant, problems that can be put right reasonably quickly and simply. There is no single simple answer to questions of relative priority.

An evaluation of the relative importance (or the effect on reliability) of remedial works and improvements can act as a guide in the decision making process and form the basis of a programme of work. This can take the form of a simple reliability model of the dam flood discharge facilities, such as that constructed for BC Hydro after an extensive survey of spillway gate installations.

The fault tree model was quantified by generic data on failure rates of systems and components. Although relatively coarse, this model yielded useful information about the order of magnitude of reliability that can be achieved by components and systems. There is no analytical method of adjusting reliability data to reflect the degraded state of equipment due to age and wear.

As an example, a reliability assessment of three 50 year-old spillway gate installations identified the following significant contributors to potential failure:

- Failure of the gate electric motor or associated equipment~30%

- Failure to raise a gate because motors are overloaded
  due to seized rollers (bearings or debris)   ~20%

- Failure of electric power supply to the gate   ~15%

- Failure of gate mechanical drive   ~15%

- Failure of gate control equipment   ~10%

PLANNING CONSIDERATIONS

Unless regular refurbishment is carried out over the years, the scope of work and costs involved in upgrading an old spillway gate system to meet modern reliability standards can be very significant.   As part of the asset management process, the dam owner may have to evaluate the need to invest money in gate refurbishment against other priorities, and consider the potential consequences of deferring or failing to take action.

<u>Factors for consideration when prioritising improvements</u>

Function of the gate system within the dam system
What is the role of the gate installation? Is it for flood routing, peak shaving (pre-spilling) or providing fish release? If for flood routing, what proportion of the flood does it discharge relative to other available discharge facilities, if any?

Consequence of gates failing to operate on demand
If a gate system fails to operate in demand, will it lead to dam failure (e.g. by overtopping) or an environmental incident (e.g. fish stranding)? Scenarios involving gates opening inadvertently due to faulty controls, flawed protocols or miscommunication should also be considered. The possible threats to river users, communities and environment, both downstream and upstream, need to be evaluated.

Reservoir and operational characteristics
A single gate system regulating frequent discharges from a reservoir situated upstream of a population centre but operated with minimal freeboard will, arguably, require higher reliability than a vast remotely located reservoir with multiple discharge outlets.

Interim control
If a gate fails to operate on demand or tends to operate without demand, is there a capability for effective intervention? Even in the case of a large reservoir with long lead time to spills, complacency must be avoided.

Maintenance personnel may assume that there is ample response time during which gate problems can be resolved, while operators – for economic or other reasons – may tend to wait until the last minute before implementing a spill, allowing no leeway for repairs.

When setting up a gate refurbishment program over a portfolio of dams, a prioritisation tool can be developed based on the above considerations (Hartford, 2005).

Improving the reliability of a gate installation is not a one-off investment. Dam owners must recognise that a recurrent investment of resources (both money and labour) will be required to upkeep an improved standard of reliability in the long term.

CONCLUSION

More stringent standards for design and greater demands from the public require a high level of reliability from gate installations, which are often critical parts of the dam system. Old installations which may have inherent design defects or may not have been properly commissioned, increasing challenges in allocating maintenance resources, and infrequent operation can lead to dam safety and environmental problems. Gate refurbishment is only the first stage in upgrading reliability. A continuous commitment to regularly maintain and test the gate system must ensue.

ACKNOWLEDGEMENTS

The authors would like to acknowledge Mr. R.A. Stewart, Director of Dam Safety at British Columbia Hydro & Power Authority, for his permission to publish this paper.

REFERENCES

Ballard, G.M. & Lewin, J. (2004). Reliability Principles for Spillway Gates and Bottom Outlets. *Proc. British Dam Society Conference, Canterbury*, pp. 175–186.

Foster, M., Fell, R. & Spannagle, M. (2000). The Statistics of Embankment Dams Failures and Accidents. *Canadian Geotechnical Journal*, vol 37, no. 5, October, pp. 1000–1024.

Hartford, D.N.D. (2005). *Personal Communications.*

Hinks, J.L. & Charles, J.A. (2004). Reservoir Management, Risk and Safety Considerations. *Proc. British Dam Society Conference, Canterbury*, pp. 220–231.

Lewin, J. (2001). *Hydraulic Gates and Valves*, 2nd Edition, Thomas Telford, London.

Naudascher, E. & Rockwell, D. (1994). *Flow-induced Vibrations: An Engineering Guide.* A A Balkema, Rotterdam.

United States Society on Dams (2002): *Improving Reliability of Spillway Gates.*

Investigation and Rehabilitation of Chardara Dam Spillway

J H MELDRUM, Mott MacDonald Limited

SYNOPSIS.   A reported vibration problem with 40 year old spillway outlet gates turned out to be a hydraulically poorly designed structure with cavitation and hydraulic impact problems requiring alterations to the structure as well as replacement of the gates. Vibration surveys, hydraulic model testing and diving surveys were all employed successfully to investigate the cause of the problem and to assist with the design of the remedial works. Dewatering of the structure was required to carry out the remedial works and as there were no provisions to do so in the existing structure a caisson gate was designed that could be floated into the site and then positioned by flooding watertight compartments.

INTRODUCTION

Chardara Dam, which was built in the 1960s, is located in the south-west of Kazakhstan and it forms a reservoir of about 4.6 billion m^3 on the Syrdarya River. The reservoir is used primarily for irrigation. It has a 80 MW hydropower station generating from irrigation and surplus releases, and it also provides flood relief to 1000 km of river valley between the reservoir and Aral Sea.

There are a number of problems with the dam and it is now being rehabilitated. This paper will deal with the investigation and design work on the four low level spillway outlets, one aspect of the rehabilitation. The station operators have considered it is not safe to operate the service gates beyond about 40-50% opening owing to vibration, whereas 80% opening is required to pass the design discharge of 1,300 m^3/s.

DESCRIPTION OF THE SPILLWAY OUTLETS

The layout of the outlets is shown in Figure 1. The four outlets are located in 2 pairs, one each side of the turbines in a combined structure. Each outlet has a 6 m high by 5.5 m wide hydraulically operated vertical lift service gate to regulate releases. There is provision for upstream stoplogs and there is a single 3 section gantry crane operated stoplog gate. There is no provision to close off the stilling basin at the downstream end.

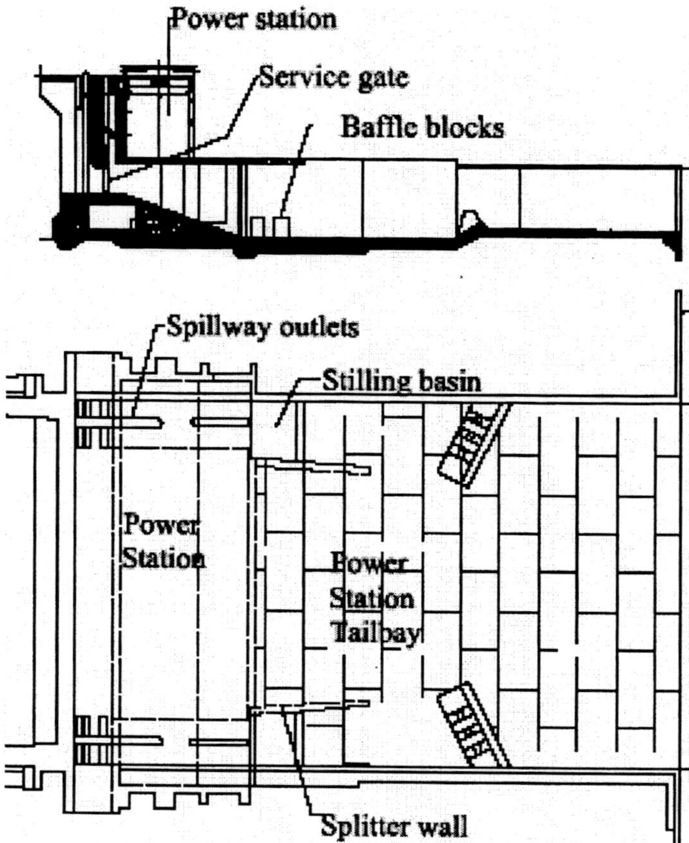

Figure 1: Longitudinal Section and Plan of Outlet Works

Significant aspects of the layout are:

- 5 m downstream of the gates the floor drops 10 m at a constant slope of about 1 on 3 (H:V) into a stilling area;
- there are three 5 m high baffle blocks shared between a pair of outlets immediately downstream of the sloping section;
- there is a partial splitter wall and two columns between the pair of outlets;
- there is a further partial wall between the stilling basins and the turbine outlet channel;
- the sloping section and baffle block parts of the basin are covered by a slab that forms the generator hall service bay and takes the station access road.

The turbines are in use throughout the year and with no plant outage they pass between 550 and 800 m^3/s depending on the level of the reservoir. The reservoir goes through an annual cycle, storing water from the snow melt releases in the spring from the reservoirs further up the Syrdarya River and releasing the bulk of the water during the summer irrigation period, although it discharges throughout the year. This regime gives a low flow period of about 5 months each year when the outlets are not necessarily required and works may be undertaken.

THE PROBLEMS

It had been previously reported in a feasibility study that remedial works were required on the service gates as they were vibrating. However, it became apparent from discussion with the operator that the problem was different from that previously reported, although the gates did prove to require replacing owing to general deterioration, but not owing to vibration. The station operator had noted damage to the upper part of the slope when examining the gates during a low flow period when the tailwater levels were low. He had employed a diving company to examine the structure further. They had found cavities, particularly at the top of the sloping section and to a lesser extent around the gate slots and elsewhere. The diving company had carried out some underwater repairs.

Inspecting the basin in operation was difficult owing to the covering slab. In addition the operator was reluctant to operate the gate beyond about 40% opening owing to the vibration as they were concerned that they would damage the structure further. It was also impossible to inspect the condition of the structure without the use of divers as the basin could not be dewatered.

Original drawings of the structure were available and these showed the configuration of the basin with its sloping section and baffle blocks. There was no design information so hydraulic calculations were carried out based on the dimensions shown on the drawings and assumed gate discharge coefficients. It was concluded that the cause of the damage was most likely to be cavitation. On the top of the sloping section of the basin, where there was an abrupt change in the floor profile, negative pressures would occur, and there was no steel liner to provide protection around the gate slots.

The cause of the vibration was not so obvious. The basin did not follow any conventional design. For a hydraulic jump basin it was too shallow for the sequent depth of a jump to be suppressed by the tailwater at design discharges, but conversely the baffle blocks were too large, which suggested they were being used to force the jump. No evidence could be found as to the events that occurred during design and construction and we are left to postulate whether there were cost, construction or other reasons for the unconventional design. It was also considered that the partial splitter wall and columns would cause flow disturbance prior to the hydraulic jump.

A number of causes of the vibration were thought to be possible, namely:
- strong turbulence in the basin due to the large baffle blocks;
- impacting of flow on the baffle blocks and deflected flow from the blocks onto the adjacent side walls;
- flow impacting on the slab over the basin due to the jump being forced on the upper part of the slope by the large baffle blocks;
- cavitation;
- gate vibration due to general deterioration, poor design or construction.

A further issue was the degradation of the downstream river bed, although not a cause of the problems that had been experienced. Since commissioning the river the tailwater levels had dropped by 1 m, and with no immediate downstream control the degradation was clearly going to carry on. Any solution would need to take the future degradation into account.

DESIGN

The complexity of the problems with the spillway had not been envisaged, and the project programme did not allow for investigations before the remedial works contract was let and there was a need to progress other remedial works which were urgently required. It was decided to include those works on the outlets that were most obvious in the contract and to make provision for other works that may be considered necessary after further investigation and testing, which would also be included in the contract.

The works that were obvious were:
- remodelling of the sloping section to provide positive pressures;
- infilling of the gaps between the partial splitter wall and columns to improve hydraulic conditions;
- provision of a steel liner around the service gate slots to protect the area from cavitation damage;
- replacement of the service gates and rehabilitation of the hydraulic operating system;
- provision of an additional upstream stoplog gate so works could be undertaken on both outlets in one pair at the same time;
- provision of a caisson gate to enable the basin to be dewatered (the gate was also to be given to the operator so that they had the means to carry out future maintenance).

The works for which provisions were made were modifications to the baffle blocks. An outline of the modifications to the outlet structure is shown on a longitudinal sectional elevation in Figure 2.

INVESTIGATIONS

The following investigations were considered as being necessary and included in the remedial works contract:

- vibration survey to try to identify the cause of the vibration;
- hydraulic model test to determine the optimum baffle block arrangement, and also to confirm the assessment that an abrupt change in the stilling basin floor slope was the cause of damage in that area;
- underwater survey to obtain information for the remedial works.

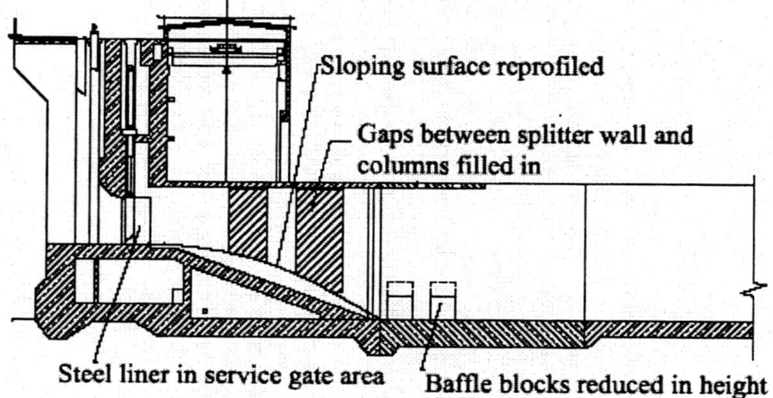

Figure 2: Modifications to Outlets

<u>Vibration Survey</u>

The vibration survey was carried out using a vibration level meter. This measured on three axes simultaneously. The readings were taken at 11 points around the gates and basin. Each reading was taken over a 60 second period with amplitudes being recorded every second. Initial readings were taken with the gates closed, to pick up background vibration from the power station, and then at 0.5 m incremental openings to 4.0 m, at which point significant long period vibration occurred on the gates and the operator refused to open the gates further. This opening was, however, greater than had been achieved during earlier inspections, and it enabled additional observations to be made.

The principal observations from the readings were:

- there was no significant variation in gate vibration with increasing gate opening except at 4.0 m, when there were distinct irregular peaks (Figure 3);

- there was a distinct increase in vertical vibration of the slab over the sloping part of the basin from a gate opening of 2.5 m (Figure 4);
- there was noticeable increase in lateral vibration adjacent to the baffle blocks (Figure 5).

The additional observations made during the survey were:
- that there was a distinct pressure fluctuation in the gate operating area at 4.0 m opening;
- that it was not possible to see a depression in the flow line after the gate on the upper part of the slope and the flow was hitting the underside of the slab from about 2.5 or 3.0 m opening.

It was apparent from the survey and observations that:
- the hydraulic jump was forming from the upper part of the sloping section (probably due to baffle blocks being too large);
- at 4.0 m opening the venting airflow was being closed off by the jump meeting the slab causing pressure variations behind the gate with resulting vibration;
- flow was impacting on the basin walls after being diverted off the baffle blocks and this was likely to be causing vibration.

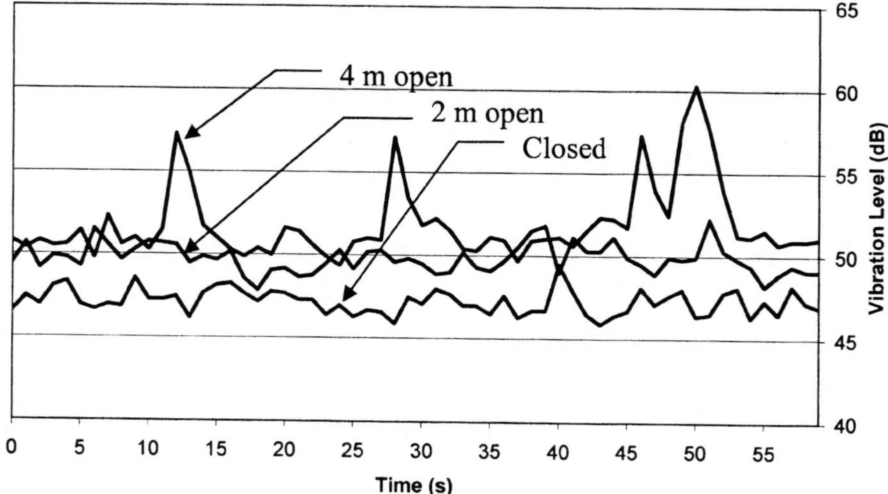

Figure 3: Vibration Survey Gate vibration Y-axis (flow direction)

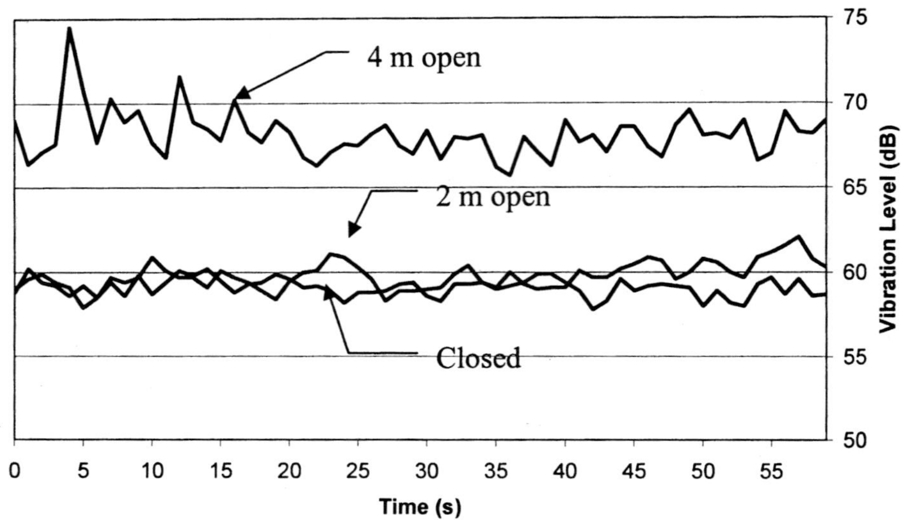

Figure 4: Vibration Survey Slab vibration Z-axis (vertical)

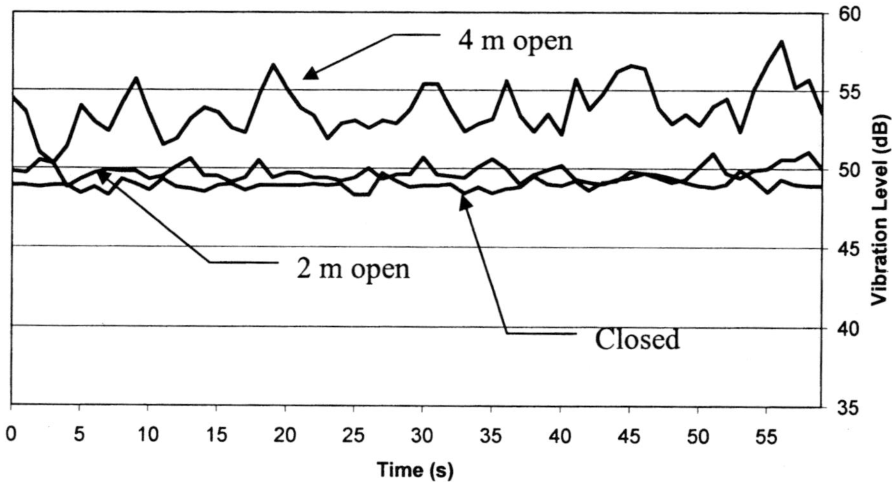

Figure 5: Vibration Survey Wall vibration next to baffle block X-axis (perpendicular to wall)

<u>Hydraulic Model Test</u>
A 1 to 40 scale physical model (Photograph A) was built and the following configurations were tested:
- the existing arrangement;
- a positive pressure profile sloping section, with the gaps between the central partial splitter wall and columns infilled, and with the existing baffle blocks;

- the modified sloping section and walls with 10 variations of the baffle blocks, involving reducing their heights; changing the layout; and profiling the front of the blocks;
- testing the arrangement with current tailwater levels and also with potential future tailwater levels after further degradation.

The model has confirmed/shown that :
- negative pressures occur at the top of the sloping section;
- hydraulic conditions are improved with re-profiling the sloping section and infilling the walls;
- an adequate and less intense jump forms further down the basin with reduced height of the existing blocks;
- the water surface will overtop the side walls of the structure with reduced (future) tailwater levels, if the baffle blocks are not reduced in height.

Photograph A: Model Test

Longitudinal hydraulic profiles for 5 m (existing), 4 m and 3 m blocks for current and future tailwater conditions are shown in Figures 6 and 7 respectively.

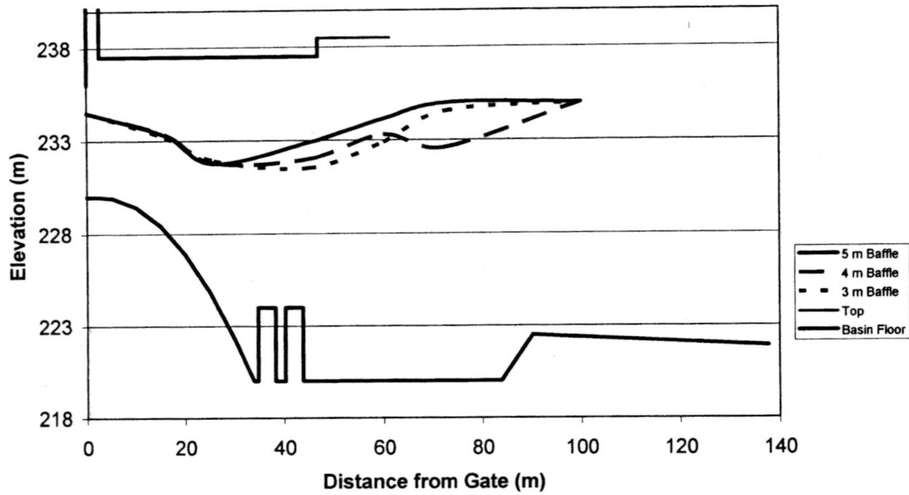

Figure 6: Model Test Results. 75% gate opening and current tailwater level

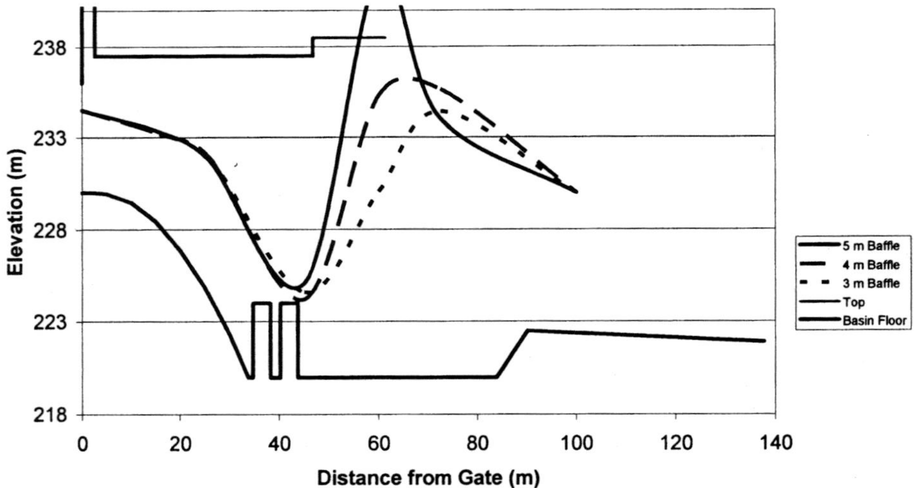

Figure 7: Model Test Results. 75% gate opening and future tailwater level

<u>Underwater Survey</u>
The diving survey was undertaken by a specialist diving company from Kazakhstan. Their survey consisted of taking a video as well as taking physical measurements of the defects and the intended location of the caisson gate.

The survey showed:
- significant cavities in the floor of the basin at the top of the sloping section with reinforcement exposed;
- cavities in the adjacent walls;

- irregularities in the concrete profile adjacent to the service gate bottom sealing plate and small cavities;
- local cavities elsewhere in the sloping section, at the base of the splitter wall and at the base of the baffle blocks;
- irregularities of up to about 300 mm in the profile of the stilling basin wall at the caisson gate location.

<u>Conclusions from Investigations</u>

As a result of the investigations it was confirmed that:

- the profile of the stilling basin floor was causing cavitation and the proposed positive pressure profile was appropriate;
- infilling of the partial splitter wall and columns improved hydraulic conditions;
- the existing (5 m high) baffle blocks needed to be reduced in height as they were forcing a hydraulic jump/causing too much disturbance in the upstream part of the basin resulting in the flow hitting the slab above the basin and causing gate vibration through pressure variations behind the gate as the air flow is temporarily occluded;
- adequate stilling could be achieved in the basin if the blocks were reduced to 3 m in height;
- the impacting flow on the side walls from deflection off the baffle blocks was causing vibration of the side walls, and this could be reduced by changing the profile of the outer part of the block face.

CAISSON GATE

To undertake the works in the outlets it is necessary to dewater the basin. As there was no existing stoplog provision for closing off the basin from the downstream end, and as coffer-damming was impractical, a new downstream gate was required. This needed to be located downstream of the baffle blocks, yet within the area bounded by the splitter wall between the outlets and power station tailbay. At this point the basin is about 14.5 m wide. The water depth may be up to 13 m during the remedial work period. In addition to the new gate, bearing and sealing surfaces also needed to be provided before the basin could be dewatered.

A solution was proposed in the tender documents and the contractor was required to provide the detailed design. The solution was to provide a gate that could be floated into the basin with water-tight chambers that could be flooded to sink the gate into position. The proposal for bearing and sealing was to bolt a suitable arrangement onto the surface of the structure.

The contractor adopted the floating gate proposal. Providing a suitable bearing sealing arrangement proved far more problematic. The contractor was concerned about bolting onto the pre-cast concrete panelled surface so he opted for a supporting system consisting of props bearing on the baffle blocks and removable arms bearing onto plates cast into holes cut into the concrete walls. In his original design the sealing was to be rubber seals placed on the concrete. This was subsequently modified to a grouted 'sausage' held in an adjustable channel when the irregularity of the concrete surface was realized from the diving survey. The gate was manufactured in 5 sections to facilitate transporting to site, and put together in an earth bunded drydock adjacent to the downstream river channel (Photograph B).

Photograph B: Caisson Gate

CONCLUSIONS
1. The vibration survey, hydraulic model test and diving survey all proved essential for investigating the outlets, in particular:
    - The vibration survey showed that with the existing configuration unstable conditions were occurring in the outlets below the design discharge;
    - The hydraulic model test showed that the water levels in the basin would exceed the levels of the structure walls in the future if the basin was modified without reducing the baffle block heights;
    - The diving survey showed that the structure surface was unsuitable for a conventional rubber seal.
2. The absence of any provision for dewatering the basin has made the remedial works far more complex and difficult. It is recommended that at least either grooves or anchorage provisions be provided for suitable stop-logging arrangements in new build where dewatering it not otherwise practical.

Application of Modern Grouting Technology to Remedial Works on Dams

AK HUGHES, Director of Dams & Water Resources, Atkins Ltd
CT KETTLE, Principal Engineer, Bachy Soletanche

SYNOPSIS Grouting to form cut-offs and to seal leakages has been used for many decades. In the past the process of grout injections were difficult to control and did not always produce the desired results.

This paper seeks to give a historical background to the grouting process and identify the need for change. The paper outlines the GIN process and goes on to illustrate the rapid changes in the grouting process based on computer packages.

To conclude the paper illustrates the work recently carried out in Wimbleball Dam in the South West of England.

<u>Historical Background – The Need for Change</u>
Over recent decades hundreds of dam cut/offs have been constructed in rock and alluvium by the injection of self-hardening grouts. The plant, equipment, grout formulations, and injection procedures were very simple, and slow to evolve. Systems and practices developed in parallel with increasing understanding of grout rheology, rock conditions, and the mechanism of grout injection. However, the new procedures could be very difficult to control in the field, where the restrictions of current technology and site practice could not always deliver the desired results.

In the early days grouts were designed on the basis of locally available materials, and often applied where there was only limited knowledge of fissure widths, patterns, and spacings.

Improvements in reservoir construction, operation and maintenance, Thomas Telford, London, 2006, 315–329

Grouts were batched by hand or by means of basic weigh batching equipment, mixed in simple low speed mixers and pumped by simple reciprocating pumps, frequently over large distances to the point of injection.

At the point of injection the pressure control of the process was achieved by manual observation of the manometer, hand control of the return line flow to the storage agitator (where one was used), or by field telephone to the pump operator. Borehole pressure gauge readings were adjusted for line loss from the point of injection, generally taking into account the additional head of grout in the borehole.

Limiting volumes of grout to be injected were controlled by manual measuring of the volume pumped from agitator holding tanks, and grout absorptions were ultimately expressed either in the form of litres/linear metre of borehole, or per discrete 'stage' (discrete length of the borehole injected, usually 2-5 m). More generally were expressed as dry weight of cement and other materials per linear metre, or per stage, since this since this formed part of the basis of payment

Monitoring of the progressive reduction in mass conductivity was made by comparing the results of water testing in un-grouted holes prior to injection, with post-grouting tests in subsequent series of boreholes and/or in separate control boreholes. Analysis and evaluation of the grouting was based upon comparing the final pressures achieved at the end of each injection, and by comparison of the total dry weight of material injected per linear metre for each stage. The results for successive phases of injection compared the results achieved between adjacent boreholes and stages, and the global average values for each phase of injection.

Results were plotted each day by hand on scale drawings of the grout curtain. These drawings carried only basic information, as they soon became crowded and difficult to read. Areal analysis of weak zones was by visual inspection of grout takes and pressures recorded on the drawings, and more complex analyses were often never seen fully plotted up until the final record drawings were submitted - too late to influence the execution of work carried out on a localised, hole by hole, stage by stage, or phase by phase, basis.

Although much excellent work was carried out by experienced and professional specialist contractors using traditional methods, there was clearly potential for many errors of execution or interpretation. The following is a summary of the key areas of risk :-

o   At the mixing station grout batching could be very inaccurate, requiring 'reconciliation' at the close of each shift between the dry weight of material delivered and used, the volume of grout mixed, and the injected volume measured from observation of the agitator tank.

o   Errors in mixing grouts of differing water/cement ratios, errors in measuring the volumes prepared in the mixer, errors in measuring the volumes pumped from the agitator, all compounded to produce large discrepancies. Site teams became adept at covertly balancing out such errors on paper, but were often unable to accurately extrapolate the amount of grout actually injected into a given stage. A performance analysis made simply on the basis of absorbed quantity of grout could therefore be flawed, and could fail to highlight stages or zones where grout takes were high and/or additional work was required, thereby leaving inadequately treated rock in situ.

o   Poor mixing practice, inefficient mixers, and the practice of starting with 'dirty water' style mixes of water cement ratios as high as 10:1, even 15 or 20:1, all combined as factors in allowing the injection of thin, unstable mixes. This frequently resulted in premature blocking off of grout lines or boreholes (sometimes undetected or unrecorded) due to segregation and/or sedimentation, or to the premature blocking off of fissures due to pressure filtration (the process of squeezing out the water in the mix). This was exacerbated in some cases by the absorption of water into dry formations, leading to poorly hydrated grouts which achieved much slower rates of gain of strength and lower ultimate compressive strengths.

o   The pressure of injection is related to conditions in the ground and the effective rate of pumping into the borehole. Similar pressures could be generated in a number of ways, by manipulating the pumping speed, and the control of the valves on the grout line and the return line at the borehole. Specified 'refusal' conditions at the end of an injection could be simulated by pumping too quickly, or by adjusting the valves. For the engineer, there was always a problem interpreting the 'snapshot' picture of the results recorded at the end of an injection, and a difficulty of verifying this information. This often made it difficult often to be confident of the data recorded on site.

o   Poor control of pressure at the point of injection arose due to inadequately damped pressure and flow from reciprocating pumps, non-continuous observation of the manometer and/or control of the return line valve, and inaccurate, poorly calibrated, manometers. As a result, limiting hydro-fracture pressures were often exceeded, and readings of oscillating pressures at the gauge were inaccurate. The resulting analysis on the basis of the final injection pressure achieved took no account of the time / pressure relationship, and often did not identify or highlight premature blocking off of the fissure or borehole. Unscrupulous contractors were able to falsify the apparent successful completion by deliberately increasing the pump speed, and/or reducing the return line flow, to effectively plug off the borehole too early, leaving the adjacent rock inadequately treated.

o   A related problem was that of poor control of hydro-fracture in the formation, or breakout of the grout at the surface, due to poor observation of the fall in grout pressure under these conditions. This lead to a potential weakening of the formation, an excessive volume being injected for the given location, and to the execution of subsequent supplementary injections due to an apparently high absorption.

o   Frequent changes from one mix to another (water/ cement ratios by weight often following a pattern such as 20:1, 10:1, 8:1, 5:1, 2:1, 1:1, and 0.5:1) added hugely to the confusion on site and the difficulty of analysing results. The amount of grout specified for each mix was often quite limited, so that sometimes the injection terminated before the new mix in the grout line arrived at the point of injection. The quantity and/or type of the final mix types for a given injection were often incorrect, leading to an inaccurate picture of the closing mixes, and a consequent mis-calculation of the dry weight of material injected. Where this value was part of the performance specification, such errors could distort the general picture, leading to either too much or insufficient quantity of grout being injected, and could result in a situation where the final grout injected was of considerably lower strength than the characteristic strength of the final mix indicated on the record sheet, and/or too low to ensure long term durability.

The above is just a brief summary of some of the major problems associated with controlling the injection process. Added to problems in controlling drilling practice and accuracy of drilling, and the whole field of water testing control and design, it can be seen that there was much the industry needed to do to ensure good design, best practice, and adequate control.

The inherent inaccuracies combined to give rise to inadequate or uneven treatment, reduced grout strengths, and reduced longevity of the grout curtain. Where grout curtains remain fully effective, it is in many cases a function of over-design, where inefficient and uneconomically high quantities of drilling and grouting were executed.

Of course, responsible contractors tried to handle all the problems of unwieldy specifications, the constraints of semi skilled local labour and long pumping distances, the limitations of the available plant and materials, such that despite all the risks, many fine examples of work were completed. However, for the designer and client the end results were often inconclusive, not demonstrably cost effective, and frequently undermined confidence in the result, the process, and the contractor.

<u>Progressive Change</u>

Over the years many incremental practical improvements have been made to the manual control which have helped significantly in improving the quality of work.

These have included:
- Introduction of automated batching plants
- pressure damping devices
- improved pumps
- high speed grout mixers of improved efficiency
- improved quality and reliability of manometers
- introduction of pressure recording rotary graphs
- introduction of flow meters, and pressure relief valves
- reduction in water/cement ratios
- improved grout materials and additives
- improved control of injection by simplification of the specification
- the introduction of dynamic refusal criteria controlling the volume, pressure, and flow rate in the final stages of injection.
- better quality, more controlled regulating valves
- pre- hydration of bentonite in slurry form prior to its introduction to the mix
- better understanding of the theory and practise of grouting
- the introduction of routine Lugeon testing as a simple, robust guide to rock characteristics and the effectiveness of the grouting programme.
- introduction and controlled use of better methods of on site testing, including the mud balance, pressure filtration device, marsh cone, and plate cohesion test

Not withstanding these improvements the end result was dependent entirely on the skill, observation, diligence, and honesty of field personnel, and the quality of management and planning of the site operations by the Contractor.

<u>Rock Fissure Treatment - GIN method</u>
In the late 1970's and early 1980's a fundamental change in philosophy of rock grouting was developed and introduced by Lombardi, Deere, and others who introduced a new concept – the **GIN** process. (GIN = Grout Intensity Number).

The GIN system, which involves the application of a constant factor P x V (pressure x cumulative volume) for each injection, allowed for a more accurate and dynamic treatment of fissured rock and a better engineering evaluation of the injection process. The method sets a limiting pressure envelope which decreases according to the volume injected, until the limiting volume (V max) is reached. This procedure is very useful in formations where grout absorptions are expected to vary significantly, such as karzstic limestone, in weak formations susceptible to hydro-fracture, and in very open formations where excessive grout travel is likely.

Apart from applying the GIN value, the system proposed the use of a single grout mix of low water cement ratio, which by the use of deflocculents eliminated the loss of mix water due to pressure filtration. This, together with the use plasticisers and the pre-hydration of bentonite slurry resulted in a 'stable' mix of greatly improved fluidity, penetrability, strength, and durability. The dry weight of material can be ignored as a calculation other than for payment, as interpretation and measurement was now directly related to the volume injected.

The further advantage of this method, arising from the fact that the grout physical and rheological properties remain constant throughout each given injection, and from phase to phase throughout the injection programme, is that it allows individual injections to be compared dynamically. By maintaining a relatively constant injection rate, the constant and stable physical properties of the grout allowed the key parameters (pressure, cumulative volume, pressure/time, and volume/time parameters) to form an accurate basis for comparison of individual injections and of successive phases of injection. The injection had become, in effect, a water test with grout.

The GIN process, though recognised immediately by engineers in the field as a major breakthrough, was for almost 10 years a difficult process to implement on site. The boundary curve had to be 'modelled' by instructing a series of decreasing steps of constant injection pressure, each linked to a limiting volume, creating a series of injection domains which were touching the boundary curve. All this had to be controlled with great diligence on site by very experienced operatives and supervisors. At the start of the works, trials were made on site to verify the appropriate single mix and the GIN value for the formation being injected.

The following Figures 1 - 3 are examples from an early reference work by Lombardi describing the GIN theory, and indicating how, at a constant rate of pumping, the evolution of the PV plot might appear for differing rock conditions. Once the PV intercept is reached the injection rate was adjusted manually in an attempt to follow the curve until the injection rate reduced to a practicable minimum (Qmin), ie to 'refusal', or until the limiting volume (Vmax) was injected, at which point the injection is deemed complete. However, it is evident that if Q was poorly controlled on site, P would vary accordingly, and the graphical plot might be meaningless since the PV plot could be made to indicate a wide range of conditions. Add to this the requirement for manually recording P and V on site, then plotting the results by hand for every injection, every day, before they could be analysed, and the scale of the practical difficulty for the grouting engineer can be imagined.

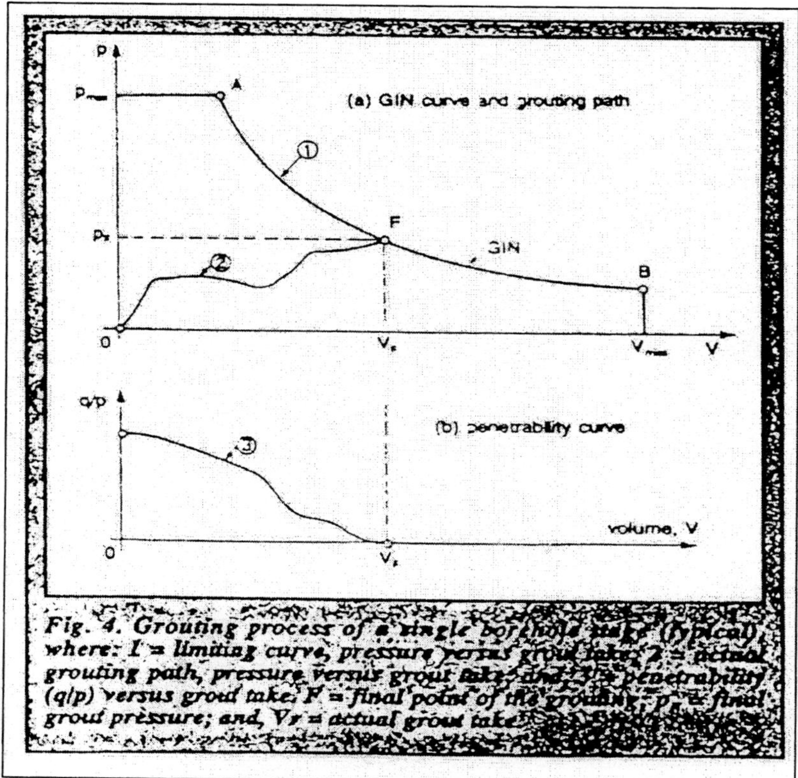

Figure 1. Illustration of an early GIN profile with a typical grouting path.

The solution to this difficulty came with the availability of the personal computer. The next fundamental improvement has been the availability of reliable computers and software to collect injection data, record, analyse, and present results in a clear visual format. This capability, coupled with the development of accurate pressure transducers and slurry flow metres, allows key parameters to be measured automatically and continuously throughout each injection. This is of huge importance and assistance to the grouting engineer, allowing the real time display and analysis of results, enabling him to make more valued assessment of the work in progress and the improvement in the foundation properties. For the first time, the entire history of a single injection could be reviewed over the period of injection, allowing a quantum leap in the understanding of the way in which the foundation fabric had been treated. Events such as hydro-fracture, fissure dilation, premature blockage, break out of grout to the surface or elsewhere, can be observed and analysed by engineers from the graphics of P, V, and Q against time.

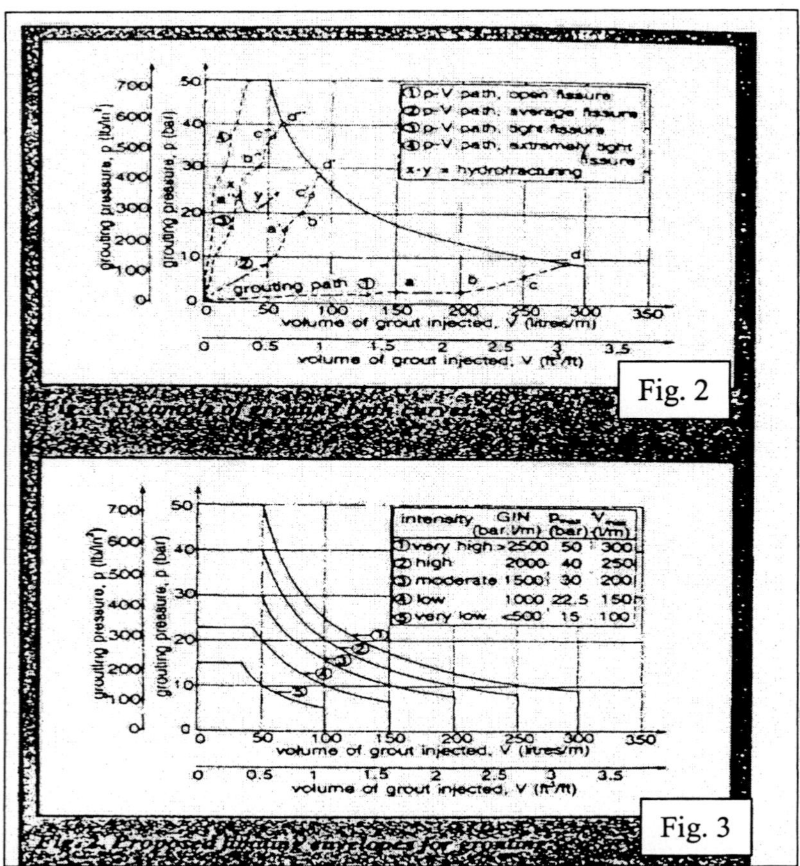

Figure 2. Comparison of different PV plots assuming the same mix and injection rate, and the same PV value

Figure 3. Indication of a range of GIN boundary curves for rock of different ground conditions, expressed as **bars.l/m**

Similarly, the constant flow rate can be both observed and maintained until the limiting PV curve for the stage is reached, and controlled / measured until the final refusal stage of the injection.

The final, obvious, but fundamental step, was to use the new computing technology not only to record and analyse results, but to dynamically control the injection pumps in such a way that the pressure and injection rate can be automatically adjusted throughout the injection according to the design parameters. With this capability, the advantages GIN process can be fully realised, as the process can be perfectly controlled throughout the injection, so that the PxV GIN value can be maintained within close tolerances, and the theoretical PV curve can be followed until either the flow rate falls below the specified minimum, indicating refusal, or until the limiting volume is injected, at which point the injection can be automatically terminated.

<u>Current practice on a fully automatic injection station - an example</u>
Major specialist contractors have either used or developed their own equipment to exploit the advances in theory, technology, and practice. Bachy Soletanche have over many years been developing their plant, equipment, and software to produce a complete, integrated, design-to-final drawing control of all types of grouting process, whether alluvial or rock grouting, particulate or chemical grout. The following example is given to illustrate what can be achieved with modern technology, and the particular benefit of this kind of approach to GIN grouting for rock formations.

Grouting is a somewhat unique activity in that everything happens out of sight below ground, and visual inspection is impossible while work is proceeding. The only proof of success is the full-scale test when the excavations have been dug or the reservoir is filled with water, too late for any modifications to be made to the design. Quality control during the course of the grouting work is therefore vitally important for the safety of the structure being built, yet the sheer volume of data collected is almost unmanageable and threatens to quickly overwhelm the user - an average-size grouting project involves several hundred boreholes, each one generally comprising up to 20 discrete stages which may themselves be grouted in multiple phases. The total soon represents several thousands of individual grouting operations.

This was a challenge which prompted the leading specialist companies to equip themselves with electronic aids which quickly became standard tools in dealing with the requirements of major grouting works. They are present at every level of operations, from the initial design for setting out borehole patterns and calculating grout quantities per injection, through on-site control of the grout pumps and data collection, to verification of the quality of the finished programme . This has necessitated not only an enormous amount of research and development into software programmes capable of modelling, controlling, and recording the injection process, but has required the development of sophisticated pumping equipment which is capable of responding to the instructed parameters with the required level of sensitivity. Such pumps today are capable of continuously measuring, and responding to, pressure readings taken every $1/100^{th}$ of a second.

The software package which has been developed by Bachy Soletanche consists of a suite of independent but compatible, interactive programmes

- ◆ ENPASOL is a package of sensors, read-out device, and software mounted on the drill rig to collect data such as torque, rate of advance, bit pressure, flush pressure, hydraulic feed pressure, etc in real time as drilling progresses. The data is presented as a continuous graphical profile of these parameters plotted against depth. These plots, which are visible to the drill operator during drilling, can be later combined by the software in different ways to highlight variations in ground density & cohesion, giving a measure of the competence and consistency of the ground. These plots are used in planning the injection programme.

- ◆ CASTAUR is a powerful design tool which creates a 3D model of the site (soil layers, structures, grouting zones, physical obstructions, etc); optimises the location of grout holes and produces graphical plots and tabulated instructions for the setting out and drilling crews, defines and calculates the theoretical volume of ground to be treated based on the soil or rock characteristics, and established a 3-D structure for the SPHINX database within which the grout injection data can be analysed and presented spatially.

- ◆ SPHINX organises information into a powerful database which processes data on all aspects of the grouting works (borehole geometry, grout types, injection parameters, grouting records) and provides all the formats for day-to-day management of this huge amount of data. More specifically, it provides for the automatic collation of grout injection data, the preparation of daily production reports in graphical and tabulated format, and the analysis of results and presentation in numeric, graphic, or spatial CAD plot.

- ◆ SPICE is the software which physically controls all grouting operations from a PC computer installed at the grouting plant, and is linked to a specialised programme which servo controls the electro hydraulic grouting pump. SPICE utilises grouting instructions regularly updated in SPHINX, and transferred to the site injection plant computer on floppy disk at the commencement of each work shift. SPICE automatically controls grouting flowrate to keep values of grout pressure and grout volume within predetermined limits. Grout injection pressure, flow rate, and volume for each pump are displayed in real time on the injection station monitor, and stored to hard disc. At the end of each individual grout injection a total of 50 categories of data, summarising the complete injection process, is stored in the database.

♦ SCAN 3-D : this highly configurable software produces sectional CAD drawings which display the grouting results and boreholes in the correct spatial position for easy analysis by the grouting engineer.

Special care was taken to design the human interface of this software package to be as approachable as possible. This is peculiarly true for the SPICE software, which is used on site by the plant operators, and therefore designed to require the minimum possible site training, of 1-2 days.

This system has been employed on a large number of dam grouting projects, ranging from very large projects such as Piedra del Aguila dam in Patagonia, Argentina (1992) which employed over 100 grout pumps and 6 computerised grouting plants, with 18 PCs for technical supervision, to very small grouting works which require only 1-2 pumps.

The main advantages of such a system, apart from considerations detailed above, are

♦ pre determined grouting procedures and parameters are pre programmed into the software,

♦ local, semi skilled personnel can be quickly trained to an operational level, requiring the presence of only a limited number of grouting specialists to run the system.

♦ computerisation enables optimisation of design, control, quality control, and production efficiency.

♦ reporting is fast, automatic, and accurate. Easy-to-read analytical diagrams can be produced and updated as frequently as necessary. These features are essential in creating an efficient work team which combines the expertise and resources of the grouting specialist contractor, the client, and the engineer.

In the end the desired effect must be to free the specialist engineer from hours of routine data processing, allowing him to employ analytical tools rapidly, thereby giving him quality engineering time to apply skills and experience to interpret the data and to refine the grouting parameters and programme for maximum effect and efficiency.

<u>Wimbleball Dam - Supplementary Curtain Grouting Employing Latest Technology</u>

Wimbleball is a 63m high buttress dam owned by South West Water and built in. During its working life it had been noted that seepages/leakages were increasing through the grout curtain on the left hand end of the dam.

A recent project requiring application of many of the most recent technical developments was the supplementary grouting to reduce long-term seepages through the grout curtain of the left bank at Wimbleball Dam in Devon. In order to carry out the grouting works a number of significant problems had to be overcome

- The dam geometry was extremely complex.
- The reservoir could not be drawn down during the works
- Injection work could only be carried out from the downstream backfill between the downstream buttresses, from up to 20m below top water level, requiring all drilling to be carried out through blow-out preventers
- Known bedrock fissure geometry necessitated boreholes to be inclined across the face of the dam, and access from the downstream fill necessitated boreholes to also be inclined upstream
- Dam geometry, combined with the required borehole orientations, required drilling through the slender face concrete with extremely high accuracy to avoid 'day-lighting' into the reservoir
- Bedrock fissures widths and rock quality varied significantly, with some very fine fissures requiring a highly penetrating grout, and some very open fissures allowing extensive grout travel both through and across the face of the original grout curtain. Leakage through the existing curtain was via a network of irregular but discrete seepage paths
- The dam was highly sensitive to hydrostatic uplift, making it essential that the extensive and complex pressure relief well system was not compromised or reduced in efficiency at any time. The pressure relief wells were in direct hydraulic connection with fissures both upstream of the curtain, and traversing the grout curtain. This necessitated that grout injection take place upstream of the existing curtain, but under conditions of extremely close control of pressures and volume
- Environmental constraints of the National Park and fish farms 2 km downstream meant that no significant seepage could be allowed to pass into the river down stream.

To meet the above constraints, an extremely high degree of care and control was required in the planning and execution of the works. Measures adopted included

- Preparation of a highly detailed 3-D CAD model of the dam, the excavation profile, the original curtain geometry, and the foundations
- Computer aided design of the grout holes with CASTAUR software, generating instructions for setting out precise borehole geometry
- Careful surveying of initial core holes which were used both as guide holes and to allow blow out preventers to be sealed into the downstream face of the dam
- All drilling was executed via blow out preventers
- Use of C3S - a stabilised & fluidified grout to designed avoid premature blocking of fissures and ensure grout travel and penetration
- Continuous monitoring of all relief well flows for pH and conductivity, with physical inspection, to observe for grout connections & seepages
- Provision for back-pumping of individual wells at critical periods of grout injection to protect the well from grout contamination
- Radio links between injection plant and drain monitoring crews
- Use of SPHINX and SPICE equipment and software for real time computer control of injection parameters - pressure, cumulative volume, flow rate - throughout each injection, with computer analysis of the results in CAD and SCAN-2D format to assist in interpretation of results
- Physical measures down stream for filtering any particulate material, grout or fissure infill, arising during injection
- Provision of facilities for citric acid dosing to correct pH imbalance arising from cement contact with seepage flows

The outcome of the works was a controlled investigation and reduction of seepages, without compromise to dam safety or environmental constraints.

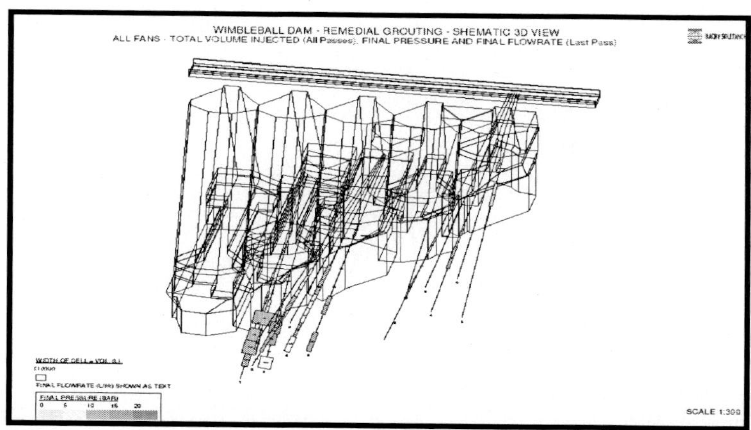

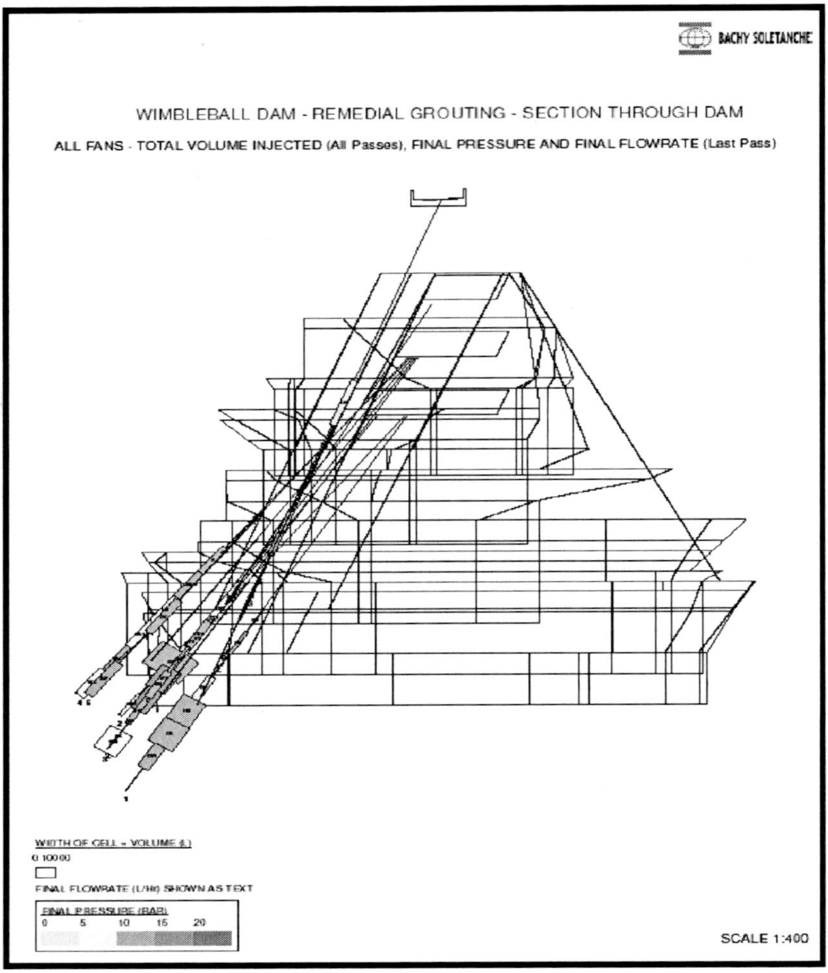

Figures. 4 & 5. SCAN -3D plots of Dam geometry, grout holes and grout injection parameters

Figure 6

CONCLUSIONS

The process of grouting injection at dams has progressed greatly in the last few years with the development of computer software to help with the control of grout injected. This coupled with the 3D software available has made the whole process much more reliable and satisfactory from all views – the contractor, the designer and the client.

ACKNOWLEDGEMENT

The authors wish to thank South West Water for their permission to publish this paper.

Emergency Underwater Rehabilitation of the Poti Main Diversion Weir, Georgia

LJILJANA SPASIC-GRIL, Jacobs, Reading, UK

SYNOPSIS.   The Dam Safety Project (DSP) in Georgia, which is a part of the Irrigation and Drainage Community Development Project (ICDDP) financed by the World Bank, was implemented by Jacobs during the period August 2003–July 2004. The DSP project included four irrigation dams as well as the main diversion weir on the river Rioni at the town of Poti, the third largest town in Georgia and currently the busiest Georgian port on the Black Sea.

The diversion structure, which discharges flows of 4,000m^3/s through 10 identical openings, was constructed in the 1950's to protect the town of Poti from frequent flooding by the river Rioni. Scour problems related to the downstream pool of the diversion structure have been known for almost 50 years. Currently the structure operates at a verge of breakdown endangering the safety of the population in the town of Poti and nearby villages, as well as the important transportation links (Poti port, motorways and railways) and sustainability of the Black Sea coastal zone. Failure of the diversion structure would cause flooding of the Rioni-Khobi Drainage Scheme which has already been rehabilitated under the IDCDP.

As a part of the safety assessment, emergency rehabilitation works have been developed. The implementation of the emergency works by the IDCDP started in October 2004 and were completed in February 2006.

INTRODUCTION

Location and use

Poti is the third largest town in Georgia and is currently the busiest Georgian port on the Black Sea. The diversion structures across the Rioni river were constructed to protect the town of Poti from frequent flooding. They comprise a left bank regulator weir and the main weir, both located about 2km from the Black Sea. The diversion structures were designed by 'GidproVodKhoz' Design Institute from Moscow between 1948 and 1951.

Construction of the structures started in 1952 and continued until 1959 when the structures were commissioned.

Since then, the main weir has also been used as the only road bridge across the Rioni river that leads to Poti. During the last decade the traffic across the bridge has intensified and the bridge, also known as the "Euro – Asian bridge", has been used for transport of heavy goods to and from the Port of Poti.

<u>Description</u>
The diversion comprises two independent weir structures (see Figure 1), namely:
- the left bank regulator which discharges $400m^3/s$ into a canal that goes through the town of Poti and
- the main weir which contains ten openings that discharge up to $400m^3/s$ each, and divert the river flood flow towards the Black Sea, away from the town of Poti.

The left bank regulator weir is placed at an angle of 120^0 to the axis of the main diversion structure. This weir is 80m long and has 20 sluice openings. Each opening was designed to discharge $20m^3/s$ and is regulated by a vertical lift gate.

The main diversion structure is approximately 180m long and is located across the Rioni river channel and was designed to discharge flows of up to $4,000m^3/s$. The structure consists of 10 identical sections; each section is 17m wide and 19 m long. The main piers (11 Nos) which support the superstructure are 3m wide and are located at 17m centres. Each opening (see Figure 2) is 14m wide and contains a radial gate 14m x 4m high. Maximum upstream water level is at +4.21m and minimum downstream water level is at –0.61m with respect to the Black Sea datum. The spillway slab is 2.2m thick and 14m by 19m in plan and spans between two piers. The spillway slabs have energy dissipators in the form of a row of "teeth". There is a 20m wide and 1.4m thick stilling slab downstream, followed by a 40m long downstream apron consisting of 2m square slabs each 0.4m thick. Upstream from the weir there is a concrete apron 15m long and 0.6m thick.

The structure was constructed in the dry with the river being diverted. To ensure that the foundation remained dry, a timber piled wall was installed on the downstream face of the structure and a steel sheet pile wall was installed on the upstream face. The construction was carried out in 17m long and 19m wide sections and it appears that there is no structural connection between the section with main joints on the axis of the piers.

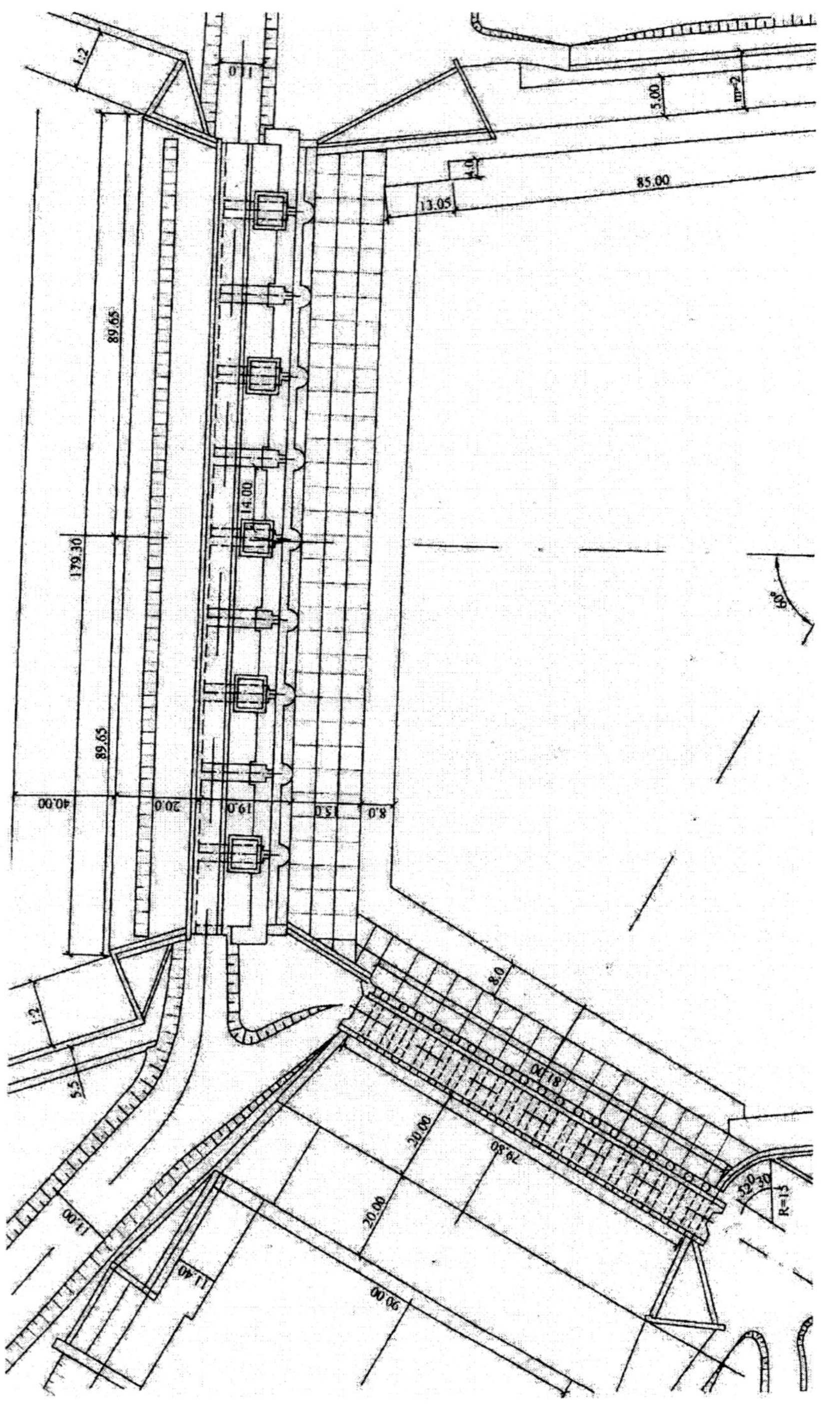

Figure 1: Plan of Diversion Structure

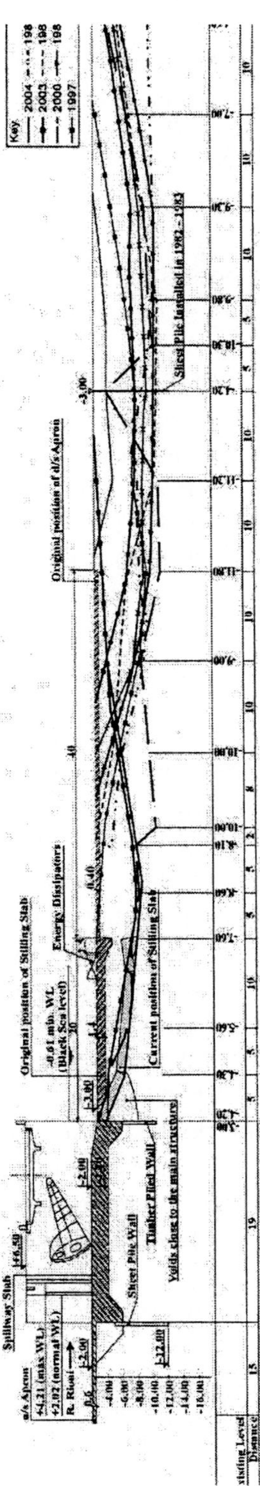

Figure 2: Section Through the Central Opening and Development of Scour

History of scour problems

Problems related to the main diversion structure have been reported for almost 50 years. The Institute of Water Industry and Engineering Ecology of the Georgian Academy of Science (hereinafter IWIEE) was commissioned in 1957 to carry out underwater observation of the diversion structure. It was reported by IWIEE that scouring of the downstream pool started immediately after the facility was put into operation. Figure 2 shows the progress of erosion of the downstream pool, and the river channel deformation at the central opening of the main weir.

At the end of November 1959, immediately after completion of construction, the first flow was passed through the left bank regulator (400 m^3/s) and into the Poti city canal. This flow damaged the downstream apron of the regulator. Scour of the river bed, just downstream of the apron of the main weir was also recorded. During the July 1963 and July 1966 floods the flow in the Rioni reached 3,000-3,400 m^3/s. The flood caused up to 6m deep scour downstream of the apron of the main weir, and by the end of 1967 the scour depth reached elevation -10m and was getting closer to the downstream apron.

In the late 1970's, under flood flow of 2,500m^3/s, the scour in the river channel immediately beyond the stilling slabs of the main diversion structure came very close to the stilling slabs (see Figure 3). The condition was considered to be an emergency, and in 1981, following the IWIEE recommendations, "GruzGiproVodKhov" (now Georgian Water Project (GWP)) from Tbilisi developed a design for the rehabilitation of the downstream apron. The design comprised the installation of a steel sheet pile wall across the river channel, at a distance of 80m from the weir, and backfilling of the eroded space with large size riprap to the original elevation of –3m. The rehabilitation works were implemented in 1982-83. However, it became apparent in the following years, that these measures did not improve the condition of the downstream apron. On the contrary, they enhanced the scour and further deterioration of the downstream pool (see Figure 3). Data from surveys carried out over the period 1986-1992 showed that the scour continued to increase. The scour was also fostered by incorrect gate regulation during floods of the river Rioni, whereby the river discharges were continuously passed only through the central sections.

Observations made in 1998-1999 revealed that the scour was progressing further. The scour depth was 6-10m, and extended to the edge of the stilling slabs.

In 2001 GWP was commissioned to carry out the design of the emergency rehabilitation measures. The design was put on hold several times due to a lack of funds. The survey carried out in July 2003 showed that the condition of the downstream pool had worsened: the apron was completely demolished over the whole width of the weir and the stilling slabs were demolished in the four central sections (see Figure 3).

Under the DSP in Georgia, which is a part of the Irrigation and Drainage Community Development Project (IDCDP) financed by the World Bank, Jacobs carried out safety evaluation of the main structure during the period August 2003–July 2004. An underwater survey was undertaken as a part of the safety evaluation and the results showed that the foundation condition had deteriorated rapidly since the last July 2003 survey, in that that the erosion had now progressed close to the main structural foundation of the central piers. Emergency rehabilitation works were recommended under the DSP and the implementation of the emergency works was carried out from August 2004 till March 2006.

INSPECTIONS AND SURVEYS

The following inspections and survey were undertaken under the DSP:
- Inspection of the hydromechanical and electrical equipment of the main diversion structure and the left bank regulator
- Inspection of records of geodetic survey
- Underwater survey by divers and the bathymetric survey
- Supplementary geotechnical investigation

Due to aged hydromechanical and electrical equipment and a lack of maintenance funds, many parts of the equipment were out of operation; i.e the main regulator could only discharge river flows through the central gates.

An underwater survey by divers and the bathymetric survey were carried out from January–March 2004. The survey was first carried out when all the gates were closed and it was subsequently repeated after the gates had been open and the sediments had been flushed. Conclusions from the survey were:
- It was found that the downstream erosion line was highly dependent on the operational regime of the gates and the flow of the sediments, related to the volume of the discharged water.
- Scour was taking place at the edge of the main structural foundation in the area of the central sections and pier No 5 where 3-4m deep voids were encountered (see Figures 2 and 4).
- The stilling slabs had been completely demolished in the central sections and partly in the remaining sections.

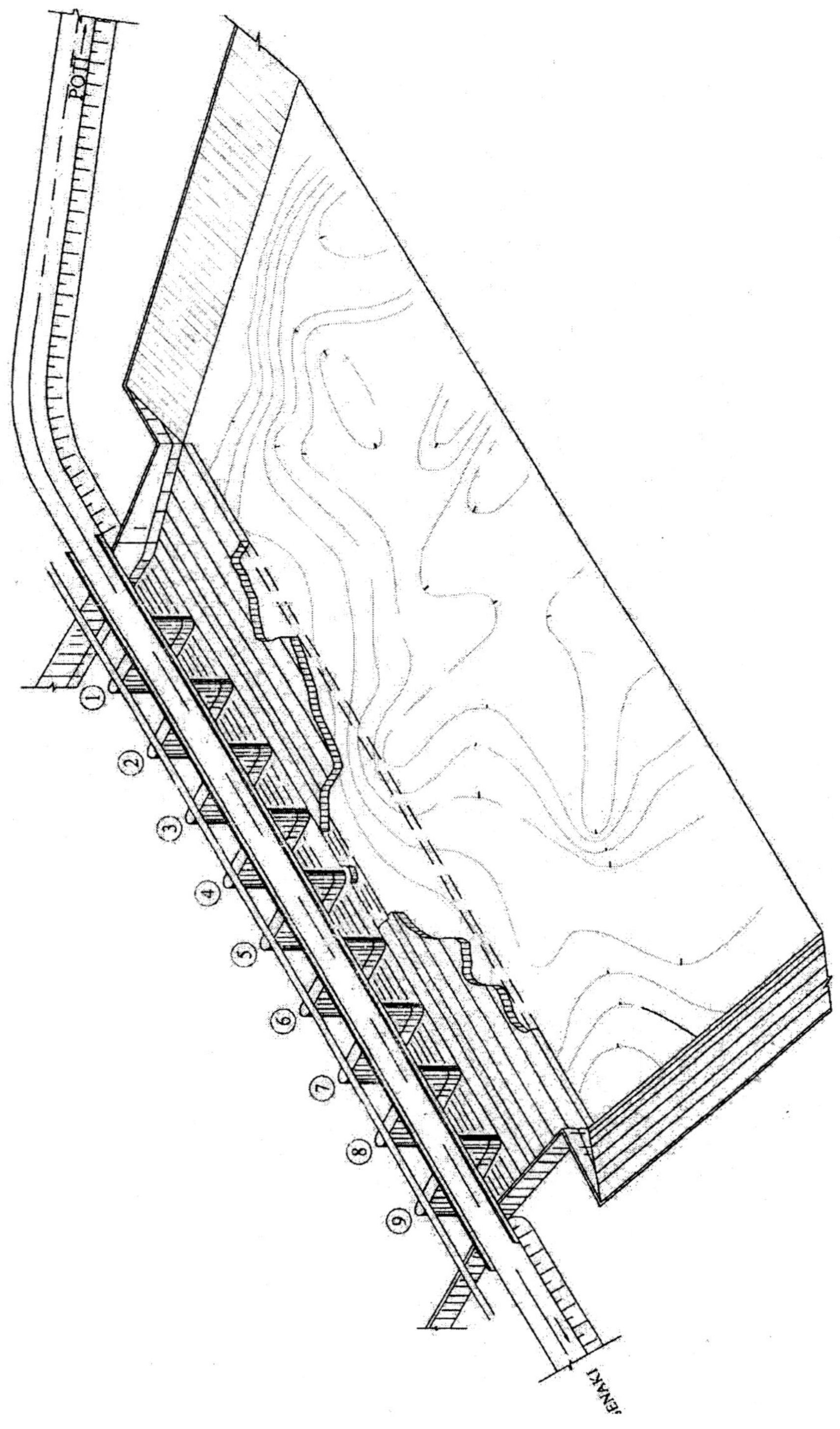

Figure 3: Extent of Damage of the Downstream Pool

- The downstream apron had been completely damaged and largely disappeared
- The upstream apron had been damaged in the central part; the last 5m of the apron slab had been damaged over a length of 100m

A small scale supplementary ground investigation was also carried out under the DSP. The investigation included drilling of one 60m deep borehole on the left bank of the river Rioni, with in-situ permeability and penetration testing and laboratory testing.

SITE CONDITIONS

<u>Hydrological Conditions</u>
The river Rioni has a catchment of 13,400 km^2 which covers about 50% of the whole western Georgia. About 68% of this catchment is on the southern slopes of the Great Caucasus Mountain. The river is fed by both rainfall and snowmelt.

The maximum flow usually occurs in May-early July and is typically 2,300-2,400 m^3/s. However, a maximum flow of 4,850 m^3/s was recorded during the flood in January 1987 which was associated with rainfall causing early snowmelt.

The lowest flows in the river Rioni are during the months of August, September and October when the typical flows are 280-320 m^3/s.

The main diversion weir was designed to discharge 4,000 m^3/s and the left bank regulator discharges 400m^3/s. However, the canal leading to the town of Poti was silted up and could only discharge 250-300m^3/s.

<u>Geological and Geotechnical Conditions</u>
The geology of the Poti area, situated in the Kolkheti depression is dominated by sedimentary deposits from the Holocene period which are of lacustrine origin. These sediments form the entire central part of the lowland and reach several hundreds meters in thickness. Lithologically, these formations consist of clays, silts and peat, whereas the underlying strata include sands, silty sands and silts.

The main founding strata of the diversion structure are the silts with occasional lenses of sand and peat.

Seismic conditions

The Kolkheti depression which, from the tectonic point of view, is the western end of the Georgian depression is bounded by major folds and active faults.

A Probabilistic Seismic Hazard Assessment (PSHA) to determine site specific seismic design parameters, namely the macroseismic magnitude (I) and the peak ground accelerations (PGA). The results of the PSHA showed that the maximal credible macroseismic intensity in the Poti area is I=8, with an average recurrence period of 1,624 years. The maximum credible bedrock PGA = 0.22g, with an average recurrence period of 8,673 years.

Due to the very weak and unfavourable foundation soils at the site a special concern was the amount of the seismic acceleration amplification that can take place through the soft foundation deposits. A site specific modification of the bedrock PGAs was carried out using the computer software which models one dimensional propagation of the seismic waves. The results of the in–situ SPT tests were used to establish the model for the seismic wave propagation. The analyses showed that the bedrock PGA of 0.22g would be amplified to 0.60g.

ANALYSES CARRIED OUT

Based on the results of the visual inspection, underwater survey, supplementary site investigation and the site specific seismic assessment, three main areas of concern were identified related to the safety of the main structure. They are as follows:
- Can the structure withstand the hydraulic forces?
- Can the structure withstand the seepage forces?
- Can the structure cope with a potential loss of bearing capacity due to the 3-4 m deep voids under the foundation and increased hydraulic gradients?

All these areas of concern were analysed and are discussed in the sections below.

Hydraulic Studies

In order to understand the mechanisms which caused the failure of the downstream apron and the stilling slabs, the original design was studied and compared with good hydraulic engineering practice. The impact of the operational regime of the weir was also studied as well as the rehabilitation measures that were implemented in the early 1980's.

The causes of failure of the downstream apron and the stilling slabs could be grouped under three main headings:
- inadequate original design

- inappropriate operation of the gates
- inappropriate design of the rehabilitation works implemented in 1982-1983

Inadequate original design
The following are the main points identified:
- The spillway slabs and downstream apron are located at high level (-3m) causing high velocities and erosion
- The stilling slabs and downstream apron were of inadequate thicknesses. The slab thickness was determined on the basis of seepage uplift pressures, as measured in a model for the following design conditions, which gave a factor of safety against uplift of 1.47:
- Normal water level in the upstream reach is at +2.02m,
- Minimum water level in the downstream reach is at -0.61m (Black Sea level)
- The design condition did not take into account the following:
    - during discharges of less than 300 m^3/s the lowest surface level of the jet flow is -1.03m (i.e. below the sea level of -0.61m). In this condition the head at the upstream end of the slab increases by some 20% and reduces the factor of safety to uplift pressure from the design value of 1.47 to 1.17
    - influence of the dynamic loading on the slab was not taken into account
- A row of energy dissipator blocks such as teeth do not work very efficiently

Inappropriate gates operation
The stilling slabs were originally designed in such a way that their stability and safety strongly depended on the correct gate operation regime. However, the operation regime of the gates has frequently been violated, mainly due to a lack of automatic gate controls, failure of gate hoists and shortages in power supply.

It was shown that the uplift pressures increased when flows of less than 300 m^3/s are passed through one or two gates, in the case when the gates are opened before the water levels downstream are steadied and the flow volumes are equalized. In such a case the shallow depth downstream appears to be insufficient for "drowning" the jump, so that the jet flow drives away the downstream waters, causing decreased pressures over the slab and increased uplift pressure. With the gates open, this condition can cause the slabs to lose their stability and start vibrating between the neighbouring slabs, indicating that the design slab thickness of 1.4m is not adequate.

Inappropriate design of rehabilitation works implemented in 1982- 1983
The main inadequacy of the design of the rehabilitation works is that the design did not take into account the erosion processes in the river channel. The design provided for the riprap backfill to come up to the original level of -3m, and without the layer being extended to the line of the sheet pile wall. Due to the inadequate rehabilitation and absence of bedding layers underneath the riprap backfill, the river bed under the riprap was soon scoured and the stone blocks, that had themselves fostered the flow turbulence, sank into the weak foundation soils.

The sheet pile wall installed 80m downstream from the structure increased the scour depth by 25-28% and encouraged the scour line to move closer to the main structural foundations.

<u>Seepage Analyses</u>
It was noted that in the original design, seepage analysis was undertaken assuming isotropic soil conditions which is an unsafe assumption bearing in mind that, due to their nature of deposition, the sedimentary soils are known to have a significant anisotropy. Therefore in the seepage analysis careful consideration was given to estimating hydraulic gradients at the point of exit.

The seepage analysis was undertaken for the following design conditions:
1. Isotropic condition; horizontal permeability coefficient, k_x, is equal to vertical permeability coefficient, k_y, with upstream apron and stilling slabs present but with no downstream apron
2. Isotropic condition, $k_x = k_y$, with upstream apron only
3. Isotropic condition, $k_x = k_y$, with no stilling slabs, and no upstream apron
4. Same as Case 1 but anisotropic condition, $k_x = 10\,k_y$,
5. Same as Case 2 but anisotropic condition, $k_x = 10\,k_y$
6. Same as Case 3 but anisotropic condition, $k_x = 10\,k_y$

The analyses showed that the actual hydraulic gradients are likely to be higher than the critical hydraulic gradient against piping failure for cases 5 and 6. This logic indicates that some internal erosion of the foundation is likely to have taken place in the areas where the downstream apron and the stilling slabs are damaged leading to their progressive destruction.

<u>Structural Bearing Capacity</u>
Structural bearing capacity was checked in order to estimate:
- The influence of the voids close to the structural foundation
- The influence of increased hydraulic gradients due to the damaged downstream pool

Structural bearing capacity checks found that, due to the voids and increased hydraulic gradients, the actual bearing capacity of the soil close to the pier No 5 could be expected to be reduced by 20%.

<u>Conclusions of the analyses</u>
It appears that the initial and the main causes of the events that damaged the stilling slabs and the downstream apron of the main diversion structure are related to inadequate hydraulic design of the downstream structural elements (the stilling slabs and the apron slabs were too thin and placed at a high elevation). These problems were however enhanced by irregular operation of the gates and installation of the sheet pile wall, as a part of the 1980's remedial works. For many years the water was discharged through the central gates only, so the damage to the stilling slabs and the downstream apron is extensive in the central area.

As a consequence of the damage of the downstream apron and the stilling slabs in the central section, the water seepage path has been shortened and consequently the exit hydraulic gradients have increased.

It was also shown that the erosive processes were likely to affect the bearing capacity of the soil underneath the main foundation of the pier No 5.

RECOMMENDED EMERGENCY REMEDIAL WORKS
Following the conclusion of the analyses, emergency rehabilitation works were recommended under the DSP. As it was not possible to divert the river away from the weir, the emergency rehabilitation works had to be carried out under water, but during a complete closure of all ten gates. The flow of less than 400 m^3/s had to be diverted via the left bank regulator into the Poti canal. The following were the recommended actions:

1. Prohibit the use of the diversion structure by heavy traffic until emergency rehabilitation works are implemented. This was recommended by the DSP and was endorsed and implemented by the Georgian Government in March 2004.
2. Rehabilitate hydromechanical and electrical equipment on the main regulating structure and the left bank regulator
3. Clear the Poti canal so that flows of up to 400 m^3/s could be discharged through the left bank regulator
4. Backfill the voids (approximately 500m^3) close to the structural foundation, in the areas of the central sections and the pier No 5, that are associated with erosive processes and are likely to have impact on the structural stability (see Figure 4). This was carried out by placing sand-cement-clay mix underwater through tremies installed through voids identified in the broken stilling slabs, close to the structure. Prior to the

backfilling a site trial was carried out to check the exact mix and the pumping pressures. Quality of backfilling works was checked by drilling and sampling of the backfilling.

5. Reinstatement of the stilling slabs (approximately 1,600m^2) and the downstream apron (approximately 7,200m^2) in the areas where the slabs and the apron are damaged (see Figure 4). This was carried out using Maccaferri gabion baskets (2m x 2m x 1m) which were assembled on the river bank, dropped into position from barges and subsequently interconnected underwater by divers. A geofabric was placed along the outer sides of the gabion baskets to prevent upward migration of the fines from the foundation soils into the gabions. Prior to placement of the gabions, a bedding layer of sand and gravel was placed over the scoured foundation to level the area. In places where the gabions were to be placed over broken slabs close to the main structure, smaller and more flexible cylindrical gabions were used. The cylindrical gabions were 1m long and 0.65m in diameter. They were also used at the ends of the "gabion structure" to provide a transition between the gabions and the natural river bed. Divers were employed to ensure that the gabions were properly placed and interconnected. An independent team of divers was employed by the implementing agency, Project Coordination Centre of the World Bank, to control the quality of the underwater works.

Detailed design of the emergency rehabilitation works was carried out by GWP under guidance and review by Jacobs. The total cost of the works was estimated by the GWP to be GEL 4M (approximately US$ 2M). It was anticipated that the rehabilitation works would be completed within 140 days, between September 2004 and February 2005, when the flow in the river Rioni is the lowest. A local Georgian contractor was commissioned to carry out the rehabilitation works. Personnel from Maccaferri's Moscow office provided specialist advice on placement and connection of gabions underwater.

The implementation of emergency rehabilitation works started in September 2004 with rehabilitation of the hydromechanical and electrical equipment on the main regulating structure and the left bank regulator, and clearing of the Poti canal. These works were delayed and had impact on the programme of placement of the gabions. Between February and March 2005 the works were undertaken with full closure of the gates, except for a short period of three days at the end of February when the flood flow was about 1,400m^3/s and gates 1, 8, 9 and 10 were opened. After passing of the 1,400 m^3/s flow at the end of February, a bathymetric survey was carried out.

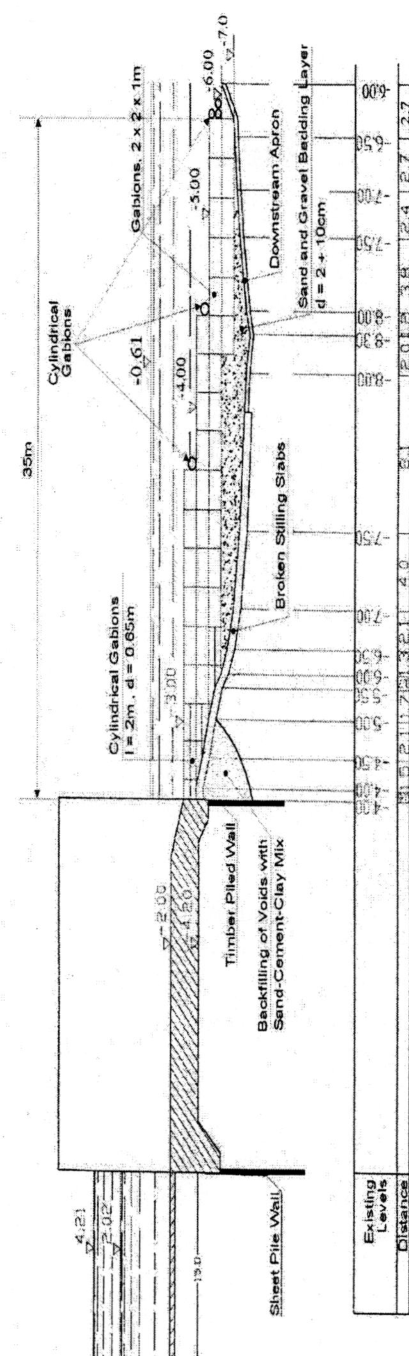

Figure 4: Emergency Rehabilitation Works- Longitudinal Section Through the Central Opening

It was found that the bedding layer and the gabions installed had not been affected by flood discharge through gates 1, 8, 9 and 10.

The rehabilitation works had to be temporarily stopped on 15th April 2005 due to high flows in the river Rioni. By the 15th April backfilling of the voids, placement of the bedding layer and installation of gabions were completed in the central sections up to an elevation of −6m. The flood period lasted till mid August, with the maximum flood of 3,000m^3/s being discharged through the weir at the end of April 2005. Underwater survey was carried out after the flood which showed that the gabions were unaffected by the flood releases, and that the river bed in the areas downstream of the placed gabions was not subjected to additional scouring. The works have since resumed and were completed in February 2006.

CONCLUSION

Inadequate design, inappropriate gates operation and a lack of maintenance fund over a prolonged period have brought the 50 year old main diversion weir on the river Rioni to the verge of collapse.

Rehabilitation measures implemented in the 1980's were not adequate and had encouraged further scour of the downstream pool. An other attempt to rehabilitate the structure at the beginning of 2001 led for the design to be put on hold several times due to a lack of funds.

Under the DSP financed by the World Bank, Jacobs carried out a safety evaluation of the main diversion weir during August 2003–July 2004 and proposed implementation of emergency rehabilitation measures. Detailed design of the emergency rehabilitation works was undertaken by GWP from Tbilisi under guidance and review by Jacobs with specialist input by Maccaferri from Moscow.

Implementation of the rehabilitation works started in September 2004 and was due to be completed by the end of February 2005. The implementation works were delayed due to a slow start of the contractor, lack of local experience with placement and connection of gabions underwater and requirements that an independent team of divers constantly checks the works. Works also had to be temporarily stopped for four months during the flood season when all the gates on the weir had to be open to discharge a flow of 3,000 m^3/s. However, this flow did not pose any damage to the works installed. The emergency rehabilitation was completed in February 2006.

Barrow Compensation Reservoir Grouting Works

A L WARREN, Halcrow Group Ltd
C HUNT, Bristol Water
M ATYEO, Bristol Water

SYNOPSIS. Barrow Compensation Reservoir has a long history of leakage and remedial works. The 12m high embankment dam, constructed in 1863, was eventually emptied in 1882 after many attempts to stem leakage. The reservoir has normally remained empty since this time, but the reservoir can substantially fill under flood conditions. Some doubts that were raised in relation to dam safety under flood conditions led to the development of a comprehensive grouting programme to seal both the puddle clay core of the dam and its rock foundation in 2004-2005. The paper summarises the history of remedial measures at the dam, and describes the site investigation works carried out to scope and measure the effectiveness of the remedial measures. The paper discusses the effectiveness and limitations of the remedial works carried out and the implications for the future monitoring.

INTRODUCTION

The great majority of the UK's reservoirs are viewed as an asset by their owners. This paper concerns recent remedial work to a reservoir which falls in the minority group. Barrow Compensation Reservoir, owned by Bristol Water, is situated south-west of Bristol. The other three Barrow reservoirs serve important water supply functions for the city, but the Compensation reservoir has served no function for the owner for well over a hundred years.

Built in 1863, the reservoir served to provide a regulated water supply to mill owners on the River Land Yeo. The dam comprises a 160m long embankment, up to 12m high with a central, narrow puddle clay core wall and cut-off trench. Problems are documented both with the stability of the 1:2 upstream face and with leakage. The history of the problems and early remedial works to the dam are described have previously been researched in detail (Watson Hawksley, 1982). In 1870, the core wall was completely removed and replaced with fresh puddle clay. The width of the core was widened to 5 feet at the foundation level and the depth of the cutoff trench increased. However, problems with leakage continued and attempts to maintain the reservoir full were finally abandoned, leading to the enabling act of Parliament being repealed in 1882. From that time to the present day, the reservoir has normally been almost empty. However, the reservoir can substantially fill during extreme storm events over the Mendips, as occurred in 1968. The only practical function that the reservoir now serves is flood protection for the villages downstream. There is a drawoff weir at elevation +81m leading to a concrete-lined drawoff tunnel and an 18-inch drawoff pipe that passes through the dam. Flow can be controlled by a valve located in a tower on the line of the core wall. The spillway crest is set at +86.9m and the dam crest is at about +89.5m. A clay blanket was placed over the upstream face of the dam in 1982 in an attempt to reduce leakage but this appears to have been unsuccessful. In 2000, localised grouting around the perimeter of the drawoff tunnel and below the valve tower also failed to significantly reduce leakage.

A ground temperature survey of leakage patterns through the core of the dam in 2002 confirmed that leakage was present near the line of the 18-inch drawoff pipeline and valve tower. In addition, dye tests indicated very rapid passage of water from the reservoir area to the drainage basin located near the downstream toe, suggesting the presence of open joints or fissures within the dam body or through the foundation.

A statutory inspection of the reservoir in 2003 recommended that either the dam watertightness should be improved to improve the stability of the downstream shoulder during flood events, or that the reservoir should be discontinued.

Studies were carried out by Bristol Water and Halcrow to compare the costs of 'notching through' the embankment (discontinuance) with the cost of remedial works. The studies concluded that the most economic solution was to improve the watertightness of the dam and foundation by grouting. In particular, the costs associated with flood protection measures for the downstream villages and environmental protection measures were prohibitive. Grouting had already proven successful in sealing leakage at Barrow 3 Reservoir in 1996 (St John, Nicholls and Senior, 1998).

GROUND INVESTIGATIONS
A comprehensive site investigation was carried out in 2004. The scope of the investigation, excluding laboratory tests, is summarised in Table 1 below.

Table 1: Scope of Site Investigation

Item	**Purpose**	**Number/Notes**
Cable tool and rotary boreholes	Drilling through core wall, shoulder material and rock foundation for installation of piezometers, soil sampling and in-situ testing.	21 no. boreholes
Standpipe piezometers within the core	For critical pressure tests to determine the susceptibility of the core to hydraulic fracture	3 standpipe piezometers installed.
Vibrating wire piezometers and common datalogger	To provide accurate measurements of pore water pressure before and during the impoundment testing.	15 vibrating wire piezometers were installed, seven of which with de-airing facilities.
In situ permeability tests	To determine the permeability of the core and embankment fill before and after grouting works.	Lugeon tests were carried out at six locations within the rock foundation.

Vibrating wire piezometers were favoured for this embankment as they have a fast response time appropriate for the rapid rise in reservoir level during floods, and would not malfunction if grouted during the remedial works.

The uppermost part of the shoulders was found to comprise a firm to stiff orange brown gravelly clay. The lower part of the shoulders is formed of softer blue grey clay. It was noted that the boundary between the blue and orange clay closely matched the piezometric profile, suggesting possible oxidation of the uppermost clay.

The foundation of the dam was found to comprise (in descending order) the following beds, with a shallow dip to the north-east:
- Interbedded limestones and mudstone, weathered to clay in places (the Wilmcote Limestone);
- A thin (<10cm) band of orange gritty calcareous sandstone;
- A white silty limestone, approximately 2m thick, with thin mudstone bands (the Langport Limestone); and
- A dark grey mudstone, weathered to clay (Cotham mudstone).

The initial lugeon tests in the rock foundation yielded values consistently over 100. In some cases it was not possible to attain sufficient water pressure to provide a result due to very high ground permeability. The band of white limestone, in particular, was suspected of contributing to this high foundation permeability. A geological plan of the dam area is shown in Figure 1 and a geological section through the dam and the foundation is shown in Figure 2.

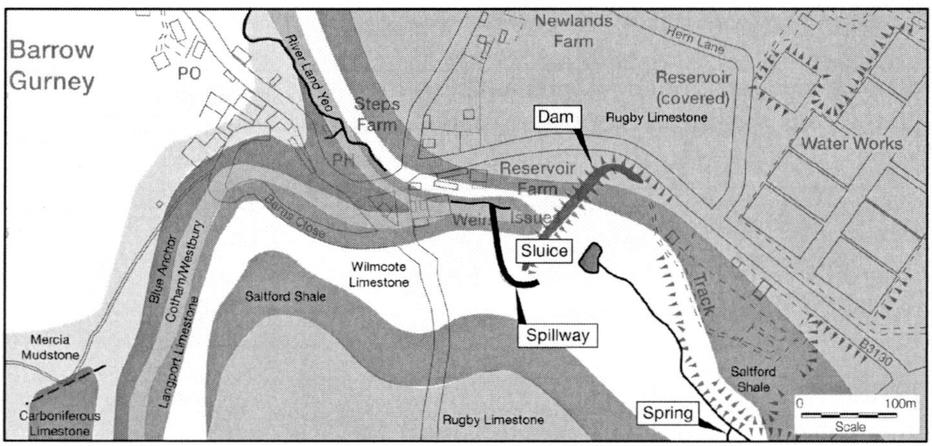

Figure 1:   Geological plan of the dam area

The ground investigation found that the puddle clay core had a permeability value in the order of 10^{-6} to 10^{-7} m/s. This indicated the likely presence of open fissures within the core wall as one would normally expect a permeability in the order of 10^{-9}m/s or lower for puddle clay.

The phreatic surface was found to be significantly higher in the downstream shoulder than in the upstream shoulder (see figure 2), particularly adjacent to the left abutment. This was believed to be a result of groundwater inflow from the valley side and was clearly detrimental to stability of the downstream shoulder.

DESIGN OF REMEDIAL WORKS

Remedial works were designed to enhance the stability of the downstream shoulder during flood events by:

- significantly reducing leakage through the foundation by grouting works;
- significantly reducing the leakage through the dam core wall by grouting works;
- reducing the permeability of the clay fill upstream of the core wall by grouting to effectively widen the core (Vaughan, 1987) and reduce the risk of hydraulic fracture re-occurring in the future.

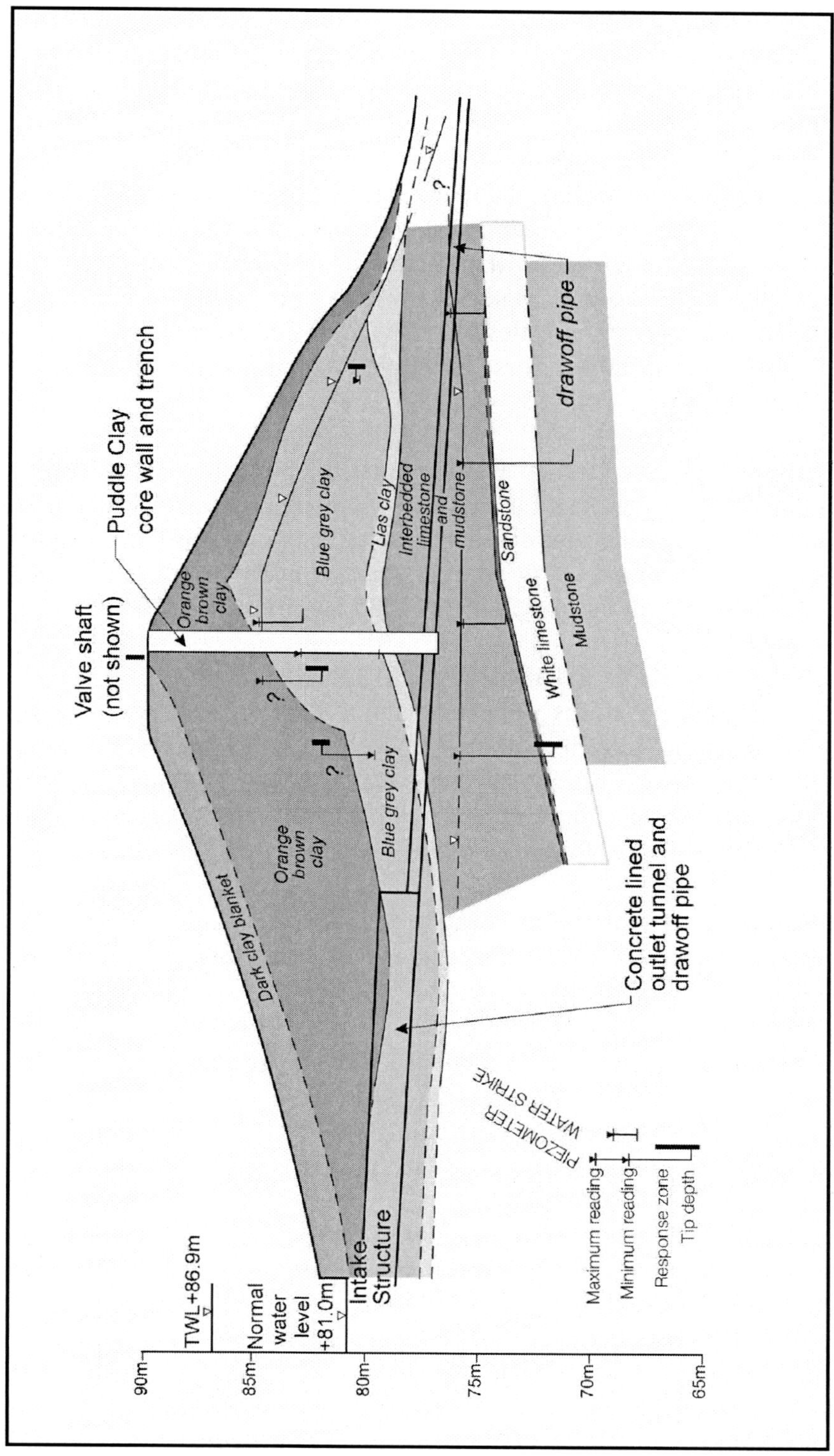

Figure 2: Geological section through dam embankment and
foundation   on the line of the drawoff works.

- providing a deep French drain in natural ground adjacent to the downstream shoulder on the left abutment to assist in draining the downstream shoulder and to intercept groundwater flowing towards the embankment.

Grouting of the core wall and foundation was deemed necessary over the full length of the embankment. Tube-à-manchette (TAM) grouting was proposed for the core wall – this technique had previously been applied successfully at Barrow 3 reservoir and a general description of the equipment and technique is described by St John *et al*, 1998. Additional TAM tubes were proposed for the periphery of the valve tower to reduce water ingress.

Further rows of grout holes were proposed upstream of the core wall, extending through the clay fill and into the rock to a varying depth below the foundation level approximately equal to the height of the embankment (to as low as elevation +67m over the central section of the dam). These were aimed at sealing fissures in the clay fill and rock joints.

IMPLEMENTATION OF REMEDIAL MEASURES
Before the grouting works commenced, the drawoff valve was closed to allow the reservoir level to rise by 3m to elevation +84m and the reservoir was held at this elevation for two weeks. This provided a baseline performance in terms of seepage and piezometric response which would serve to evaluate the performance of the remedial works.

The detailed design of the grouting works was developed in consultation with the proposed grouting contractor, Norwest Holst Soil Engineering (NWH).

TAM sleeve tubes were proposed at 1.5m longitudinal centres on the centerline of the core wall. Grouting ports were proposed to have 0.5m vertical centres with the tubes extending over the full height of the core wall/trench.

Two grouting techniques were considered for the line of grouting upstream of the core wall. With either method it was proposed to use two rows of grouting holes, located 1.25m and 2.25m upstream of the centerline of the core wall, with the holes staggered at 2m spacings.

1. Pressure grouting of the foundation below packers and subsequent TAM grouting of the clay fill upstream of the core wall.

2. A simpler 'open hole' approach whereby grout would be injected into sacrificial tubes extending into the foundation with perforations in the tubes over the length within the rock foundation. With this method, grout would also travel up the annulus between a slotted plastic casing (used to support the hole) and the clay fill, thereby filling voids within the clay fill.

The first method was favored technically, as grouting of the rock foundation could be completed using higher grout pressure. TAM grouting of the clay fill would also provide a greater degree of control in the grouting process, provide a detailed record of grout takes, and the grout mix could be varied to suit.

The advantages of the second method were in terms of cost and programme. This would arise from the shorter amount of time that critical equipment would spend at each hole in supporting the packers. With this method, which was proposed by the contractor, the grouting pressure is limited to the hydrostatic pressure available from the dam crest elevation.

It was agreed that the second method should be trialled and lugeon tests conducted to check for the effectiveness in sealing the foundation and reducing the permeability of the clay fill immediately upstream of the core wall/trench. This trial resulted in two foundation permeability results of zero lugeons in the foundation and therefore the 'open hole' method was adopted for the remainder of the work.

The grout mix used for the TAM grouting used bentonite and OPC in equal measure by weight and a water:cement ratio of 10:1. This mixture was designed to have a shear strength (when fully set) similar to and less than that of the clay core.

Following initial trials for the foundation grouting to determine the optimum ratio of water to OPC, a 3:2 water:cement (by weight) mix was selected. This mix was used for all of the foundation grouting except in cases of limited acceptance where a 3:1 (by weight) mix was used.

Grouting Performance
Falling head permeability tests in the clay fill and water pressure tests in the rock foundation were carried out to assess the effectiveness of the grouting works. The results are summarized in Table 2.

Table 2: Grouting Results

Location	Permeability prior to grouting	Permeability after grouting
Puddle clay core wall	1E-06 m/s based on three tests.	2.5E-07m/s based on five test results
Clay fill on plane immediately upstream of core wall	1.2E-05 m/s based on two tests. Taking into account tests on the downstream shoulder fill, the fill permeability is typically in the range of 1.0E-05 to 1.0E-06 m/s.	1.9E-06 m/s based on six test results.
Rock foundation	Five borehole water pressure tests: two results of >100 lugeons; unable to obtain sufficient water pressure in the other three boreholes.	**Result (lugeon)/Formation** 60/ Wilmcote 40/ Wilmcote/Langport >100/ Langport 0 / Langport 0 / Langport 70 / Wilmcote.

The reduction in permeability in the core wall/trench and the clay fill upstream of the core wall was considered a success. The foundation results were disappointing, given the promising trial results (the two results of zero lugeons). However, overall the results indicated that the permeability of the foundation had been significantly reduced, if not to the standards normally associated with new dams. Discussions with the inspecting engineer concluded that the improvement was sufficient to warrant a trial impoundment of the reservoir to assess the effectiveness of the remedial works. The option to pressure grout sections of the dam foundations remained available at this time, but it was decided to reserve this option in the event that the impoundment trials indicated that the works had been ineffective.

The cost of the grouting works completed at the dam was approximately £400k.

<u>Trials</u>
To test the effectiveness of the remedial works, a programme for raising the reservoir level to top water level and monitoring the response of the

embankment was prepared. The reservoir level can be raised simply by closing the 18-inch drawoff valve. The timing of the test impoundment was dictated by the requirements of a badger license and inundation of the sets in the reservoir area could not commence until the summer months. A further constraint was the proximity of an adjacent storage area owned by the Undertaker which is inundated at reservoir top water level. This area had to be cleared of equipment before the test could commence.

The strategy for testing the remedial works was as follows:

1. Close the drawoff valve and slowly raise the reservoir level to +84m and hold at this level for two weeks with the valve cracked open.
2. Compare the results with the partial raising completed before the remedial works. Assess the level of seepage flows and the piezometric response.
3. Close the drawoff valve and slowly increase the reservoir level from +84m to +86.9m, continuously monitoring seepage and piezometric pressures. This phase was implemented over a period of over three weeks.
4. Hold the reservoir at top water level for a period of at least 2 weeks and assess the results.
5. Crack open the valve and slowly draw the reservoir level down. The reservoir was drawn down at a rate not exceeding 300mm/week to guard against instability of the upstream face.

In November 2005, after the reservoir had been maintained full for nearly three weeks with no problems arising, the remedial work was declared successful. In practice, the reservoir would be full for a shorter period of time in the event of an severe flood.

It was observed that the seepage rate from the reservoir when full (+86.9m) was significantly less than had been observed prior to the grouting works when the reservoir was empty (+81m).

The piezometric responses were generally in line with expectations. The piezometers located upstream of the core responded more markedly after the remedial works in response to the partial raising, indicating reduced permeability of the core. The piezometers in the foundation beneath the downstream shoulder and in the downstream shoulder material itself generally displayed a more subdued response to the impoundment.

FUTURE MONITORING
The performance of the dam will continue to be monitored using both the vibrating wire piezometers and an improved seepage monitoring weir in the

tailbay channel immediately downstream of the dam. These provisions will enable the Undertaker and the Supervising Engineer to effectively monitor any deterioration in the performance of the remedial works.

CONCLUDING REMARKS

This project describes the latest and most successful attempt to reduce seepage and improve the safety of Barrow Compensation Reservoir; a dam which has proved troublesome ever since its construction in 1863 due largely to a highly permeable foundation.

It is an example of where expensive remedial works have been completed for the purposes of reservoir safety despite the reservoir having no useful function for the owner.

In practical terms the project provides a further example of how puddle clay cores can be improved using TAM grouting. Despite the relatively crude approach adopted for the foundation grouting, a significant reduction in permeability was achieved.

ACKNOWLEDGEMENTS

The authors wish to thank Bristol Water for permission to publish this paper.

REFERENCES

Charles JA and Watts KS, 1987. The measurement and significances of horizontal earth pressures in the puddle clay cores of old earth dams. Proceedings of the Institution of Civil Engineers, Part 1, Vol. 82 pp123-152.

Vaughan PR, 1987. Discussion. The measurement and significances of horizontal earth pressures in the puddle clay cores of old earth dams. Proceedings of the Institution of Civil Engineers, Part 1, Vol. 87 pp1249-1255

Skempton A W, 1989 "Historical development of British embankment dams to 1960". Keynote address, Proc. Conf. on Clay barriers for Embankment Dams, London, pp15-52. Thomas Telford.

St John T, Nicholls R A and Senior K W, 1998. Grouting the puddle clay core at Barrow No 3 Reservoir, Bristol. Proceedings of the tenth conference of the BDS, Bangor, pp255-264. Thomas Telford.

Watson Hawksley, 1982. Review of the History of Barrow Compensation Reservoir.

Reservoir Safety and Refurbishment Works at Severn Trent Water's Howden, Derwent and Linacre Reservoirs

S.A.ROBERTSON, Jacobs Babtie, Glasgow, UK.

SYNOPSIS.
Severn Trent Water (STW) initiated a contract for essential reservoir safety and refurbishment works following a routine 10 year inspection, under Section 10(2) of the Reservoirs Act 1975. The required works were varied and included, valve refurbishment, additional valves, spillway repairs and general maintenance work.

At Howden and Derwent Reservoirs there was concern that operation of 9no. 30 inch diameter Blakeborough scour valves was causing cavitational damage and some valves could not be operated through their full travel. It was recommended that further investigations were carried out 'in the interests of safety'.

At the Linacre Reservoirs (Upper, Middle and Lower) the works to be carried out, some 'in the interests of safety' and others for refurbishment, included repairs to spillway walls and floors, restoration of a drainage facility, installation of a scour guard valve and re-lining of the Lower Reservoir draw-off scour pipeline.

This paper reviews the works involved from the early feasibility investigations through to the construction works. The construction period was August 2005 to March 2006.

HOWDEN AND DERWENT RESERVOIRS

Howden and Derwent Reservoirs are situated in the Upper Derwent Valley area of north Derbyshire, within the Peak District National Park, approximately 20km west of Sheffield.

The reservoirs are impounded by concrete/masonry gravity dams, approximately 35m high and 330m long, which were constructed during the early part of the 20th century. Overflow from Howden Reservoir spills, via a

Improvements in reservoir construction, operation and maintenance, Thomas Telford, London, 2006, 355–367

central crest weir (ref Figure 1 : Howden's central crest weir) , into the head waters of Derwent Reservoir. A similar weir structure on Derwent Dam discharges flood water from Derwent Reservoir into the headwaters of Ladybower Reservoir. The three reservoirs are used as a source of raw water for Bamford Water Treatment Works, which serves a population of about a million customers throughout Derbyshire, Nottinghamshire and Leicestershire.

Figure 1 : Howden's central crest weir

Inspecting Engineer, Mr Alex Macdonald, of Jacobs Babtie (JB), carried out a routine 10-year inspection of Howden and Derwent Reservoirs, under Section 10(2) of the Reservoirs Act 1975. Whilst confirming that both reservoirs had adequate provision for emergency draw-down, he was concerned about the effects of cavitation on the duty scour valves and the inability of some of the valves to be operated through their full travel.

Howden Dam
Howden Dam incorporates six pairs of in-line 30 inch (750mm) diameter Blakebourgh guard and duty scour gate valves, three pairs in the east valve chamber and three in the west.

These 'duty' scour valves are hydraulically operated. Hydraulic pressure to power the actuators is provided by the reservoir head, through tappings off the pipework on the upstream side of the three duty valves in the west chamber. The pressure is then boosted by a compressor.

Derwent Dam
Derwent Dam incorporates three pairs of in-line 30 inch Blakebrough guard and duty scour valves (ref Figure 2 : Derwent Valves), situated within the

west valve chamber of the dam. The outlet pipes leading from these valves discharge into a tailbay stilling pool.

The scour valves are similar to those at Howden. Hydraulic head to operate the valve actuators is provided by means of a 2 inch diameter asbestos cement and uPVC pipeline from Walkers Clough, a small reservoir situated high up on the eastern hillside overlooking the reservoir.

Figure 2 : Derwent Valves

<u>Feasibility Investigations</u>
During the feasibility investigations it was considered that renewal of the duty valves would have been considerably more expensive than refurbishment. The valves were also partially encased, to half barrel, in concrete. Complete removal of the valves would have incurred a significant amount of effort as well as cost.

Options were therefore considered for repair, refurbishment and renewal of various components of the scour valves at Derwent and Howden. The valve components were; body, bonnet, wedge, stem and nut assesmbly, operating cylinder and piston, control valve, pressure relief valve, bolts, gaskets and ancillary items such as indicator plates.

The investigation at this stage of the works was restricted to bringing together a limited amount of factual information. A clearer understanding of the extent of refurbishment could only be ascertained when the valves were dismantled. However a significant element of the data available was a recent CCTV survey carried out by STW. This clearly showed that there was evidence of cavitational damage to some of the valves. It was also clear that where new sleeves had been installed in two of the Howden valves that this recent refurbishment work had been effective. The situation was similar at Derwent where one valve had been repaired with a metal filler which appeared to be performing satisfactorily.

Due to the apparent success of the previous valve refurbishment, it was recommended that repairs to the areas affected by cavitation be carried out.

To communicate the extent of repairs to Norwest Holst (NWH), STW's framework contractor assigned this work, a detailed specification and scope of work was compiled listing the valve components. The specification also elaborated on the timing of the works ensuring that emergency drawdown would not be affected by the construction works.

The information, with respect to valve condition, was limited. Therefore the specification stated that after stripping the valves down on site and before any refurbishment / fabrication work was carried out, a valve condition report would be prepared for each valve.

The feasibility review of the work also encompassed health and safety issues and environmental impact of the works in a National Park. Good communications with the park Rangers assisted access and egress issues.

<u>Extent of Valve Works</u>
The specific requirements for the valve work were extensive, but listed below are some particular items of interest;

Valve Body and Bonnet Repairs
It was specified that both the body and bonnet would have guides dressed, minor damage repaired and damage to any corroded retaining pins replaced. If sustained damage had occurred to the guides, in particular at the body/bonnet interface, a filler material would be required to allow full refurbishment.

Wedges
The wedges once removed completely from the body of the valve were to be shot blasted and painted with their faces dressed and any minor damage repaired. The associated new bush, pin and roller sets would be manufactured and installed to suit body guide sizes or if shoes, these would be dressed and minor damage repaired.

Operating Cylinder and Pistons
Damage to the operating cylinders would be machined and new pistons, piston guides (if required) and seals were to be manufactured to suit the oversized cylinders.

Pressure Relief Valve
The pressure relief valves were to be fully refurbished and re-set to prevent the valve wedge and control piston from experiencing excessive forces.

Construction Works

In July 2005 Norwest Holst commenced construction work with subcontractor Blackhall Engineering. A comprehensive valve condition report was issued by NWH for each valve as it was dismantled. The pretender assessment of proposed works were found to be a good estimate and refurbishment was initiated at Blackhall Engineering's factory. Typical valve condition found following stripping can be seen in Figures 3 and 4 : wedge cavitation and scored channel guides. Work during refurbishment can be seen in Figure 5 : view showing the wedge and guide brushes

Figures 3 and 4: Wedge Cavitation and Scored Channel Guides

Figure 5 : View showing the wedge and guide bushes

At feasibility stage it was considered that removal of the scour valves would be difficult as there were no existing provisions in the valve chambers for removal. In addition access was very difficult as there were no direct routes that plant could travel to the valve chambers. However, Blackhall Engineering designed and constructed temporary gantries to allow safe removal and the work proceeded very efficiently.

Following refurbishment the valves were tested and commissioned. Tests were carried out under the available reservoir head and proved that the valves would operate smoothly through the full range of travel. During the final tests, adjustments to the pressure relief valves were made making sure that the wedges were not fully embedded onto the bottom of the valve body. If this were to occur, there would be a risk of the valve jamming shut.

The works 'in the interests of safety' had identified that cavitation was a significant concern. Although no physical works were implemented to alter this hydraulic phenomenon it was recommended to STW that the scour valves should only be used when opened to a position which was outside the cavitational zone (20% to 70% open). During the final testing the ease of closure was noted as was the reduction in noise by those having witnessed the operation of the old valves. Operator confidence to move the valves through the zones has now increased.

LINACRE RESERVOIRS

Linacre Reservoirs consist of three cascading reservoirs appropriately named, Upper, Middle and Lower. They are situated on the Holme Brook, approximately 5 kilometres west of Chesterfield in Derbyshire. They are impounded by earth embankment, clay core, dams, which were completed in 1854 (Lower), 1864 (Upper) and 1911 (Middle). All three Linacre Reservoirs are currently non-operational and are allowed to fill up to spill level.

Inspecting Engineer, the late Mr. John Beaver, of Halcrow Group, had carried out a routine 10 yearly inspection of Linacre Reservoirs, under Section 10(2) of the Reservoirs Act 1975. His report made a number of recommendations. Those referred to in this paper are;

- refurbishment works to spillway walls and floors,
- restoration of a drainage facility,
- installation of a scour guard valve,
- re-lining of the Lower Reservoir scour pipeline.

The Works

This project, like Howden and Derwent, complied with the Construction (Design and Management) Regulations 1995. Early involvement of the Planning Supervisor assisted in the application of the Regulations throughout the project. During the feasibility stages of work the design risk assessments were compiled, defining the significant hazards that NWH would later consider and develop in their Health and Safety Plan for the construction works.

Of the more general but significant risks identified was the potential for the working areas, such as spillways, to be inundated with water. Planned draw down of the reservoirs were successfully implemented during construction by STW Operations and NWH.

Refurbishment works to spillway walls and floors
The works for this element of the project were considered at first to be relatively straight forward. The scope of work required general repairs in the spillway channel retaining walls and repairs to damaged concrete and joints in the bed of the channel. A site visit early in the project with STW, JB and NWH identified the extent of work referred to in the Inspecting Engineer's Report. (Ref Figures 6 and 7 : Typical floor joints prior to repair and spillway channel)

Cracks on the surfaces of the concrete walls and floors were to be repaired with a concrete waterproof repair mortar. The joints were to have the existing material removed and replaced with a gun grade sealing compound. Once on site NWH's suspicions regarding the existing joint filling material were aroused. The material was easily broken when removed. All work in spillways channels was stopped and a sample of the joint filling material taken away for analysis. It was found that the material contained approximately 15% asbestos!

As part of the developing risk assessment of the planned work it was considered that leaving the asbestos in place could be safer than removal. However it was decided that the risk of leaving the problem asbestos to others in the future was not an acceptable solution and a specialist and licensed contractor was employed to remove the 156 linear meters of jointing material before work could re-commence on the spillway channels.

In the case of these works the specialist contractor notified the Health and Safety Executive of the planned asbestos work complying with The Control of Asbestos At Work Regulations 2002.

Work was successfully completed on this section in December 2005.

Figure 6 and 7 : Typical floor joints prior to repair and spillway channel

Restoration of a drainage facility
The purpose of the drainage facility located adjacent to the Middle Reservoir embankment and at the top of the Lower Reservoir was uncertain and historical construction records were not available.

A small burn flowing down a 'closed off valley', in the direction of the reservoir was intercepted by a chamber, measuring approximately 4 metres by 2 metres. It contained a depth of silt of approximately 2 metres.

The Inspecting Engineer requested that the silt and other obstructions be removed to restore the drainage facility and the outlet investigated. Various attempts had been made in the past to find the outlet of the drainage facility without success.

During the feasibility stage, environmental issues prevailed. An Environmental Assessment was carried out indicating a number of issues that could influence the method of working;

- Badgers – protected under the Badgers Act 1992. A sett was found close to the chamber. It was anticipated that restricting the working area would only cause some short term disturbance to the badger's territory.
- Plants – Impact on wood barley and woodland ground flora was a significant issue. The working area was confirmed prior to entry and access/ egress routes defined.
- Nesting Birds – The working area was checked prior to entry for nesting birds and access routes again defined.

The initial proposals considered a prolonged programme of manual working but with NWH's input balancing the limitations of manual working with available plant, a method of working was evolved without detriment to the environmental issues. The planned 'duck board' paths into the valley were replaced with scaffolding extending into the 'closed off valley'. Pulleys replaced potential endless journeys using wheelbarrows. Unfortunately the silt from the chamber had still to be removed by spade.

This work was completed in November 2005. Not long after the chamber was inundated with water during a heavy downpour and the burn outflow into the top of the Lower Linacre reservoir became clearly visible.

Installation of a scour guard valve
The provision of a scour guard valve at the Upper Reservoir was requested by the Inspecting Engineer as there was no 'secondary' valve on the scour outlet.

Figures 8 : Upper Linacre valve tower

The feasibility review of the required works considered various alternative locations. With the tower being relatively small in diameter and with numerous pipes and valves already accommodated a suitable location at higher level was opted for. Figures 8 : Upper Linacre valve tower shows the internal and external views of the tower.

As indicated, the logistics of providing the guard valve within the existing valve tower chamber was difficult, with confined space access and manual handling problems the prevalent risks. Scaffolding, again, was adopted as the best option for safe access, being constructed from the bottom of the chamber up. Confined space entry was rigorously employed with the Local Fire Services taking the opportunity to use a site visit as confined space entry practice!

Once the reservoir water level was sufficiently drawn down, work commenced on cutting the existing pipework for the addition of the valve and spindles. Actuators were provided and located at the top of the valve tower giving STW Operations a safer means of operation.

Relining the Lower Reservoir scour pipe
It was reported that during a site inspection water was seen to be issuing from the bulkhead wall upstream of the supply culvert. The Inspecting Engineer examined the flow of water and was of the opinion that the scour pipe had either suffered through-wall corrosion or the joints had sprung a leak. Leakage could therefore have been taking place along the pipe/clay interface and into the culvert downstream of the bulkhead. In advance of construction works the scour valve downstream was opened to mitigate the effects of leakage. The Inspecting Engineer, at the time of his inspection, indicated that lining the scour pipe would be an acceptable solution.

The scour work also included the feasibility and construction of an operable scour guard valve. The existing valve, refer to Figure 9 : Lower Linacre existing guard valve, was currently only accessible by divers as it was located in 9m depth of water. Reasons for this arrangement were only speculative.

During the feasibility stages of the works JB identified that there could be significant risks of flooding or even damage to the dam, if during proposed construction works, the reservoir remained full of water. Before any lining works could be carried out the existing scour pipe would have to be cleaned. This work could exacerbate any cracks and leakage. It was therefore agreed to empty the reservoir to mitigate any flooding and limit other risks while at the same time providing safe and relatively dry access for the removal of the scour guard valve.

In turn, the implications of emptying of the reservoir became a critical environmental issue since the timing of the work could have a significant impact on the local wildlife. The Linacre environmental report had identified that the bird nesting season could be disrupted.

The discharge waters, containing silt, were considered to have a potential impact on the downstream river life if uncontrolled. Early interaction with the Environment Agency by NWH assisted the smooth progress of the works. Many options were considered including bringing to site stilling tanks, with the addition of coagulating chemicals, to the more practical option of using the existing spillways channels to removed any silt. The latter solution was developed, by NWH, into a workable method. The reservoir waters were pumped into a series of lagoons which were constructed at the top of the spillway channel (ref Figure 10 : Lower Linacre lagoons) before flowing down the channel through a series of straw bales. This method was adopted and implemented successfully for the removal of the final 2 metres depth of reservoir water which contained a large amount of silt.

The lining work itself underwent a feasibility review. At this stage early discussions with a specialist contractor and NWH assessed very quickly that installation of the lining would be difficult due to access problems to the location of the scour pipework. There were no access roads or paths on which plant could be taken and moving any equipment would probably have to be done manually.

Figure 9 and 10 : Lower Lincare , Existing Guard Valve and Lagoons

To compound issues there were no existing record drawings of the Lower Linacre reservoir and pipework in STW's archives. This proved to be significant. The scour pipe was considered to be continuous through the dam but, following a CCTV survey to assess the condition of the scour pipe, it was discovered that the pipe stopped near the centre line of the embankment, possibly in the area of the clay core and bulkhead, and continued to the upstream side of the dam and scour valve via a 500mm x 500mm square masonry culvert as shown in Figure 11 : section through Linacre Lower Reservoir. A further CCTV survey was carried out to assess the condition of the culvert from the upstream side. Although the reservoir was empty at the time of the survey there was evidence to suggest that water could percolate through the structure. The reasons for pipe leakage was not through pipe wall corrosion but through the culvert. Leakage through the bulkhead wall remained unresolved. After much consideration it was agreed with the Inspecting Engineer that the 300mm diameter cast iron scour pipe and culvert would be lined with a 250mm diameter continuous polyethylene pipe. The annulus between pipes and culvert was filled with grout.

Calculations estimating the drawdown capacity assisted in arriving at the lining solution. Comparing a new 250mm diameter polyethylene pipe to a 300mm diameter heavily encrusted cast iron pipe confirmed that there would be an improvement in the current drawdown capacity.

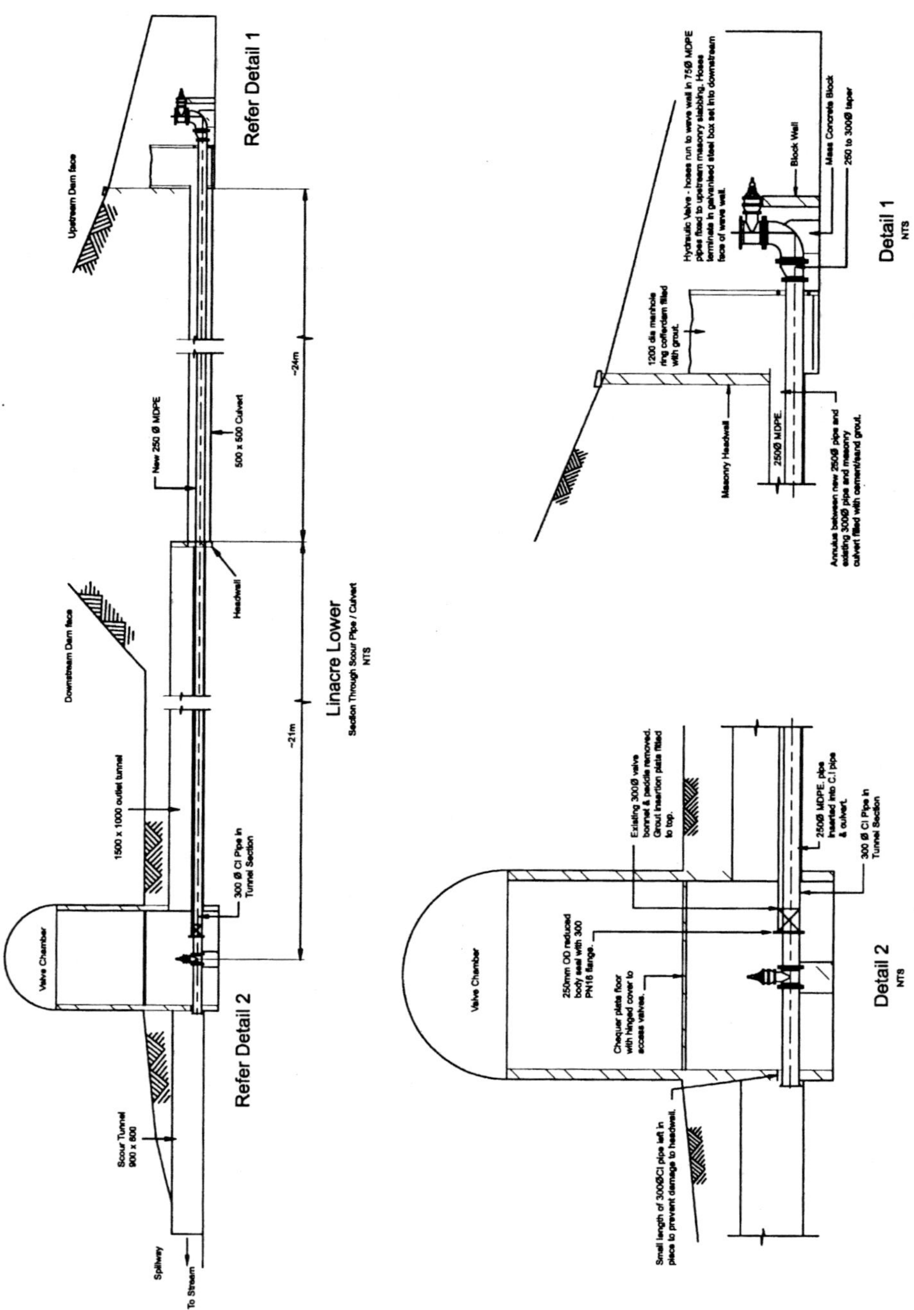

Figure 11 – Section through Linacre Lower Reservoir

Despite early pigging problems clearing the heavily encrusted 300mm diameter cast iron pipe, the polyethylene pipe was pulled through, tested and grout fill injected. Work was completed in early February 2006.

CONCLUSION

Some of the works carried out at the Howden, Derwent and Linacre reservoirs were in the 'interests of safety' and others general maintenance works.

At Howden and Derwent the 9no. 30 inch diameter, scour valves were stripped and refurbished without the need for complete replacement and therefore major costs.

At Linacre, like Howden and Derwent, the works including repairs to spillway walls and floors, restoration of a drainage facility, installation of a scour guard valve and re-lining of the Lower Reservoir scour pipeline were jointly progressed between STW, NWH and JB. Significant health and safety and environmental issues were addressed and overcome to complete the works.

ACKNOWLEDGEMENTS

The author would like to express thanks to Messers Neil Williams and Stewart Harwood of Severn Trent Water for their assistance during the course of the works and to Severn Trent Water for their permission to publish this paper.

The Management of Siltation at Hillsborough Dam, Tobago

D A BRUGGEMANN AND J D GOSDEN, Jacobs Babtie (formerly KBR)

SYNOPSIS

Hillsborough Dam is situated on the Hillsborough East River, 4 km north of the village of Mount St George on the island of Tobago and is owned and operated by Trinidad and Tobago Water and Sewerage Authority (WASA). The dam, commissioned in 1952, forms one of the main sources of water supply for the island of Tobago.

WASA have been concerned for a number of years that the volume of Hillsborough Dam was decreasing due to the deposition of sediment in the reservoir. In 2004 Inter-American Bank funding was obtained by the Government of Trinidad and Tobago to let a contract for the feasibility study and detailed design for desilting and rehabilitation works at Hillsborough Dam, Tobago. The project was carried out between January and August 2005.

The paper describes the investigations that were carried out to determine the extent of siltation, discusses the causes of siltation, describes methods proposed for desilting the reservoir, outlines the environmental legislation and discusses issues related to the disposal of silt. The paper concludes with proposed methods for future reservoir and catchment management.

INTRODUCTION

The island of Tobago is approximately 42km long and 12km wide at its greatest width. The island lies 32km north-east of the island of Trinidad and 120km north-east of Venezuela in South America.

Hillsborough Dam is situated on the Hillsborough East River, 4 km north of the village of Mount St George on the island of Tobago and is owned and operated by Trinidad and Tobago Water and Sewerage Authority (WASA). The dam forms one of the main sources of water supply for the island of Tobago. It is believed that first excavations began in 1944 and the dam was

Improvements in reservoir construction, operation and maintenance, Thomas Telford, London, 2006, 368–380

officially commissioned in 1952. The dam is about 18m high and comprises an earthfill embankment with a concrete core wall which was keyed into the foundations. The draw-works comprise a 450mm (18 inch) diameter draw-off pipe positioned on the upstream face of the dam, with draw-off valves at three elevations. A masonry cascade spillway (Figure 1) with a crest length of 30m is provided on the right abutment. The design capacity at maximum retention level is 1.03Mm^3 (about 227 million gallons).

Figure 1 – Masonry spillway in need of rehabilitation

The catchment is about 5km long with an area of 5.2km^2. The valley side slopes are generally steep and the average gradient of the river is about 12%. The average rainfall at the dam is just over 2,200mm per annum.

Soil erosion was known to be a problem in some parts of Tobago as a result of deforestation and poor agricultural practices. Initially, it was not believed to be a significant problem in the Hillsborough catchment because most of the catchment was covered by rainforest which has had forest reserve status for over 100 years with no development permitted since the commissioning of the dam.   Nevertheless WASA was concerned that storage at Hillsborough Dam was being lost as the result of siltation of the reservoir.

In 2004 Inter-American Bank funding was obtained by the Government of Trinidad and Tobago and a contract was let for the feasibility study and detailed design for desilting and rehabilitation works at Hillsborough Dam, Tobago. The work was carried out between January and August 2005.

GEOLOGY OF THE STUDY AREA

The island of Tobago is composed mainly of Mesozoic igneous and metamorphic rocks which evolved in an oceanic island arc. Hillsborough Dam is located in the Bacolet Formation which is principally composed of pyroclastic deposits comprising tuffs and agglomerates of labradorite and andesite with intercalated lava flows of similar composition. The catchment area is underlain by rocks of the mid-cretaceous plutonic suite comprising diorite-gabbro over the southern half and ultramafic rocks over most of the northern half. The geology changes from the Bacolet Formation to diorite-gabbro about halfway up the reservoir.

Brown et al (1965) noted that the soils (sandy loams) derived over the deeply weathered friable diorite erode readily leaving a thin cover of soil, or in some cases bedrock.

EXTENT OF SILTATION

The extent of siltation was difficult to determine because the datum used to derive the original Elevation – Volume curve was unknown, there was little published information on soil erosion studies for Tobago, bathymetric survey was hampered by dense vegetation (see Figure 2) which prevented access to the edge of the lake and the most recent aerial photography was more than 10 years old.

The first report which acknowledged that siltation might be taking place was prepared in 1996 (Howard Humphreys, 1996) for an inspection carried out following the principles of Section 10 of the UK's Reservoirs Act, 1975. The original reservoir Elevation – Volume curve was compared with a curve prepared from a hydrographic survey carried out in 1986. This suggested that there had been a significant reduction in the volume of water stored in the reservoir since commissioning in 1952. However, in 1996 it was recognised that the elevation datum used for each curve was different and therefore no firm conclusions were drawn at that time.

Figure 2 – View of lake shore showing dense vegetation

<u>Investigations</u>

In order overcome the difficulties identified above it was necessary to estimate the extent of siltation using several methods and these included the following:

- Desk study to evaluate existing data including published information on soil erosion in Trinidad and Tobago
- Bathymetric survey of the reservoir using the Medusa system
- Silt depth probing
- Reconnaissance of the lake and catchment
- Examination of turbidity measurements at the treatment plant

Information examined during the desk study included the original Elevation-Volume curve and a hydrographic survey carried out in 1986 which formed the basis of the Elevation – Volume curve included in the 1996 Inspection Report. These curves suggested that about 180,000m^3 of reservoir capacity had been lost to siltation over 36 years.

Little information has been published on soil erosion in Trinidad and Tobago. Ahmad and Breckner (1974) investigated the rate of erosion of three soils in Tobago without any vegetation cover i.e. bare soil, thus the data was not directly applicable to the forested catchment of Hillsborough

Dam.   A search on the internet, (TriniNetwork.com, 2000)) provided information on soil erosion from a study carried out in the Northern Range in Trinidad.   There was no indication of the source study, but the information indicates erosion yields between 50 and 5,000 tonnes/km^2/year for forested and bare soil respectively.

The data provided by Ahmad and Breckner suggested that the sediment yields for bare soils in Tobago could be twice those for the Northern Range in Trinidad. The sediment yield was estimated on the basis of a weighted average assuming 90% of the catchment was forested and 10% was bare soil on the basis of the Northern Range data for forest adjusted for Tobago soils. Using a silt density of 1.3tonnes/m^3, the weighted average silt volume was estimated to lie between 236,000m^3 and 307,000m^3 over 53 years.

A diving survey carried out in 2002 indicated silt depths of up to 2.4m (8 feet) thick. This survey estimated that there was about 85,000m^3 of readily dredgible silt.   This estimate was considerably less than the volume suggested by the shift in the Elevation – Storage curves but was based on relatively few silt depth measurements.

A bathymetric survey was carried out in February 2005 using a Global Positioning System in conjunction with the Medusa system. This system comprises a probe dragged over the floor of the reservoir gathering information on water pressure, background gamma radiation and reservoir bed roughness which are translated into water depth, chemical composition and physical characteristics.   Fifteen sediment samples were taken for chemical and physical testing to calibrate the Medusa readings.

The results of the bathymetric survey were used to build a digital terrain model (DTM) of the reservoir floor and derive an up to date Elevation – Volume curve. The Elevation – Volume curves determined over the life of the dam are shown in Figure 3 below which indicates an ongoing loss of storage volume.

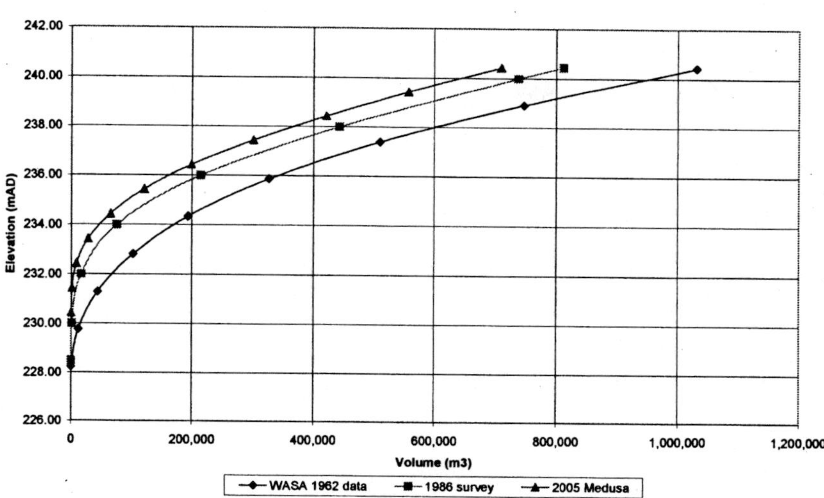

Figure 3 – Elevation – Volume Curves

The extent of siltation was also investigated by probing the depth of silt using a 19mm diameter pipe with a 100mm diameter foot plate. The pipe with foot plate attached was lowered from the boat until it rested on the surface of the silt and the depth of the water was recorded. The foot plate was then removed and the pipe lowered to the surface of the silt and then pushed into the silt until refusal. It was estimated that the pipe refused when the consistency of the silt was about 130kPa i.e. in the stiff range (BS5930). Estimates of the volume silt in the reservoir area up to the refusal depth were between 100,000m^3 and 150,000m^3.

The head of the reservoir where the Hillsborough East River discharges into the lake was accessed from the reservoir side. This visit revealed that the river outlet area was completely silted up. Silting up of the tributary river outlets was also observed. The silting up of the reservoir inlets also has an impact on bathymetric surveys as the reservoir surface area at top water level has decreased over time.

This behaviour is typical of reservoir siltation where silt deposition starts at the upstream end of the reservoir and migrates towards the dam. Additionally throughout the reservoir the extent of islands shown on the old mapping had also increased substantially through deposition of sediment in shallow water. The volume of silt deposited in these areas was estimated at about 60,000m^3. A typical silted up inlet is shown in Figure 4

Figure 4 – Typical silted up river inlet

In order to determine a reasonable estimate of the volume of silt to be dredged, the sediment yield was calculated for a range of methods as shown in the Table below.

Table 1 – Sediment Yield Estimates

Source	Volume (m^3)	Sediment Yield (tonne/km^2/year)
Silt volume, catchment yield estimate (published data)	115,000-220,000	550-1,050
Comparison of 1962 and 1986 Elevation – Volume Curves	218,000	1,600
Comparison 1962 and 2005 Elevation – Volume Curves	320,000	1,500
Silt volume, 2002 estimate	>85,000	>400
Silt volume (readily dredgible), 2005 estimate (current study)	160,000 – 210,000	750 – 1,000

Examination of the above Table shows that the comparison of the Elevation – Volume curves give relatively consistent sediment yields of 1,500 t/km^2/yr and 1,600 t/km^2/yr. These estimates are also of the same order as that estimated from the published data on soil erosion rates and the readily dredgible volume from the current investigation. On this basis the volume of silt deposited in the lake was estimated at 320,000m^3 from the Elevation – Volume curves. The silt depth probing calculations estimated a volume of between 160,000m^3 and 210,000m^3 of silt. The probing penetrated soft to stiff deposited silt and therefore it was postulated that there was about 110,000m^3 of very stiff to hard silt that was not penetrated by the probe. The presence of compact layer of sediment is credible as the reservoir is drawn down to minimum levels frequently and thus cyclical increased effective stresses in the silt deposits over 50 years could create a highly consolidated and thus compact layer of silt.

CAUSES OF SILTATION

A reconnaissance of the lake and catchment was carried out by boat and on foot. The reconnaissance revealed that the catchment was not completely undisturbed and the remnants of an old public road (Mt. St. George to Castara) were still present as well other tracks understood to have been constructed when limited timber extraction was permitted for a short period. Maps showed that the old public road ran along the whole length of the western edge of the catchment boundary.

Erosion features were observed along the route of the old public road and on some of the secondary tracks. Where the track has been cut into the side of the valley, numerous localised slope failures were noted in the face created by the cut. These slips were of significant size, often up to 4m long by 3m high. It was also noted that vegetation did not re-establish itself in the slipped material or on the exposed face and thus slope failures provide a constant source of silt. Other features such as localised erosion gullies were also observed in some place along the edge of the tracks.

The reconnaissance thus revealed significant catchment disturbance and there was anecdotal evidence that illegal logging continues to be carried out thus causing further degradation of the catchment.

Morris and Fan (1997) mention that dramatic variations in sediment yield may occur, even in small catchments, when subject to disturbance. They present data for a small (4ha) catchment which showed that watershed disturbance by logging activities can cause the ratio of the disturbed to the undisturbed catchment yield to increase from 2 for minimal disturbance to 550 for the mass erosion of haul roads. Therefore, the evidence of past logging activities and the old public road in the catchment of Hillsborough Dam substantiates estimated sediment yields for Hillsborough Dam well in excess of those given for a forested catchment in the Northern Range of Trinidad.

PROPOSED METHOD OF DESILTING

Design Criteria and Operational Constraints

In determining the most appropriate dredging techniques, the duration of the works and particular requirements of the works the following design criteria and operational constraints were applied:

- The reservoir is to remain at its normal operational levels, i.e. the level cannot be reduced to facilitate the dredging works;
- Excessive suspended solid levels are to be avoided in the vicinity of the intakes;
- Dredging works are to be undertaken during the rainy season over a maximum period of 6 months (24 weeks);
- Enabling works can take place during the dry season;
- The upper reaches of the reservoir where the sediments are at or above the Top Water Level (TWL) will need to dredged when the reservoir is full or within 200mm of TWL;
- Dredging level to be to a "clean" dredge to the original reservoir bed, a maximum depth of about 12m from top water level (240.44m AOD) to include for dredging below the invert of the bottom intake;
- The reservoir is to remain operational at all times during the dredging works;
- The gross volume to be dredged is assumed to be 320,000m^3 giving a net volume of 256,000m^3 allowing for 20% being un-recoverable;
- Dredging tolerance to be +0mm to −500mm

Dredging Methods

The relative merits of various inland dredging methods were examined in the context of Hillsborough Dam and a preferred method was then recommended. The methods examined included the following:

- Backhoe from pontoons
- Dragline
- Cutter suction
- Excavators and bulldozers in a dewatered reservoir

Backhoe dredging mounted on a pontoon was recommended as the most appropriate method as it has the following advantages:

- Backhoes are readily transportable to the site by road
- The method can cope with un-decomposed matter, such as branches, which is likely to be entrained in the sediment
- An appropriate backhoe will be able to cope with the range of sediment consistency likely to be encountered in the lake bed
- The method produces the driest sediment and so minimises the amount of water that would be removed from the reservoir and subsequently transported

Prior to dredging operations, the reservoir surface would be cleared of floating vegetation such as water grass, fallen trees and bamboo which was observed during the reconnaissance. To minimise truck movements and to make transport easier the green waste will be chipped/shredded prior to transport.

<u>Sediment Transport to the Disposal Area</u>

The transport of the wet dredgings from the dredger to the selected disposal sites involves transport from the dredger to the shore and then transport from the shore to the disposal areas.

Transport from the dredger to the shore can be carried out by pumping into a floating pipeline or by mud hoppers propelled by tug. At Hillsborough Dam the length of floating pipeline could be up to 1.2km. This option was rejected at an early stage because of the likelihood of high concentrations of un-decomposed vegetation in the sediment which could cause blockages.

For disposal from the shore to the disposal areas a combination of pumping to buffer lagoons and trucking from the buffer lagoons was favoured at first. However, it was learned subsequently that the proposed site for the buffer lagoons would not be available and an all trucked option was selected.

ENVIRONMENTAL LEGISLATION

As part of its commitment to developing a national strategy for sustainable development, the Government of the Republic of Trinidad and Tobago enacted the Environmental Management Action (Act No. 3 of 2000), which created the Environment Management Authority (EMA). This is an independent body governed by a ten-member multi-disciplinary board, appointed by the President of the Republic of Trinidad and Tobago. This body assumes sole responsibility for environmental management and protection of the natural resources of Trinidad and Tobago.

The goal of the National Environmental Policy is the conservation and wise use of the environment of Trinidad and Tobago to provide adequately for meeting the needs of present and future generations and enhancing the quality of life.

The Policy recognizes the linkages among the human resource, natural systems and development processes and the competition for use of the same resources by different interests. It offers a framework for the management and use of resources to yield the sustainable benefit for the population.

In order to carry out the desilting of Hillsborough Reservoir a Certificate of Environmental Clearance is required from the EMA, with the extent of environmental impact assessment required stipulated by the EMA.

SILT DISPOSAL ISSUES

The disposal of silt was problematical as there is little flat land on the island and in particular in close proximity to the dam site. Disposal at sea was rejected at an early stage owing to the presence of a marine reserve and potential impact on beaches and tourist facilities. Moreover, the potential for spreading silt on agricultural land was also limited as there is little large scale agriculture practised on the island today. The total area under agriculture was estimated at 5,872ha in the 1982 Agricultural Census of which 70% were considered to be small holdings averaging less 2ha.

A total of 14 potential sites were identified from the 1:10,000 mapping as having the potential to receive sediment. The distances from the dam site varied from 150m to 18km and the areas from about 1ha to 2 ha.

In order to determine the best sites, a ranking system was developed on the basis of the following factors; proximity to the dam site, access to the disposal area, environmental / social impacts, area available for disposal.

Each site was ranked on the basis of a rating for each of the above factors from 1 (unfavourable) to 5 (most favourable). The maximum score possible was 18 and sites which achieved a score greater than 50% (9) were considered as suitable for sediment disposal. Three sites achieved the threshold.

However, only one of the three sites was available for the disposal of silt owing to pre-existing agreements or the proximity to a public leisure facility. Thus only one site, about 5km from the dam site was available for disposal. This site is an existing solid waste landfill site and thus silt disposal would not contribute to further degradation of the area.

Arrangements for local farmers to collect silt for use on their small holdings are intended to be implemented.

PROPOSALS FOR FUTURE CATCHMENT MANAGEMENT

Siltation will continue to reduce the reservoir capacity. In order to minimise the need for regular dredging of sediment from the reservoir, action should be taken to reduce the quantity of silt entering the reservoir. Any measures that are implemented in the catchment are likely to require continual long-term maintenance if they are to remain effective in reducing the ingress of silt. If there is no commitment to the long-term upkeep of these measures then a regular dredging operation will be required and would be the most sustainable approach to maintaining storage capacity.

Potential measures to control the ingress of silt would include catchment management, control measures at the head of the reservoir at the inlets from the various streams and implementation of erosion control measures higher up in the catchment.

The primary catchment management measure would be to prevent vigorously the establishment of any new tracks in the forest. The ban on logging should also be enforced vigorously as this activity is the chief cause of establishment of new tracks and general disturbance of the catchment.

Following dredging of the inlets to the reservoir, gabion baskets could be installed across the inlets to act as silt traps. The areas behind the gabions will fill up with silt in time and the areas will need desilting from time to time if they are to remain effective.

Erosion control measures which should be implemented in the catchment would include:

- Stabilising the surface of the slipped mass at slope failures in cuttings
- Establishing vegetation on cut slopes and the scarp faces of old slips
- Providing effective drainage on the existing tracks and remnants of the Castara Road with appropriate silt traps
- Blocking off existing erosion gullies and installing check structures along these routes.

ACKNOWLEDGMENTS

The Authors would like to thank the Project Monitoring Team established by the Ministry of Agriculture, Land and Marine Resources for permission to publish this paper. Acknowledgment is also due to the Project

Implementation Unit of WASA, Tobago Services for their invaluable assistance during the execution of the project. The authors would also like to thank Alpha Engineering and Design (2002) Ltd., without whose support the project could not have been carried out, Land and Water Remediation Ltd for their contribution to the evaluation of siltation and dredging methods, and Planning and Advisory Consulting Services Ltd for their advice on environmental issues and preparation of the application for the Certificate of Environmental Clearance.

REFERENCES

Brown C B, Hansell J R F, Hill I D, Stark J and Smith G W (1965) Land Capability Survey of Trinidad and Tobago, No. 1, Tobago

Howard Humphreys and Partners Ltd. (1996), Inspection of Hillsborough Dam, Report prepared for the Water and Sewerage Authority, October

Ahmad M and Breckner E (1974) Soil Erosion on Three Tobago Soils, Tropical Agriculture (Trinidad), Vol. 51, No. 2 April, pp 313 – 324

TriniNetwork.com, Trinidad and Tobago - Facts and Figures, November 2000

BS 5930 (1999) Code Practice for Site Investigation, British Standards Institute.

Morris G L and Fan J (1997), Reservoir Sedimentation Handbook, McGraw – Hill, pp 7.4 – 7.5

Desiccation Assessment in Puddle Clay Cores

A. KILBY, Damwatch Services Ltd., NZ (previously Thames Water., UK.)
A. RIDLEY, Geotechnical Observations Ltd., UK.

SYNOPSIS.   The embankment dams to Banbury Reservoir and Lockwood have a history of high level leakage, clay core repairs and TWL restrictions. Current TWL restrictions posed a risk of damage to the kneeler beam of Banbury Reservoir during storms, whilst on Lockwood Reservoir undermining of the kneeler beam and associated slabing due to exposure to wave action was actively occurring. In seeking relaxation of the restrictions to mitigate damage, and to recover potential storage capacity, Thames Water were requested by the AR Panel Engineer to complete a desiccation assessment. This paper describes the principles and techniques adopted for the desiccation assessment of the clay cores, including laboratory testing of high quality samples and the installation and monitoring of two arrays of GeO flushable piezometers. Visual inspection did not identify desiccation cracks within the cores, although there is evidence that the cores have previously been desiccated to greater depth. The potential of desiccation processes is highlighted, with the monitoring of pore pressures within the clay cores demonstrating the seasonal activity and depth within the clays cores to which suctions can be experienced.

INTRODUCTION

Top Water Level (TWL) restrictions to Banbury Reservoir and Lockwood Reservoir posed a continued risk of damage to the internal slabing of Banbury Reservoir during storms, whilst on Lockwood Reservoir undermining of the slabing due to exposure to wave action was actively occurring. In 2003 Thames Water sought the relaxation of the restrictions to mitigate the risk and with the added benefit of recovering raw water storage within the Lee Valley. A desiccation assessment was requested by the AR Panel Engineer in advance of any trial raising of the reservoir levels. The initial desiccation assessment was undertaken in March 2004, with monitoring of soil suctions at a selected site on each reservoir continuing until May 2005.

Improvements in reservoir construction, operation and maintenance, Thomas Telford, London, 2006, 381–390

RESERVOIR DESCRIPTION

Inaugurated in 1903 Banbury Reservoir and Lockwood Reservoir are both non-impounding storage reservoirs formed by continuous earth embankments, with a central puddle clay core keyed into the underlying London Clay Formation. Located within the flood plain of the Lee Valley the reservoirs were founded on "soft" Alluvial soils. The puddle clay cores were constructed using the underlying alluvial clays of very high to extremely high plasticity. The shoulders or "Filling" comprises a mixture "clayey Gravels" and "gravely Clays", Figure 1. Both reservoirs experienced considerable settlement, up to some 10% of the core height above surrounding ground level in the 40 to 50 years post construction.

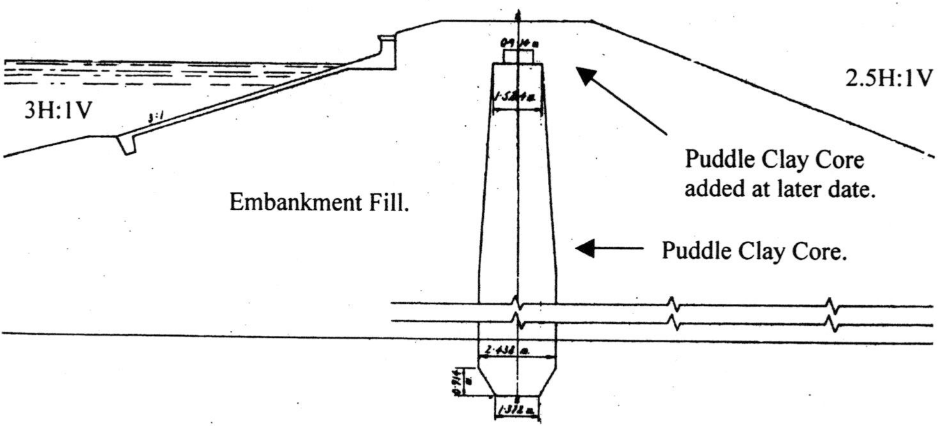

Figure 1: Representative Embankment Cross Section of Lockwood Reservoir

During the Second World War the reservoirs were kept at a lower water level, Lockwood requiring repairs at two locations due to bomb damage. It is suspected that during this period the top of the puddle clay cores may have "dried out" to the extent that leakage occurred during refilling. Raising of the clay cores was undertaken on Lockwood Reservoir during 1943 to 1945 and on Banbury Reservoir between 1957 to 1958. No specific information has been found with respect to the selection of the material for raising the cores. Whilst it is anticipated that a key with the existing core would have been provided (as the case for the raising of the clay core to the Warwick Reservoirs undertaken during a similar period) this is not shown on record drawings.

Occurrences of leakage continued throughout the 1960's to the 1980's with a series of investigations undertaken and TWL restrictions applied. A number of remedial works were subsequently completed during the 1970's to 1980's including asbestos sheet piling, grouting and core remoulding (Ray and Bulmer, 1984.)

THE ASSESSMENT OF DESICCATION

A soil is desiccated when it has either (i) soil suction in excess of that, which would be expected or (ii) a moisture deficit.

<u>Soil Suction</u>

Traditional assessments of desiccation make use of suction measurements to identify its presence. The stresses on an element of soil in the ground are made up of vertical (σ_v) and horizontal (σ_h) total stresses and the pore water pressure (u) at the depth of the element.

$$p' = (\sigma_v + 2\sigma_h) / 3 - u$$

When an element of soil is removed from the ground the total stresses are reduced to zero and the pore water pressure becomes negative (i.e a suction). If the sampling is perfect and the sample is truly undisturbed the soil suction will be equal to the effective stress in the ground (p'). If the soil is desiccated the measured suction will be greater than the expected in situ effective stress. This is evident as a bulge in the soil suction/depth plot (Figure 2).

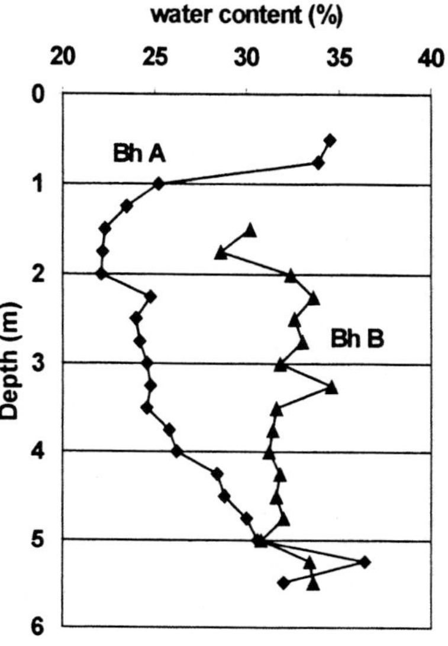

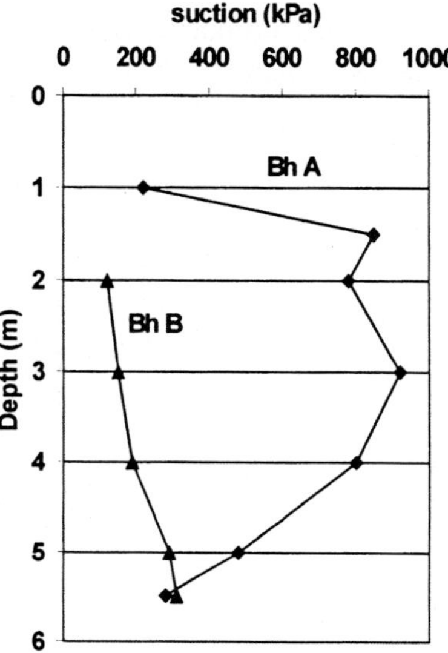

Figure 2: Typical profiles of water content and suction in a desiccated location (A) and a normal location (B), after BRE (1996.)

In an embankment the initial stress condition on an element of soil in the ground is complicated by the fact that it lies above the natural ground surface and above the natural water table. Therefore there are no significant horizontal stresses on the soil, particularly at shallow depths. Moreover the compaction, which occurred when the core of the embankment was constructed, would have introduced an inherent suction into the clay. The magnitude of the suction measured now may provide an indication of desiccation, whether current or historical.

<u>Moisture Deficiency</u>
Soil moisture deficit (in mm) is defined at the amount of water per unit surface area, which the soil surface will absorb before further precipitation cannot be stored in the profile (i.e the soil has reached it's field capacity, although it is not necessarily saturated in this state). Volumetric water content (θ) represents the volume of water in an element of soil and is defined as follows:

$$\theta = (S_r.e) / (1+e)$$

Where, S_r is the degree of saturation and e is the void ratio.(Note: In a saturated soil the volumetric water content is equal to the porosity, n)

The moisture deficit for an element of soil is the difference between the volumetric water content at field capacity and the desiccated volumetric water content. The soil moisture deficit is measured over the whole profile and is the difference between the volumetric water content profile at the field capacity and the desiccated volumetric water content profile, Figure 3.

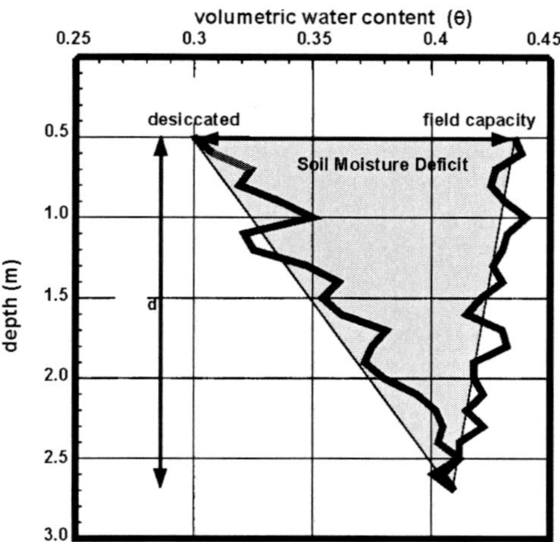

Figure 3: Illustration of Soil Moisture Deficit in the ground for a desiccated profile and a profile at field capacity.

Void ratio and degree of saturation are mutually dependent. Therefore in the absence of direct measurements of volumetric water content it is acceptable to represent moisture deficit in terms of the degree of saturation estimated from the measurements of the bulk density, water content and specific gravity (G_s).

$$\gamma_{bulk} = \cdot\, \gamma_{water} \cdot (G_s + S_r.e) / (1+e)$$

$$S_r = (w.G_s) / [G_s(1+w).(\gamma_{water}/\gamma_{bulk})-1]$$

FIELD SAMPLING AND LABORATORY TESTING
Samples of the clay core were retrieved using thin walled "Shelby" sample tubes. To improve recovery and reduce the amount of disturbance each 1m tube was only pushed 0.5m into the clay core. Samples were extruded using a hydraulic jack with a leveling platform, with the direction of extrusion consistent with the sampling direction.

Standard laboratory testing was undertaken including;

- Gravimetric Water Content
- Atterberg Limits
- Bulk Density
- Suction

In addition the degree of saturation was inferred from the measurements of bulk density and water content, using assumed values of specific gravity. Suctions were measured using suction probes adopting the technique presented by Ridley et al (2003).

The results of the laboratory testing from the samples obtained at Chainage 600 on Lockwood Reservoir are shown in Figure 4 (a to d). The results for bulk density, degree of saturation and water content all indicate that there has been some desiccation at an elevation of 13m to 13.5m. This is consistent with the bulge in the graph of suction verses depth and coincident with the restricted top water level. It should be noted that the suction measurements provide only a representation of the suction at a point in time. The investigations were undertaken in March 2004 and the suctions are only considered representative of end of winter.

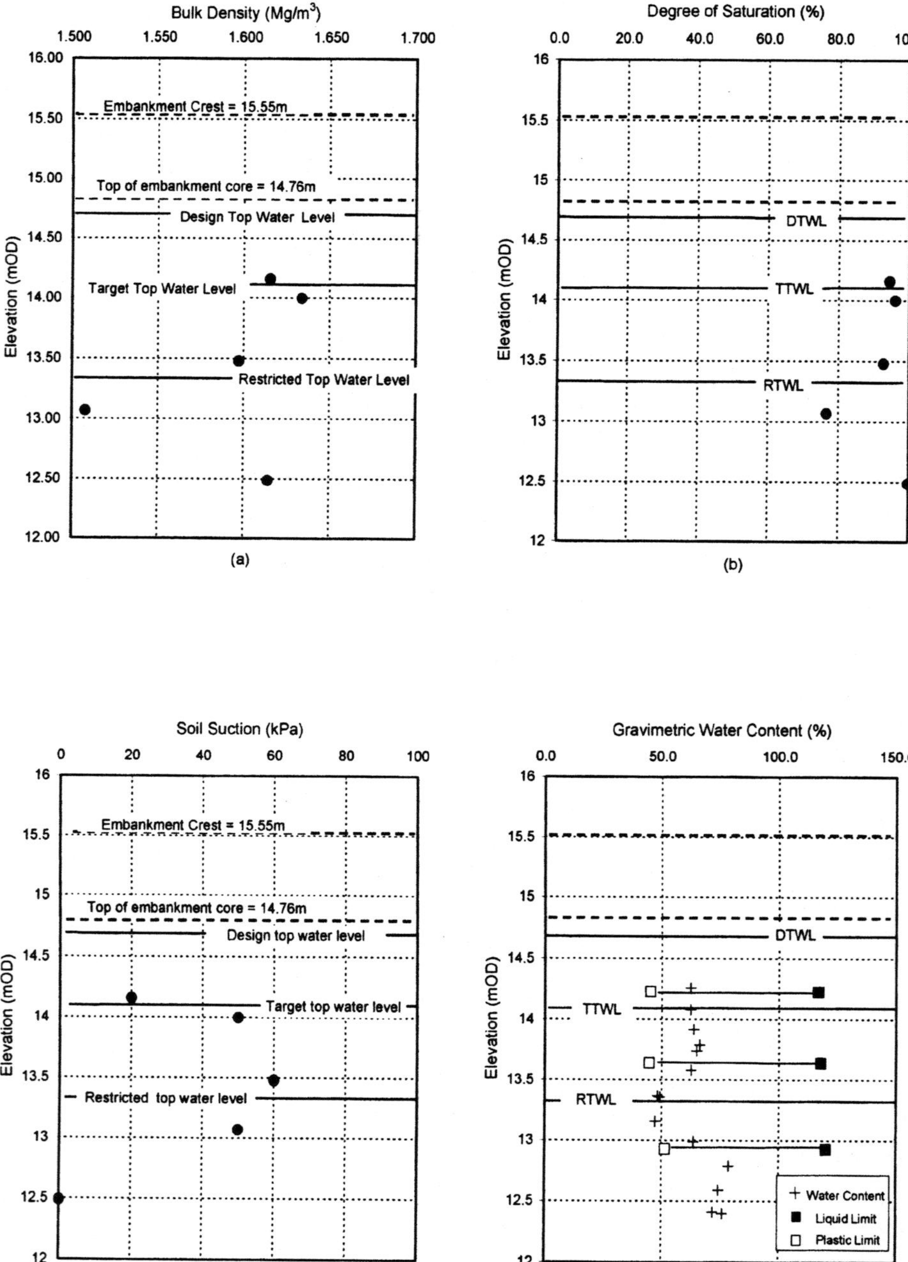

Figure 4: Laboratory Test Results on samples of clay core from Chainage 600 on Lockwood Reservoir (a) Bulk Density, (b) Degree of Saturation, (c) Soil Suction and (d) Gravimetric Water Content.

SUCTION MONITORING

With the suction measurements undertaken on the samples only providing a reference of the end of winter conditions within the clay core GeO flushable piezometer's were installed to investigate the seasonal behavior of suctions within the clay core. Although GeO piezometers are widely used in investigating embankments and slopes, this was the first application of the instruments within the clay core to an embankment dam.

GeO Piezometers were installed at Chainage 600 on Lockwood Reservoir at three depths;

- between the Design TWL (14.69mOD) and Target TWL (14.09mOD),
- between the Target TWL (14.09mOD) and Restricted TWL (13.37mOD),
- below the current Restricted TWL (13.37mOD),

This is shown on Figure 5. A similar array was also installed on Banbury Reservoir at Chainage 300. The monitoring period for the GeO piezometers ran from April 2004 through to May 2005, during which period the reservoir level's in both Lockwood Reservoir and Banbury Reservoir was maintained close to the restricted TWL's with no significant operational drawdown experienced. The results of the monitoring are shown in Figure 6 (a to c).

At all depths the GeO Piezometers show strong seasonal response, with suctions reaching a maximum at the end of summer. The response from the GeO Piezometers at 1.04m and 1.64m depth recorded suctions in excess of 90kPa. The GeO Piezometer at a depth of 2.48m, approximately 0.7m below the current Restricted TWL, approached a suction of 80kPa.

Although not presented herein the suctions of the array of GeO Piezometers installed at chainage 300m on Banbury Reservoir showed a similar seasonal response. Suctions approached 80kPa above the Restricted TWL and 30kPa below the Restricted TWL.

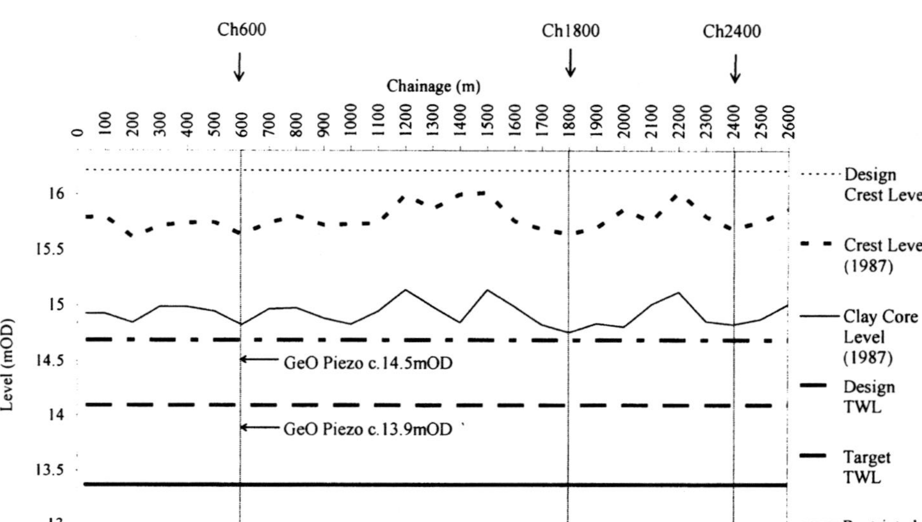

Figure 5: Long Section for Lockwood Reservoir, showing TWL, Clay Core Level and locations of the GeO Piezo's.

DISCUSSION

There is an obvious link between suction and the development of desiccation cracks. Visual inspection of cores retrieved during the investigations considered that there had been previous desiccation and this is consistent with the historical observations of leakage and remedial works.

The monitoring record from the GeO Piezometers has provided confirmation of the seasonal behavior as might be expected and has also provided an initial indication of the magnitude of the suctions and depths to which they may be experienced. Whilst monitoring results will vary from year to year, further research would be required to establish;

- to what depth are the seasonal variations in the core effective
- how much above 100kPa were the suctions experienced

The latter may be addressed by the measurement of suctions from undisturbed samples taken at different seasonal periods, i.e. at the end of winter and at the end of summer.

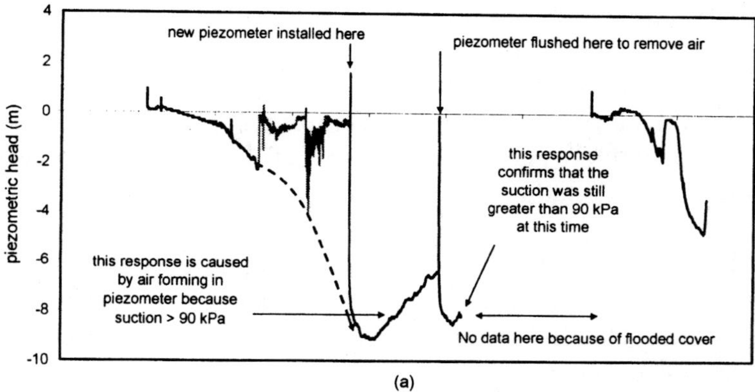

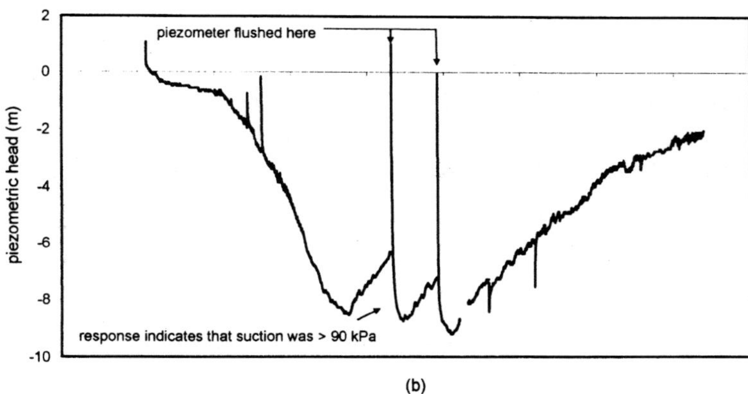

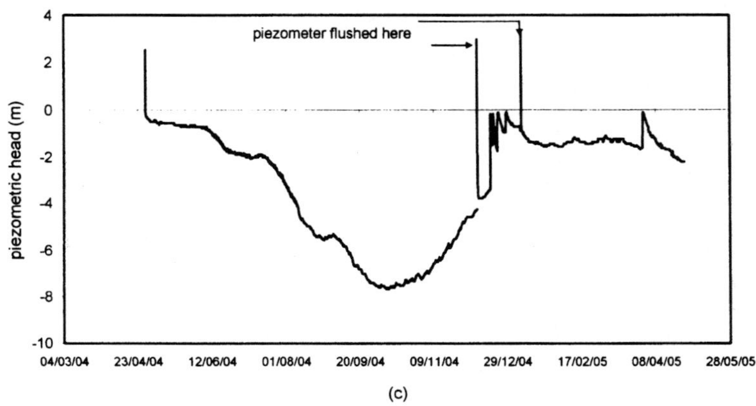

Figure 6: GeO Piezo Suction Measurements at Chainage 600 on Lockwood Reservoir (a) depth 1.04m between Design TWL and Target TWL, (b) depth 1.64m between Target TWL and Restricted TWL, and (c) depth 2.48m below Restricted TWL.

Whilst at Banbury Reservoir and Lockwood Reservoir there were no physical indications of cracking of the clay core, as would result in high level leakage, there is clearly the potential for cracking to occur as a result of desiccation. It is therefore reasonable to anticipate that high level leakage could occur in the future, particularly upon refilling after reservoir drawdown and / or during drought event years.

Owners need to be aware of this mechanism, with high plasticity clays more likely to suffer desiccation and potential settlement, and to recognize that increased surveillance is clearly very important during such periods. Whilst high level leakage occurring from desiccation may not necessarily be a threat to the safety of a dam it will restrict operational performance until the leakage has been addressed.

ACKNOWLEDGEMENTS
The Authors would like to thank Jon Green and Dr Andy Hughes for their support of the presented works. The data is presented with the kind permission of Thames Water Utilities Ltd.

REFERENCES

BRE, 1996. *Digest 412 Desiccation in Clay Soils.* Building Research Establishment Watford.

Marsland, F., Ridley, A.M., Vaughan,P.R. & McGinnity. *"Understanding Vegetation and its influence on the stability of slopes."* Proc. 2nd International Conference on Unsaturated Soils, Beijing, China. pp , Vol.1., pp 249-254. International Academic Publishers, 1998.

Ray, W.J.F. & Bulmer,T., 1982. *"Remedial Works to Puddle Clay Cores."* Proceedings of BNCOLD 1982 Conference, University of Keele, September 1982, pp.27-44.

Ridley A.M., Dineen K., Burland J.B., and Vaughan P.R. (2003*) Soil Matrix Suction – Some examples of its measurement and application in geotechnical engineering.* Géotechnique 53, No. 2, pp 241-253.

The failure of the Mostiště embankment dam

JAROMÍR ŘÍHA, Brno University of Technology, Czech Republic
JIŘÍ ŠVANCARA, AQUATIS a.s., Consulting engineers, Czech Republic

---

SYNOPSIS.   The Mostiste Dam on the Oslava River is a rockfill dam with a relatively thin inclined impervious core. The maximum dam height is 29 m and its crest length is 292 m. The Mostiste Dam was completed in 1960 and at that time it was the first compacted rockfill dam in the Czech Republic (CZ). Leakage first occurred in 1996 in the grouting gallery which indicated that its impervious core sealing might have been impaired. The results of follow up surveys carried out up to the end of 2004 demonstrated the progressive worsening of the problem. As a result, the reservoir level has been significantly lowered since the end of 2004. In May 2005, the governor of the region proclaimed a state of emergency to protect the lives and property of inhabitants downstream from the dam.

The Mostiste reservoir normally serves as a source of drinking water for more than 70 000 inhabitants. Therefore, the remedial works concept design had to respect the requirement that water supply remained uninterrupted without any dramatic reduction.

This paper provides an analysis of the failure's initiation and gives information about the remedial measures accepted. The extra-operational procedure - the gradual filling of the reservoir accompanied by careful monitoring - should be carried out during the winter and spring of 2006 depending on hydrological conditions. The authors hope to be able to summarize the results of this test operation during the oral presentation at the conference.

BASIC INFORMATION ABOUT THE DAM

The Mostiště Dam on the Oslava River has served for more than 40 years as the water supply source for waterworks at the cities of Velké Meziříčí and Třebíč. Normally, the reservoir supplies the region with drinking water.

The dam is located on the Oslava River (Fig. 1), its backwater length is 5.385 km and the catchment area is about 223 km^2.

The dam was finished in 1960 as the first compacted rockfill dam in the Czech Republic. The dam sealing is made of a thin inclined loess core. The maximum dam height is 29 m and its crest length is 292 m.

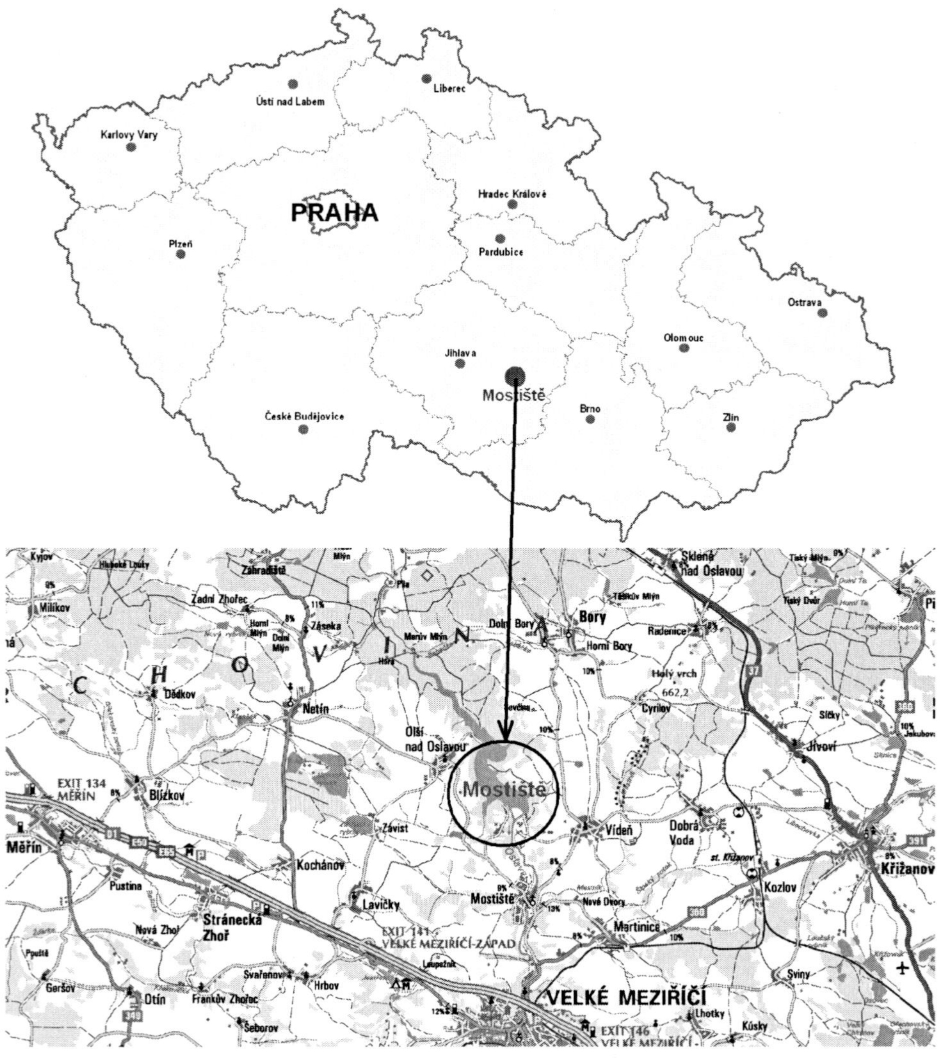

Figure 1: A map of the Czech Republic with the Mostiště Dam

The reservoir volume division corresponds to the main purpose of the scheme – the supply of drinking water. The flood protection effect of the dam is relatively small; for bigger flood wave volumes it is practically negligible.

Figure 2: Aerial view of the dam

Total reservoir volume	11.9373 M m^3
Uncontrollable flood storage	0.9437 M m^3
Controllable flood storage	0.6094 M m^3
Live storage	9.3389 M m^3
Permanent storage	1.0453 M m^3

The dam site is located upon a moldanubium granite massif, and the dam sub-base is composed mostly of high quality granite rock.

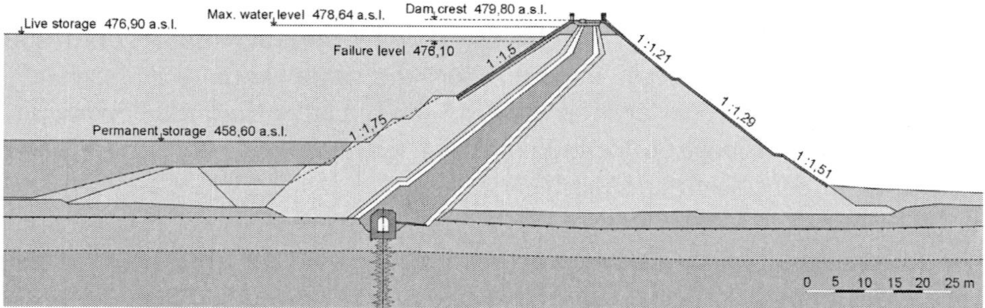

Figure 3: Typical cross section of the dam

The dam is composed of compacted quarry-rock fill with a thin loess soil core (Fig. 3). In the upper part the core is vertical; in the lower portion it is inclined close to the upstream face. The dam core is keyed to the sub-base by a grouting gallery. The sub-base was grouted along a single line of grout holes.

According to the dam safety classification system in the Czech Republic the Mostiště Dam is registered as category I, i.e. one of the most carefully observed hydraulic schemes. There are 26 such dams in the Czech Republic.

When compared with other rockfill dams in the Czech Republic, relatively small vertical and horizontal displacements have been observed at the dam during its service life. However, the set of critical comments addressed to the Mostiště Dam design concept deals with the recent state of the art in large dam engineering when compared with the state of knowledge in the 1950's. The dam has very steep slopes from 1:1.2 to 1: 1.7, meaning the reserve in dam stability is minimal. The impervious core is relatively thin; protective zones (filters) are made of crushed stone with gradation, not assuring resistance against contact suffosion. Moreover, the additional survey gave rise to doubts about the quality of the filters, their real extent and functionality. The transition zones between the "stiff" rock fill and the impervious core are completely missing. The dam was not equipped with any drainage system, which is a common feature in dams built in later periods. There is only one bypass bottom outlet (the recent Czech standards require at least 2 bottom outlets) and a small hydropower penstock.

THE DAM'S FAILURE AND ITS PROGRESSION

Technical dam safety inspections and surveillance indicated increased seepage through the dam body since 1996. The defects could not be localised due to the lack of a drainage and seepage observation system. The uncertainties over future dam behaviour called for additional surveys, focused firstly on the probable locations of potential defects. The careful investigation of records from the building site during construction showed that the dam material used was very heterogeneous, and at some locations quite improper (wooden logs, steel rods, etc.). During the completion of the dam body, suitable loess material for the dam core was probably unavailable, and its upper portion contains sandy layers from the borrow pit. Moreover, the compaction technology used was not sufficiently developed at the time of construction in the 1950's.

The failure was assessed as being a complete hydraulic fracturing of the impervious core in at least two places above the level of 476.00 m above SWL. Further on, a geological survey indicated the potential for future damage to the thin vertical part of the core at any point in the dam. Nevertheless, the systematic location of hazardous places was not possible. At the same time, the geological survey showed that the impervious core thickness varies at its top significantly, and its level does not fit the requirements of present dam safety standards.

The failure was triggered by a combination of unfavourable factors. The most significant factor was the use of heterogeneous core materials, locally improper for the sealing zone, and also implicated was poor compaction of the impervious core. This was caused by the lack of suitable material at the end of construction, and the use of poor compacting equipment (ramming plate). One of the accompanying effects was the varying levels of moisture of the core material. Moreover, at some places transition zones (filters) are missing or do not fit non-suffosion criteria.

The complex analysis was carried out with the use of FEM stability analysis. The results of the modelling documented the significance of
- unsuitable construction methods, namely the non-uniform filling of the embankment body in the cross section,
- the development of differential settlement in the core and shoulders followed by "an arch effect" and corresponding stress redistribution. The accompanying effects of such a failure are a decrease in vertical stresses in the core, water penetration to the joints, and a successive increase of soil moisture at the affected zones.

It is evident that the failure development has been gradual and over a long period of time. The total hydraulic fracturing was undoubtedly preceded by a gradual increase in seepage through the core. A more accurate assessment of the reasons for the failure is however quite difficult due to a lack of reliable data.

The geological survey of the dam and thus additional knowledge obtained showed the necessity for the complete and systematic repair of the dam sealing.

FOLLOWING ACTIVITIES AND PROVISIONS
At the end of the year 2004 the findings mentioned above led to the statement that the dam did not fulfil elementary safety requirements and could not be operated without immediate measures being taken. The main goal of the dam owner was to avoid increasing risk for the inhabited area downstream of the dam. Based on the modified manipulation order, the water level in the reservoir was lowered by 13 metres. This enabled the routing of a 20-year flood through the reservoir without exceeding the

critical water level of 476.00 m above SWL. Following operational decisions and proposed conditions for the remedial works a compromise was reached between contradictory demands for the maintenance of a tolerable level of risk in order to obtain increased dam safety, and for ensuring a continuous drinking water supply. If the summer months were dry the insufficient water storage capacity and impaired water quality could complicate raw water intake from the reservoir.

With respect to the extreme snow cover in the catchment area during spring 2005, a 7 m long section of side spillway wall was temporarily removed (Fig. 4). This part of the wall was blasted out in its entire height to release an additional discharge of 14 m^3/s without exceeding the critical water level in the reservoir.

Figure 4: Emergency spillway - partial removal of spillway wall

To partially increase dam safety, further provisional repairs were carried out at the evidently damaged parts of the dam sealing. These activities consisted of filling caverns in the core using low-pressure grouting and sand jetting into the drill holes directed to the transition zones (filters). These measures temporarily reduced the risk of dam failure to a tolerable level.

Knowledge obtained from the provisional repair contributed to a more reliable description of the state of the dam body:
- The real geometrical shape of the impervious core does not correspond to that given in the original project documentation,
- The transition zones are locally missing or are of insufficient thickness and improper grain size, which does not protect the impervious core against contact suffosion.

To precisely map the extent of the potential dam-break floodplain, dam breach and flood routing modelling was used. The calculations considered the mechanism of dam failure to consist initially of hydraulic fracturing, piping and subsequent dam breaching at the documented points where weakened zones of the core lie. This analysis was the basis for the preparation of an early-warning system in the area below the dam.

In order to accelerate the tender and contractor selection process, the governor of the region proclaimed a state of emergency to protect the lives and property of the inhabitants living downstream from the dam according to the Emergency law. This enabled immediate design activities and the preparation of tender documentation with one principle target - to finish all necessary remedial works to increase the safety of the dam by the end of 2005.

THE DAM REPAIR DESIGN CONCEPT

The design concept for the remedial works was approved by the technical committee in March 2005. From that time on, intensive design and consultation work was carried out to prepare project documentation for the construction work. Conceptually, the remedial works at the Mostiště Dam were divided into two stages:

- 1st stage – urgent safeguarding work averting the emergency state, consisting of
  - reconstruction of the dam sealing - impervious core,
  - a system improving technical safety monitoring,
  - reconstruction of the water supply main,
  - additional grouting of bedrock.
- 2nd stage – reconstruction of the dam, which comprised
  - arrangement and remedial works at the dam crest,
  - verification and completion of a monitoring system.

The 1st stage included repair of the dam core over the entire extent of the dam's length, and improvement and completion of technical safety instrumentation to enable verification of the results of reconstruction. This stage was required to be finished before the end of 2005 to enable the filling of the reservoir, and the provision of a continuous raw water supply to the waterworks and distribution network as soon as possible. However, the requirements of dam safety-related standards would be fully fulfilled after the completion of the 2nd stage of the remedial works.

Reconstruction of the impervious core

The main goal of the works was the installation of a reliable sealing element in the core, which significantly decreased its permeability in damaged and degraded portions. The raising of the impervious core crest to the standard level is also a necessary condition for remedial activities planned in the 2nd stage of the remedial works.

The repair consists of the development of a continuous inclined sealing element (seepage barrier) along the entire length of the dam (292 m) to the depth of 7 m (to the level of 473.00 m above SWL - Baltic system), close to the upstream face of the impervious core (see Fig. 5). As the sealing membrane could not be vertical (due to the inclined part of the core), it was decided to use the technology of jet grouting from the dam crest (see below). The total area of the seepage barrier was about 1700 m^2.

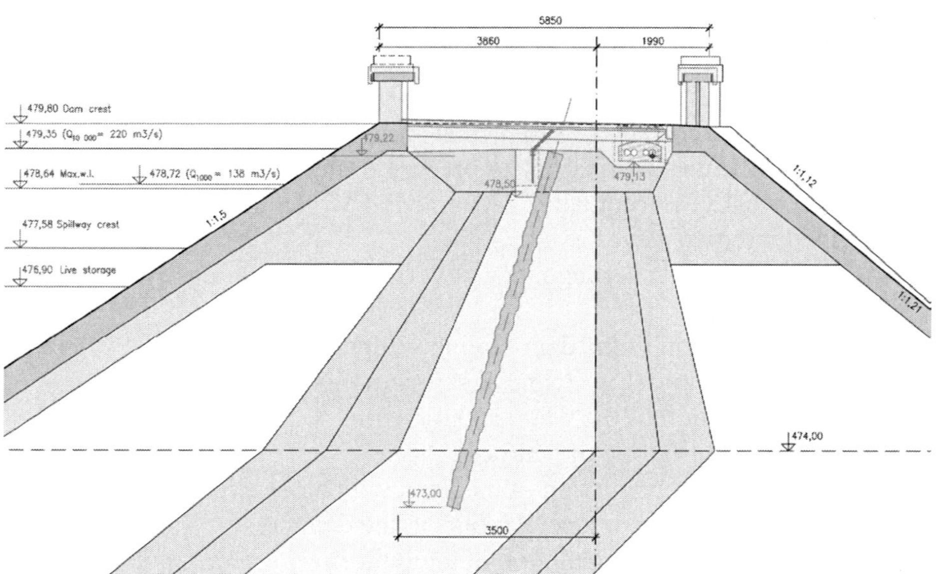

Figure 5: The sealing element in the dam core made by jet grouting technology

The system of technical safety monitoring

In the 1950's, during the construction of the dam, no drain was built at the downstream toe. Because of this, it has not been possible to reliably identify seepages and thus localise eventual failure zones. To improve this situation a new drainage system has been designed and installed at the downstream toe of the dam. This will stabilise the groundwater level and enable sensitive pressure, groundwater level and seepage measurements. The new drain is subdivided into sections to enable the spatial identification of possible increased seepages in potential failure zones. Moreover, the right bank abutment has been drained into the grouting gallery. Two automatic gauges have also been set up inside the grouting gallery to measure leakage. After the revision of the system of observation boreholes in the grouting gallery, it was decided to replace them with new ones, and the system was extended and automated. The newly installed devices give much accurate and frequently measured data, which provides the dam operator with better information as a basis for decision making within the early warning system. The automated system contains both outer and inner observation boreholes. For remote data collection new cable lines were installed in the grouting gallery and outside of the dam body. A conceptually new monitoring system will be gradually integrated into the system serving for the operation of the scheme.

<u>The water supply main</u>
The 45 year old water main taking water from the reservoir to the waterworks interfered with the newly designed toe drain. Therefore, the pipeline was preventively replaced by a new one.

<u>Additional grouting</u>
The grout curtain was finished at the beginning of the 1960s. Due to its age, it is showing local degradation, and so new grouting of the most exposed bedrock was performed as a part of the activities at the 1st stage of the reconstruction scheme. The locally damaged parts of the concrete structure of the grouting gallery (caused by the concentrated leakages) were repaired. In the right-bank part of the gallery along the downstream side, four separately controlled sections were developed by local grouting of the back side of the wall. The "closed" sections created in this way were drained to the gallery and thus enabled localisation of potential increased seepages through the dam's impervious core.

Recently, the 1st stage of the reconstruction was finished, and the trial filling procedure was started at the beginning of December 2005. During this procedure the detailed and frequent observation of dam behaviour will be carried out using the newly installed monitoring system.

The 2nd stage of the reconstruction is planned for the year 2006. It comprises dam crest repairs and the completion of the monitoring system. These arrangements will place the scheme in accordance with current standards and enable adequate control over any uncertain factors specified during the preparation works.

THE REPAIR OF THE IMPERVIOUS CORE

<u>Peculiarities of the impervious core repair process</u>
Various technical methods were originally presupposed when choosing an appropriate reconstruction method. A range of international experience from ICOLD documentation was reviewed and analysed. Finally, three possible methods were assessed in more detail, namely:
- Removal of the upper portion of the dam and replacing the impervious core with appropriate material.
- Construction of a standard vertical slurry wall.
- An inclined curtain constructed using jet grouting technology.

Finally, the relatively "young" method of jet grouting was selected using the formalised multi-criterial optimisation method. It must be stated that this method is generally used for the improvement of subsoil properties (namely bearing capacity) and that its use to such an extent in the repair of an impervious core was probably a world first.

In the impervious core, the technology of jet grouting could not be applied in the usual manner. Besides a sufficient level of impermeability, the deformation characteristics had to follow the properties of the original core material to assure the appropriate deformation behaviour of the resulting improved sealing element. Thus, the target was not a high degree of strength and stiffness for the new element, but rather long-term deformation properties fitting the deformations of the original dam core.

The whole reconstruction process has been influenced by the low stability reserves of the dam, where the safety factor along critical failure surfaces reached values of about SF = 1.275. As it was feared that the jet grouting technology could impair the stability of the dam, the detailed 3D finite element model was set up to simulate the sealing element construction process. The results of the numerical solution served as a recipe for the technological procedure, which was based on the gradual development of the seepage barrier in a set of sequences, and a permanent check on strength acquirement in individual runs. The aim was to avoid worsening the stability conditions of the dam during construction.

As there is only limited experience with jet grouting technology being applied to the impervious cores of dams, the contractor was obliged to prove the method before the start of reconstruction at the test field in the right abutment. The test showed that the technology is vital in the creation of a core-sealing curtain of the required parameters.

Experience with impervious core repair using jet grouting technology

Jet grouting technology was applied in five sequences with reference to the restricting requirements based on the results of stability calculations (Fig. 6).

Figure 6: Procedure of jet grouting in five sequences

Furthermore, the technology had to avoid damaging the impervious core, so the grouting boreholes were inclined by 16 degrees in the direction of the inclined core so as to be sufficiently far from its faces. The procedure was based on the modified single jet method proposed by Soletanche.

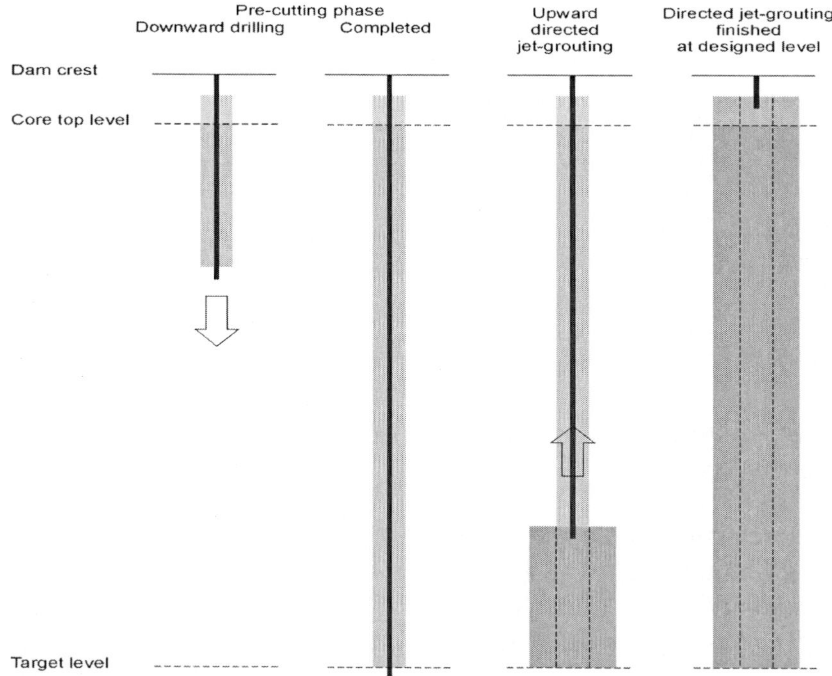

Figure 7: The scheme of jet grouting

The grouting in individual boreholes was carried out in two steps. The first one was the pre-cutting phase, which was done in a downward direction. Pre-cutting with the grouting suspension during drilling ensured a pillar diameter of approx. 0.3 m to the depth of 7 m from the dam crest. The second - upward phase was performed by directed jetting, creating butterfly–shaped lamellae with 0.4 m long "wings" in the direction of the longitudinal dam axis.

The grouting procedure was carefully monitored and maintained so as not to exceed maximum prescribed displacements of the dam crest. Those were originally assumed to be the values of 4 mm vertically and 4 mm horizontally (widening of the dam crest).

Unfortunately, the prescribed displacements were not gained during construction, and the measured deformations, both vertical and horizontal, exceeded 45 mm in some observation profiles (Fig. 8). The following stability analysis indicated higher pressures in the drills, which in combination with weakened zones in the dam core were the reason for the relatively high displacements. The additional calculations showed that the safety factor of the repaired dam could, at certain locations, have been slightly decreased to the value SF = 1.25!

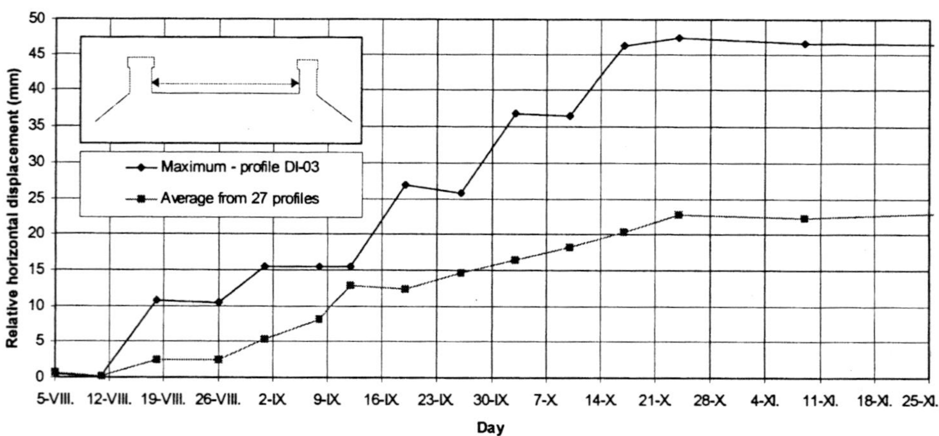

Figure 8: The development of horizontal displacements of the dam crest in individual sequences

Therefore, the recommendations for the trial reservoir filling procedure and parallel careful monitoring of displacements and seepage were set up by consulting engineers. The effect of the trial filling and the consequent sudden drop of reservoir water level on dam stability was numerically modelled in advance before the repair procedure. The modelling results enable the dam owner to control displacements and check if they are within the expected range.

CONCLUSIONS

Recently, the 1st stage of the remedial works has been finished, and, at the time of writing, the project documentation has been completed for the 2nd stage of the reconstruction. At the beginning of December 2005 the filling of the reservoir was slowly begun according to the plan for the trial procedure. Unfortunately, the filling stagnated during December due to the lack of water in the catchment.

Careful continuous monitoring was carried out from the beginning of December, when the first reference readings of all observation devices were completed. No significant displacements and seepage has appeared by January 2006. However, it is still too early to conclude and proclaim the effect and success of the dam reconstruction at the time of writing. It is hoped that an update on the Mostiště Dam's behaviour will be provided at the conference meeting.

ACKNOWLEDGEMENTS

The contents of this paper are a part of the research supported by the Grant Agency of the Czech Republic, project No.103/05/2391.

The authors wish to thank the dam owner – the Morava River Basin Agency, for providing the valuable data published in the paper.

REFERENCES

Mostiště Dam – The project documentation of the dam reconstruction, AQUATIS a.s., 2005.
Doležalová, M. et al. *Mostiště Dam – dam reconstruction. The set of dam stability assessments - Final Report.* Dolexpert Geotechnika and AQUATIS a.s., 2005.

Construction of three RCC dams forming part of the Ghatghar Pumped Storage Project in India

V.C.SHELKE, Government of Maharashtra Water Resources Department
M.R.H.DUNSTAN, Malcolm Dunstan & Associates
J.L.HINKS, Halcrow Group Ltd.
T.ZELENKA, Halcrow Group Ltd.

SYNOPSIS

The 250-MW Ghatghar Pumped Storage Scheme, now nearing completion near Mumbai in India, has three roller-compacted concrete (RCC) dams forming the Upper and Lower reservoirs. The level difference between the two reservoirs is 410 m. The two dams forming the Upper reservoir were constructed first and were useful precursors to the 84-m high dam impounding the Lower reservoir.

The Lower dam was built over two seasons with a halt for the 2005 monsoon when there is heavy rain – 640 mm fell at the site in eight hours on 28 June 2005; on the same day 940 mm fell in Mumbai – in total 7,200 mm rain fell in just over four months during the monsoon.

This paper describes the construction of the dams and also the thermal analyses used to choose the optimum placing temperatures.

GENERAL

The Ghatghar Pumped Storage Scheme, now nearing completion about 100 km north-east of Mumbai, has an underground power house in a cavern measuring 123 x 23 x 46 m in which there are two 125-MW pump/turbines. There is also a separate cavern housing the transformers.

The Upper reservoir is retained by the Upper dam, which is 15 m high, 495 m long and has a volume of 32,000 m^3, and the Saddle dam, which is

Improvements in reservoir construction, operation and maintenance, Thomas Telford, London, 2006, 404–411

12 m high, 250 m long and has a volume of 12,500 m^3. Both are RCC dams.

The Lower reservoir is retained by a single RCC dam that will be 84 m high, 415 m long and will have a total volume of 656,827 m^3. The total cost of the Lower dam is $ 35.6 million. This implies a cost of $ 54 per m^3 for the dam structure.

SADDLE DAM

The Saddle dam, which was constructed first, was considered to be a large full-scale trial for the Upper dam. RCCs with several different mixture proportions were tried in this dam. The Portland cement content varied between 88 and 98 kg/m^3 and the low-lime flyash content between 142 and 152 kg/m^3. The specified maximum placing temperature was 28 $^{\circ}$C, although the actual maximum placing temperature was only circa 24 $^{\circ}$C, and the joint spacing was 25 m. The dam was constructed in two halves, the first half, the Left-Hand Side (LHS), between 9 March and 23 March 2003 and the second half, the Right-Hand Side (RHS), between 21 April and 9 May. To date the dam is free of any detrimental cracking and has no seepage (see Figure 1).

Figure 1: Downstream face of Saddle dam

UPPER DAM

The Upper dam has a central spillway with five overtoppable radial gates measuring 12 m long by 3 m high. The remainder of the dam is constructed of RCC the great majority of which contained 88 kg/m^3 of Portland cement and 132 kg/m^3 of flyash.

The placement of RCC started just before the 2003 monsoon. At the time the joint spacing was 25 m and the specified maximum placing temperature was 28 °C (although the majority of the RCC had a placing temperature below 25 °C). The RCC was placed in two sections either side of the central spillway. The lower section of the RHS was placed between 19 and 23 May 2003 and the lower section of the LHS between 1 and 10 June. In both cases the volume of RCC placed was a relatively low proportion of the total volume. During the placements the ambient temperatures were high with maximum temperatures well over 40 °C.

During the 2003 monsoon, some uncontrolled cracks were found between the contraction joints in both sections of the dam that had been placed in May and June. Some thermocouples had been placed in these sections and it was found that the adiabatic temperature rise was circa 25 °C rather than the 17 °C that had been used in the design. A preliminary thermal study was therefore undertaken and this showed that with the new data, the Factor of Safety against cracking in the Upper Dam was close to unity while that for the Saddle dam was in excess of 1.5. Using these data the joint spacing in the Upper dam for the second season was halved to 12.5 m while the maximum allowable placing temperature was maintained at 28 °C. The remainder of the LHS was placed between 26 November and 20 December 2003 and the RHS between 2 and 16 January 2004. No further uncontrolled cracks have been found in the upper sections of the dam (other than those propagated from the lower sections of the dam). Figure 2 shows the Upper dam and reservoir.

Figure 2:   Upper dam and drawn-down reservoir

LOWER DAM

The Lower dam has an upstream face sloping at 0.141:1 (H:V) and a stepped downstream face sloping at 0.782:1. The central 58 m of the crest is a free overflow spillway.   The crest width in the non-overflow sections is 8.0 m.

In the same way as the Upper dam, the Lower dam was constructed over two seasons, either side of the 2005 monsoon. The RCC placement started on 14 December 2004. The mixture proportions initially were the same as the Upper dam (i.e. 88 + 132) but this was changed to 75 kg/m^3 of Portland cement and 155 kg/m^3 of flyash at the beginning of February 2005. As the volume of each layer of RCC was too great to be placed within the limits for a 'hot' joint (for which there is negligible treatment), the dam was spilt into two halves. One half was raised circa 3.6 m, the plant was then transferred to the other half of the dam (using a 1.8-m high mobile ramp (see Figure 3)) and then the second half was raised 3.6 m. The first season was completed on 31 May 2005 by which time 235,500 m^3 of RCC had been placed.

Placement for the second season commenced on 14 November 2005 and, at the time of writing, it is expected that the RCC placement will be completed – there is 421,250 m^3 of RCC to be placed in this season – in the second half of May 2006. Figure 4 shows the RCC placement on 20 January 2006; the escarpment on top of which is the Upper reservoir can clearly be seen in the background.

Figure 3: Plant being moved from one side of the 'split placement' to the other

Figure 4: Ghatghar Lower dam under construction on 20 January 2006

THERMAL STUDY

To confirm the assumptions made during the preliminary Thermal Study, it was decided to carry out a full thermal stress analysis for the Lower dam. This analysis was started at the end of the first season (December 2004 to May 2005) when the dam was already 27 m high and when the contraction joint spacing had already been chosen to be 15 m. The main purpose of the analyses was to determine a suitable maximum placing temperature for the second season (November 2005 to May 2006).

The modelling technique was a more sophisticated version of that employed on the Khlong Ma Dua dam in Thailand (Cordell, 2002) and at various other RCC Dams such as Platanovryssi dam in Greece (Kolonias et al, 1989) and New Victoria dam in Australia (Hinks et al, 1991). As at Khlong Ma Dua it was planned to build the Lower dam using the 'split-placement' methodology so there was a need for separate models for the LHS and RHS of the placement as well as for the spillway and non-spillway sections.

The temperature analyses, as well as the stress analyses, were carried out using ANSYS since this program now offers all of the facilities built into the program DAMHOT which has been used on previous occasions (Hinks and Copley, 1995). The use of a single program offers considerable advantages in that it is easy to transfer data between the temperature and stress modules.

Because thermocouple readings were available from the Lower dam's first season there were ample data available to check the calibration of the temperature model. In the event it was found that predicted temperatures were generally close to those measured in the dam (see Figure 5).

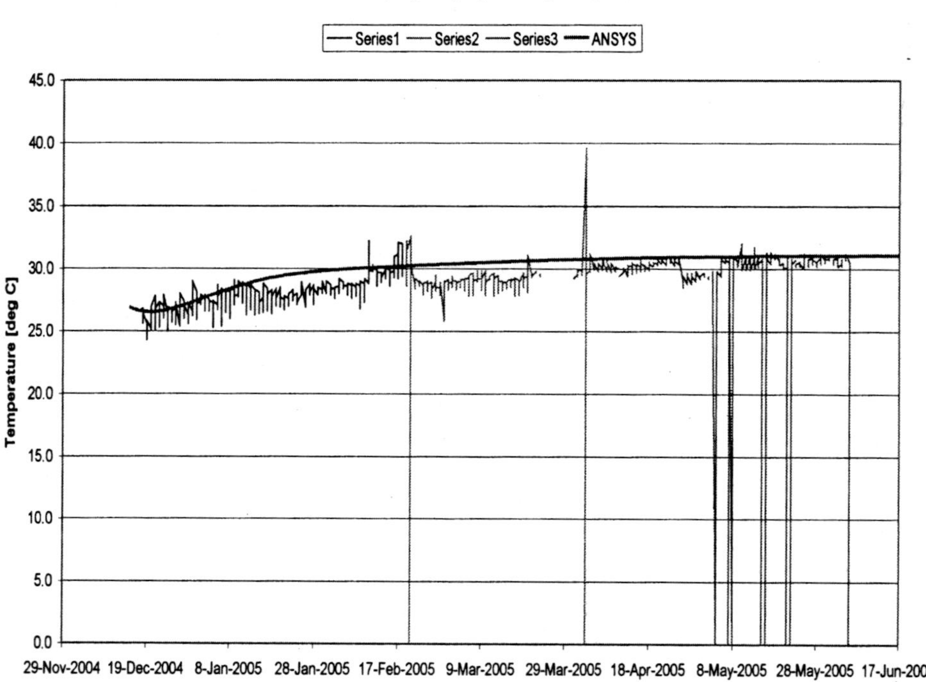

Figure 5: Comparison between predicted and measured temperatures in the Lower dam at Ghatghar

It had originally been intended to construct the spillway crest in conventional concrete at the end of May 2006. However the thermal stress analyses showed excessive stresses in the RCC immediately beneath the conventional concrete so it was decided to defer the construction of the spillway sill until the cooler weather in November 2006.

The conclusion of the thermal analyses was that it would be safe to increase the placing temperature from 22 °C (as used during the first season) to 25 °C. This increase was expected to produce a cost saving of $500,000 to $600,000. In the event, the maximum placing temperature was relaxed to 23.5 °C for the first few months of the second season (November to February) when the temperatures on the site are not particularly high (less than 35 °C) and it is possible that it will be further relaxed to 25 °C for the last few months of the second season (when the temperatures are rather higher (up to 45 °C).

ACKNOWLEDGEMENTS

The authors acknowledge the kind permission of the Government of Maharashtra Water Resources Department (GoMWRD) to publish this paper.

Principal Consultants for the scheme are EPDC (now J Power) of Japan and Tata Consulting Engineers (TCE) of India. Malcolm Dunstan & Associates are advising GoMWRD on all aspects of the RCC and Halcrow Group Ltd undertook the final Thermal Study.

REFERENCES

Cordell M.C. (2002), Designing against Thermal Cracking in an RCC Dam, *Dams & Reservoirs, London,* July.

Hinks J.L, Copley A.F. and Dunstan M.R.H. (1991), Thermal Modelling for RCC Dams, *International Symposium on Roller Compacted Concrete Dams, Beijing, China,* 6–9 November.

Hinks J.L, and Copley A.F. Thermal Analysis for RCC Dams, (1995). *International Symposium on Roller Compacted Concrete Dams, Santander, Spain.* 2-4 October.

Kolonias E, Dunstan M.R.H, Hinks J.L. and Copley A.F (1989), The Design of Platanovryssi: Europe's Highest RCC Dam, *Water Power and Dam Construction, London.* November

Lightweight fill in dam remediation - a case study

K. M. H. BARR, W. A. Fairhurst & Partners, Glasgow, UK

SYNOPSIS.   The east dam at Millbuies Loch was reconstructed following observations of crest settlement. Investigations found the embankment was founded on a layer of peat. It was concluded that there was significant potential for further settlement of the peat foundation if additional soil were added to make up the freeboard. The embankment was reconstructed using a combination of existing fill and new expanded polystyrene lightweight block fill.

INTRODUCTION

Millbuies Loch is an amenity reservoir located about 6km south of Elgin, in Moray in the north-east of Scotland. It is understood that the reservoir was originally used for water supply, but is now redundant for that purpose. The loch now forms the centrepiece of Millbuies Country Park, an important recreational asset owned by The Moray Council. The park is popular with walkers and the loch is also used as a trout fishery operated by the Council.

The reservoir is situated in a narrow valley with steep forested slopes to north and south. On the north side the ground rises steeply about 15m to a ridge and then falls away. On the south side the ground rises steeply to an escarpment and then continues to rise steadily at an average gradient of about 1 in 10. A catchment area of about $1km^2$ drains to the reservoir from the south.

The loch lies approximately east-west and is about 900m long but with an average width of only about 50m. The reservoir is retained by two small earth embankments at the east and west ends which drain to separate tributaries of the River Lossie. The location of the loch is shown on the plan in Figure 1.

The west embankment is about 5.5m high and 45m long. The east embankment is about 3 m high and 40 m long. Each dam has a narrow stone pitched overflow channel at its north abutment with the weir crests set at the same level. The date of construction is not known and no original

Improvements in reservoir construction, operation and maintenance, Thomas Telford, London, 2006, 412–418

record drawings are available. The smaller but deeper western part of the reservoir is joined to the eastern part by a narrow channel over a ridge at the original watershed. The depth of water here is less than 1m, so the reservoir divides in two if significant drawdown occurs.

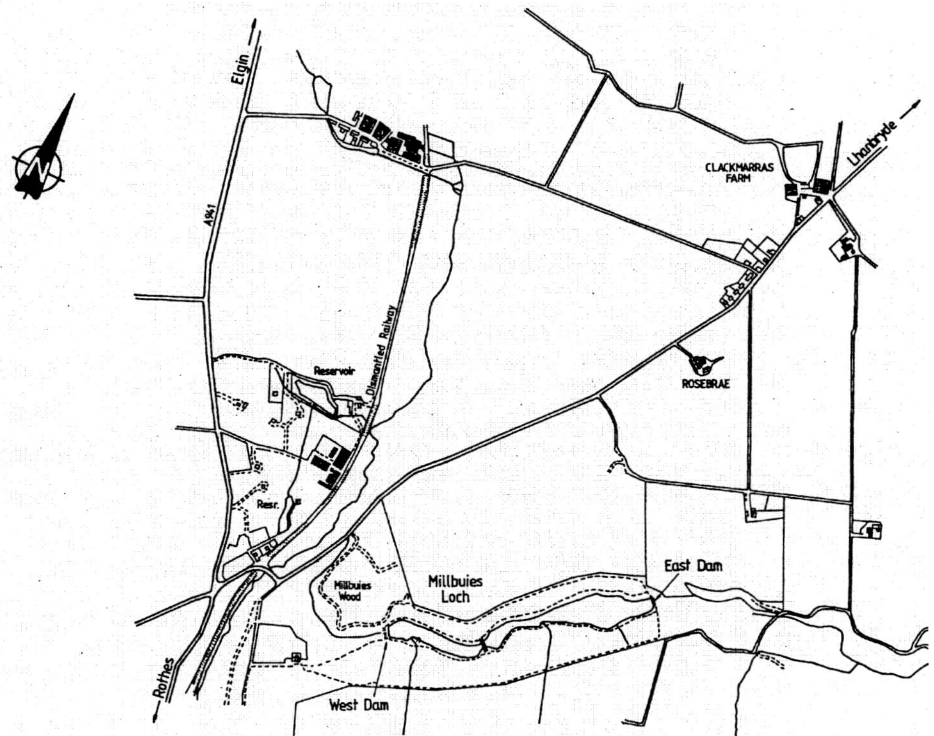

Fig. 1. Location plan

The control valve on the bottom outlet of the west dam is operated by a spindle located in the downstream slope of the embankment. This may suggest that the west dam originally retained a smaller reservoir and was raised at a later date, at the same time as the east dam was constructed.

There is poor vehicular access to the embankments at Millbuies Loch. An unsurfaced forest track runs close to the north abutment of the west dam, but at a level further up the slope well above the dam crest. At the east dam, a long length of unsurfaced forest track with steep gradients runs to the top of the escarpment above the south abutment of the dam. At this point the ground level is about 20m higher than the dam crest. Access is only possible for tracked vehicles via a cleared path through the trees directly down the slope to the dam. Access for wheeled vehicles is very difficult.

Previous remedial works

There is no record of the Loch having been inspected under the Reservoirs (Safety Provisions) Act 1930. The first inspection under the Reservoirs Act 1975 occurred in 1986 The overflow capacity and freeboard were found to be deficient. In addition, both embankments were showing signs of seepage at their respective downstream toes.

Remedial works were carried out in 1989-1990. The crests of both embankments were made up by a modest amount to improve the freeboard and 300mm high rubble masonry wave walls and stone pitching on the upstream faces installed. The two overflow channels were widened and gradients improved and new overflow weir sills installed. The existing mortared stone pitched channel linings were retained. A rockfill toe was installed at the west dam and a berm and toe drainage was installed at the east dam. A new outlet pipe was installed at the east dam to allow the reservoir to be drawn down to the level where its two parts separate.

FLOOD DAMAGE AND INVESTIGATIONS

Flooding on the Moray Firth coast often occurs during summer and autumn rainfall events and in northerly airflows associated with frontal systems. There is a long history of flooding from similar events.

On 30 June and 1 July 1997 persistent rain was driven onto the Moray coastal plain and hills by a strong north-easterly wind. Rainfall of up to 150mm was experienced over a two-day period in the band of countryside from Inverness to Banff. This severe rainfall event caused widespread flooding. Local to Millbuies, significant flooding of property was experienced in the Fogwatt and Longmorn areas, downstream of the west dam, and in the Lhanbryde area, downstream of the east dam. In the wider Moray area serious flooding of property was experienced at Elgin and Forres.

During the flood event the outlet valves were opened to reduce the water levels in the reservoir. The water level was estimated to be 450mm above the overflow crest level near the time of the peak of the event.

At the Supervising Engineer's visit to Millbuies Loch following the flood event damage that had occurred to the east overflow during the flooding was discovered. A large scour hole had formed at the bottom of the overflow channel, with undercutting of the slope of the hillside above. Cracking was evident in the south side slope of the overflow channel in the vicinity of the access bridge, suggesting settlement of the dam. There was also minor cracking of the mortar of the rubble masonry wave wall. No damage was found to the west dam. A periodical inspection was carried out the following day and a topographic survey shortly thereafter.

The survey of the flood damage revealed that significant settlement had occurred at the east embankment. The minimum crest level established in 1990 had reduced by about 300mm, effectively eliminating the additional freeboard that had been created by the earlier remedial works.

No survey of the crest levels had been undertaken between 1990 and 1997, so it is not known how much of the settlement occurred between these dates. However, the evidence of damage seen in late August 1997 suggested that at least some of the settlement was recent. It is not thought that the duration of the flood event was sufficient for the weight of the flood surcharge to have a significant effect, so the mechanism of further settlement is not fully understood. However, the elapse of time before the damage was discovered and the survey carried out may be significant. It is speculated that the additional weight from saturation of the embankment may have resulted in further settlement over that period.

Ground investigation

In May 1998 a ground investigation contractor was appointed to carry out trial pits and boreholes to examine the causes of the observed settlement. Five cable percussion boreholes and four trial pits were taken in the east dam and one cable percussion borehole was taken for comparison in the west dam.

It was found that the east dam is founded on a layer of peat. The boreholes in the centre section of the east dam showed over 5m of sandy gravelly clay dam fill material overlying 1.5m of soft fibrous peat. Below the peat there are glacial tills. In the wetland area downstream of the dam about 3-4m depth of peat was found.

It appears that the east dam was constructed by spreading dam fill material on top of an area similar to the wetland downstream of the dam. The peat foundation would have been compressed as the fill material was placed until an approximate equilibrium was reached when the weight of the fill material matched the bearing capacity of the peat layer. Settlement is likely to have continued over the life of the embankment. By 1989 the freeboard was inadequate and it was restored by making up the dam crest with fill material. Crest levels may have been made up on previous occasions although no record of this is available. The existence of the peat layer in the foundation was not known in 1989. Although a ground investigation was carried out, it was limited to relatively shallow trial pits because of the difficulty of access for drilling rigs and consequent cost.

The foundation conditions at the east dam are in contrast to the west dam, where below the dam fill material thin deposits of peaty topsoil associated with original ground level were found overlying glacial sands, silts, and gravels.

REMEDIAL WORKS OPTIONS

The first option considered was to reduce the overflow level to restore the required freeboard. However, this option would have reduced the viability of the fishery and required reconstruction of a boathouse, so the Undertaker did not wish to pursue it. It was decide to examine methods of restoring the freeboard at the Loch at a reasonable cost by raising the crest.

A number of alternative methods were considered for restoration of the freeboard. It was considered that adding more infill to raise the crest would result in further settlement and be self-defeating in the long term. Thought was also given as to whether it was practicable to surcharge the crest with additional material until the peat layer was compressed such that no further settlement would occur. It was not considered that this point could be defined with precision and the access difficulties made import of any significant quantity of material problematic. Excavation and complete reconstruction of the embankment was also ruled out on the basis of cost.

Geotechnical solutions involving some form of piling or ground improvement to increase the bearing capacity of the foundation were also considered. However, all the available methods required plant that could not easily reach the embankment or had significant establishment costs that could not be justified by the limited extent of work. The most likely candidate appeared to be jet grouting but the practice had previous experience of unsuccessful jet grouting trials in peat dam foundations and the cost was considered too high relative to the risk.

An alternative philosophy was considered involving reducing the loading on the peat foundation by removing some of the dam fill material and raising the dam using a lightweight material. In this way the dam could be raised without increasing the load on the foundation. Lightweight fill has been used starting in the 1970s in Scandinavia in road embankments on compressible foundations and is now considered an established technology. This option was chosen as representing the best value for money.

The lightweight fill material chosen was the Fillmaster, an expanded polystyrene (EPS) block system with a density of about 20kg/m^3. It is available in various grades to suit different loading conditions. EPS is an inert and durable material, the main hazards to it being fire prior to being buried and solvents such as petrol and diesel. In road embankments it is normally protected from solvents by wrapping it in a low density polyethylene membrane. The manufacturer did not consider this necessary for Millbuies, so a geotextile separator was substituted.

It was decided to retain the existing upstream face of the embankment below top water level and to reconstruct the crest without a wave wall. This required crest levels to be raised by up to about 500mm and levels on the downstream slope to be increased by over 1m in places.

The main design considerations were ensuring no significant surcharge of the underlying peat, the stability of the EPS fill and maintaining the watertightness of the dam.

Standard profiles of the block fill were derived for different dam cross-sections. These provided sufficient lightweight fill to eliminate additional loading in each part of the crest and downstream slope. EPS blocks are water resistant, but water from the reservoir could escape through the joints between the blocks so it was decided that a new low permeability element was required in the dam cross-section upstream of the blocks. The chosen design was a self-setting bentonite/cement slurry wall in a trench excavated in the existing dam crest following demolition of the existing wave wall.

To prevent uplift pressures on the base of the EPS blocks, they were constructed on a drainage layer. A filter drain and sand layer was placed before the blocks were installed. The blocks were then carefully backfilling using excavated fill material.

The downstream face of the dam was protected with an erosion control mat to prevent erosion if any occasional over-topping should occur. A perforated pipe drainage system was included to drain the filter drain under the EPS blocks.

CONSTRUCTION
Following tendering Balfour Beatty Construction Limited was appointed to construct the works in 2000.

During the early stages of construction the contractor approached the designers with a view to consideration of alternatives to the cement-bentonite slurry trench. The contractor was concerned over the financial risk associated with delays to its specialist sub-contractor in the event of adverse weather or access difficulties. A review was carried out with the contractor of the design concept and possible alternatives. Following consideration of various possibilities the contractor proposed the use of lightweight steel sheet piling installed using a small piling hammer mounted on a midi-excavator. It was considered this alternative offered advantages over the original design so the contract was amended accordingly. Mabey M11 interlocking lightweight steel sheet piling was used. The final design cross-section was as shown on Figure 2.

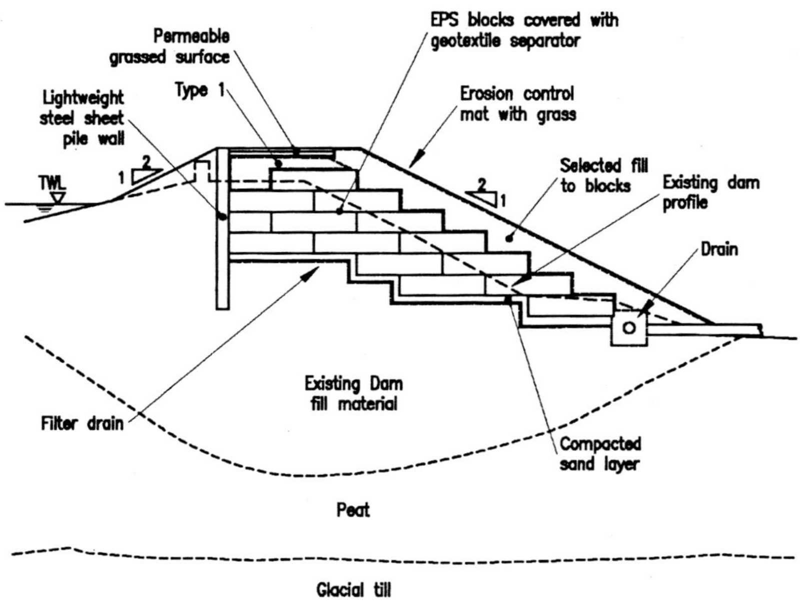

Fig. 2. Typical cross-section

Construction of the remedial works at Millbuies Loch was completed in early 2001. The reservoir was certified following a periodical inspection under the Reservoirs Act in the same year. The dam has now performed satisfactorily for 5 years since completion of the work.

CONCLUSION

Millbuies Loch east dam presented particular problems in finding a suitable method of raising the embankment crest to restore freeboard. Lightweight fill was found to be a cost-effective method in a situation where access for more conventional solutions was difficult.

REFERENCES
VR Fillmaster - Technical Information and Design Details, Vencel Resil Limited, Kent.

The Influence of Inspection and Monitoring on the Phased Construction of the Barragem de Cerro do Lobo

M. CAMBRIDGE, Cantab Consulting Ltd, Ashford, Kent, UK
M. OLIVEIRA TOSCANO, Dam Supervisor, Somincor S.A., Portugal

SYNOPSIS.   The SOMINCOR S.A. Neves Corvo mining complex is located near Castro Verde in southern Portugal. The mine commenced processing copper and tin ores in 1988, the storage capacity for the resulting residues (tailings) and process water supply being provided by the Barragem de Cerro do Lobo. This embankment, constructed between 1988 and 2005, comprises a complex zoned rockfill dam designed to store the sulphidic tailings generated during ore processing sub-aqueously to prevent acidification.   The embankment periphery has been extensively instrumented for both operational and performance monitoring, and throughout both construction and operation has been subject to independent annual inspection.   The paper demonstrates the benefits of independent inspection, combined with instrumentation and monitoring to meet the requirements of safety and stability as well as the constraints of environmental compliance and mine tailings storage requirements.

BACKGROUND
The Neves Corvo Mine is located in the southern part of Portugal on the south-western limit of the Iberian pyrite belt, and was developed from 1985 onwards by Somincor S.A., a joint venture between Rio Tinto Zinc plc and the Portuguese State Mining Company, (Real and Franco, 1990). The mine complex is located some 20km south of Castro Verde in the Alentejo region of Portugal between the villages of Neves and Corvo, Figure 1, and includes underground operations, twin process plants and, some 5km to the east of the mine facilities, a tailings management facility, the Barragem de Cerro do Lobo. Production from the mine has risen over this period from 1.6Mt/yr to 2.3Mt/yr in recent years and is targeted to increase to more than 2.5Mt/yr in future. Between 90% and 95% of the mine throughput is produced as process waste (tailings), which has a specific gravity of 4.1 and contains more than 85% pyritic materials. These tailings are pumped from the process plant site to the depository in the Cerro do Lobo basin as a slurry at

a pulp density averaging 20% with a pH of 10, (Cambridge and Coulton, 1990).

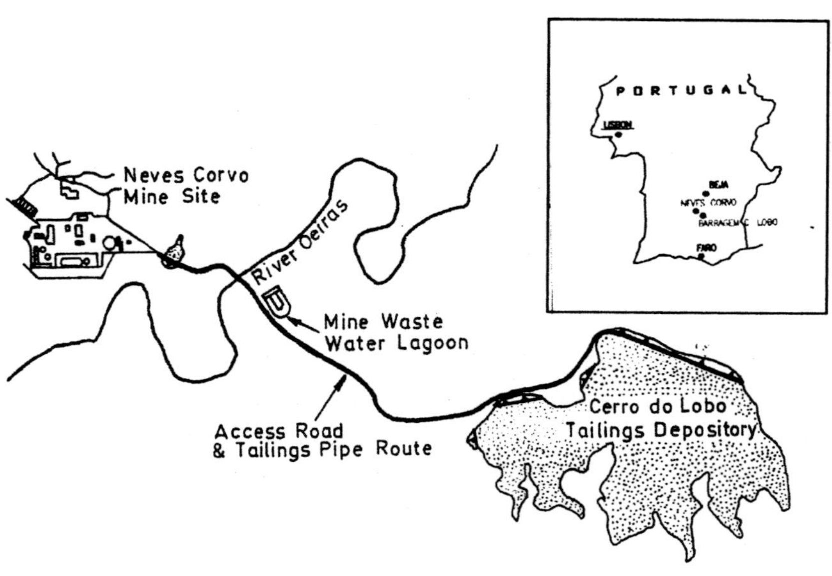

Figure 1. Project location plan

In order to store the anticipated volume of tailings to be produced throughout mine life, a stage-constructed deposition facility was required, Figure 2. This tailings management facility comprises a principal embankment dam across the Lajes Stream, a tributary of the Oeiras River and ultimately of the Guadiana River, together with three saddle dams and appurtenant works, and was constructed in four phases between 1987 and 2005. The first phase of construction was completed in October 1988 and the fourth in 2005, increasing the capacity of the facility from $6 \times 10^6 m^3$ to more than $20 \times 10^6 m^3$ and the height from 28m to 42m. Tailings were to be stored sub-aqueously with 1metre of water cover, thus preventing oxidation of the pyritic waste and, in addition, reducing the possibility of wind erosion. The geological setting for the embankments is characterised by greywackes and shales which exhibit relatively low permeability and suitability for the storage of the pyritic wastes. The embankment periphery, which now extends over 3.3km, has been extensively instrumented for both operational and performance monitoring. In addition, the facility has been subject to independent annual inspection throughout both construction and operation, and the instrumentation and monitoring data to similar scrutiny.

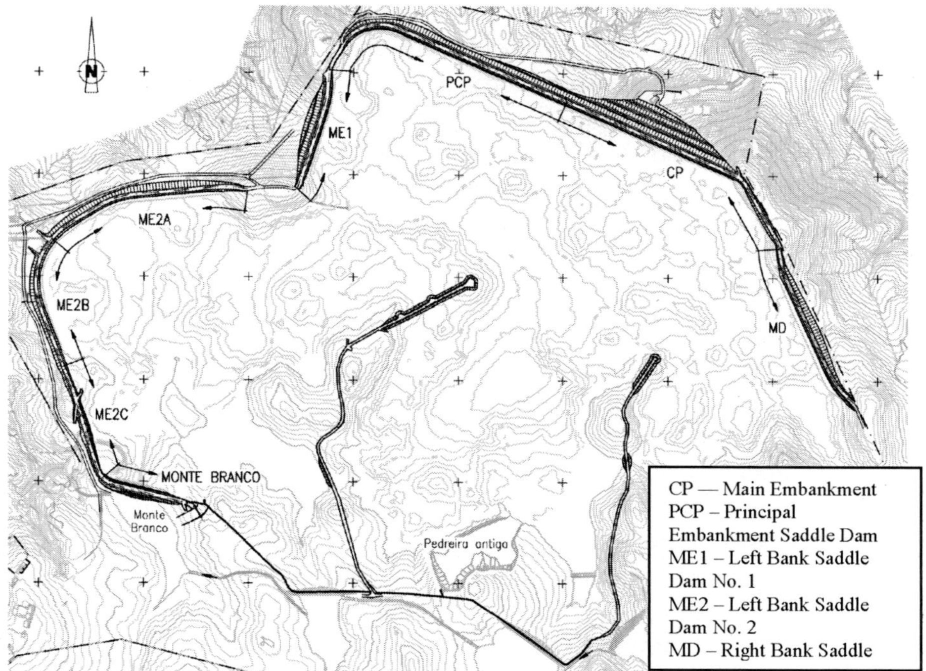

Figure 2. General plan of Barragem de Cerro do Lobo

DESIGN AND DISPOSAL

This tailings disposal facility was ultimately constructed in four phases to meet the demands not only of waste storage but also of process water supply to the plant, (Oliveira Toscano and Cambridge, 2006), as summarized in Table 1 and described briefly below.

Table 1 – Construction Phases

Construction Phase	Crest Level (mOD)	Construction Period
Phase 1	244.00	1987-88
Phase 2	248.00	1990
Phase 3	252.00	1992-93
Phase 4	255.00	2004-05

Phase 1

The Phase 1 works were undertaken under an EPCM contract and involved the construction of the main embankment, together with two left bank saddle dams (ME1 and ME2), to impound a storage volume of some 6Mm^3. The initial embankment was 28m high, had a crest length of 850m and comprised a rockfill dam with an upstream inclined clay core. Expert review of the design, with particular respect to the proposed embankment cross-section and the geotechnical characteristics and availability of suitable clays, resulted in a modified embankment cross-section, the sloping clay zone being replaced by a narrow vertical core. The revised section included the clay core, downstream filter protection and mine waste rockfill shoulders with slopes of 1:1.8 upstream and 1:1.7 downstream. All seepage from the facility was to be collected via basal blanket drains and fed into seepage sumps located at low spots throughout the periphery. These sumps were fitted with float valve controlled pumps in order to return all seepages back into the main reservoir.

Construction of the Phase 1 facility commenced in 1987 and was substantially completed in 1988 in time to receive initial tailings from the process plant at start-up. Instrumentation installed in the embankment and foundations included 19 No. piezometers, continuous flowmeters to record seepage volumes in each sump, and settlement beacons. Environmental performance was to be monitored via a series of external groundwater piezometers, together with seepage quality sampling and testing and annual environmental auditing. In addition to the statutory inspections by the Portuguese dam authority, LNEC, and the designer, Somincor instigated annual audits by the expert reviewer. These inspections and the monitoring data have been instrumental in modifying and improving both the cost-efficiency of all construction stages and also in optimising the tailings disposal operations and process water supply.

Initial Tailings Disposal
Initial tailings deposition commenced into the Barragem de Cerro do Lobo facility in October 1988. The disposal system adopted involved pumping the thickened tailings slurry via a 5km long, 500mm diameter HDPE pipe into a similar manifold system laid on the crests of the main embankment and saddle dams. A series of valved offtakes from the manifold connected to floating HDPE fingers enabled tailings to be distributed throughout the basin, thus minimising dead-storage areas. The operating requirement was to maintain a minimum of 1 metre of water above all deposited tailings surfaces and thus prevent oxidation.

The facility also provided a major source of process water to the metallurgical plants on the main mine site. The return water pumping station controlled process water supply, augmented as necessary by make up from the nearest fresh water reservoir, Santa Clara, some 40km to the south. The operational criteria were therefore to maintain even distribution of tailings across the depository, to control and store runoff from the local catchment and to ensure a continuous supply of process water for the plant. During the initial period, climatic conditions required the storage of significant volumes of process water and resulted in the reservoir level within the BCL rising faster than predicted and, therefore, to prevent any possibility of there being an untoward discharge over the emergency spillway, Phase 2 construction was brought forward. Since 1988 the timing of the construction phases has been driven by water storage requirements rather than tailings deposition. The rate of rise of reservoir with storage of tailings is shown in Figure 3.

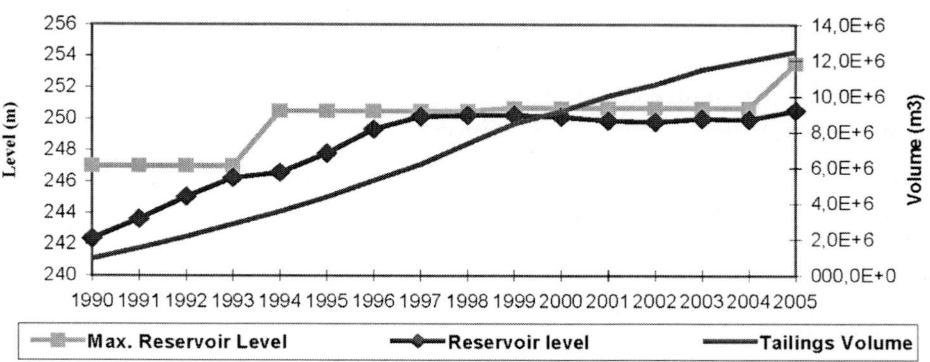

Figure 3. Rate of reservoir rise with stored tailings volumes

<u>Phase 2</u>
The Phase 2 construction was brought forward to 1990 and included raising the crest level by 4m by extending the embankment in the downstream direction, the lateral extension of the existing embankment and saddle dams and the construction of the additional right bank saddle, MD. Again, after expert review of the proposed upstream inclined extension of the central vertical core, the designer replaced the clay with an HDPE liner, (Cambridge and Maranha das Neves, 1991). The modified design section included the installation of double roughened HDPE on a filter blanket to protect against any untoward leakage through the periphery, as well as extending the embankment toe downstream using selected mine waste rockfill and upgrading the seepage control system. The overflow spillway was also relocated to Monte Branco and upgraded to meet international standards for embankment flood design. The Phase 2 construction works were commenced in 1990 under the management of the same EPCM team as for Phase 1. The embankment raise was substantially completed in the same year and included a further 12 No. piezometers, additional seepage monitoring flowmeters and replacement settlement beacons.

Phase 3

The Phase 3 lift was again undertaken ahead of schedule due to the demand for water storage at a time when mine throughput was increasing and climatic conditions in Alentejo were unusually wet. The Phase 3 construction involved a further 4m raise, with an additional extension of the embankment in a downstream direction using similar construction techniques to those employed for Phase 2. The increased crest elevation also necessitated extension of the saddle dams and the modification of the emergency overflow spillway to raise the maximum reservoir level.

Embankment construction commenced in 1992 using a local contractor supervised by Somincor, and was completed in 1993. The embankment raise was substantially completed in the same year and included a further 4 No. piezometers, additional seepage monitoring flowmeters and replacement settlement beacons.

Embankment Monitoring 1994-2003

Extensive performance monitoring was undertaken between Phases 3 and 4 with particular respect to structural stability, deposited tailings densities, water balance, piezometric data and seepage volumes and quality, (Oliveira Toscano and Fonseca, 2004). The annual expert audits offered the opportunity to regularly assess geotechnical and geochemical performance and to review the ongoing instrumentation data with respect to the need for the construction of the final permitted raise.

The water balance modelling indicated the benefits, both operationally and environmentally, of reducing the volume of run-off reporting to the reservoir. The construction of five 5m to 6m high embankment diversion dams was proposed along the southern periphery, to be interlinked by a deep overflow channel to divert catchment runoff around the facility and bypass the tailings reservoir. These dams were constructed in 1996 and reduced the effective catchment area from $4.0km^2$ to $1.91km^2$.

Tailings disposal monitoring had been undertaken on a continuous basis from early on in the project. During Phase 2, sub-aqueous tailings densities were predicted initially from a series of bench scale drained and undrained column tests. Subsequently, large-scale sedimentation tests were undertaken on site in 2m high columns. These tests indicated the long-term density of the submerged tailings to be of the order of 1.5 t/m^3 to $1.6t/m^3$, (Cambridge, 1993-97). Annual hydrographic surveys were undertaken to define the settled tailings surface and thus to assess the stored densities in the depository. The density determinations are presented in Figure 4, which shows the trend towards an average density value of $1.58t/m^3$, consistent with the laboratory data. The differences between field and laboratory data almost certainly arise from edge effects which influence sedimentation tests in small diameter columns. However, the column tests proved to be a reasonably accurate indicator of long-term sub-aqueous deposition and, together with the updated water balance predictions, indicated the need to instigate the final permitted raise of the facility in 2004 to meet predicted mine production to 2012.

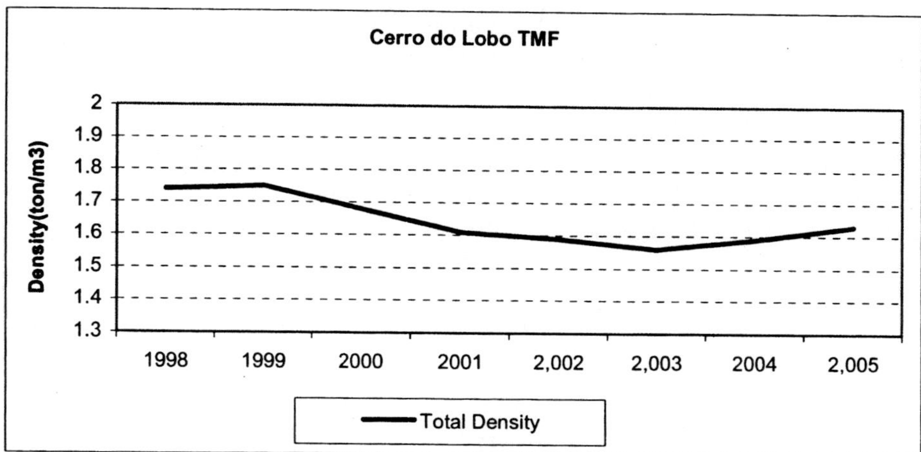

Figure 4. Density measurements for tailings deposit

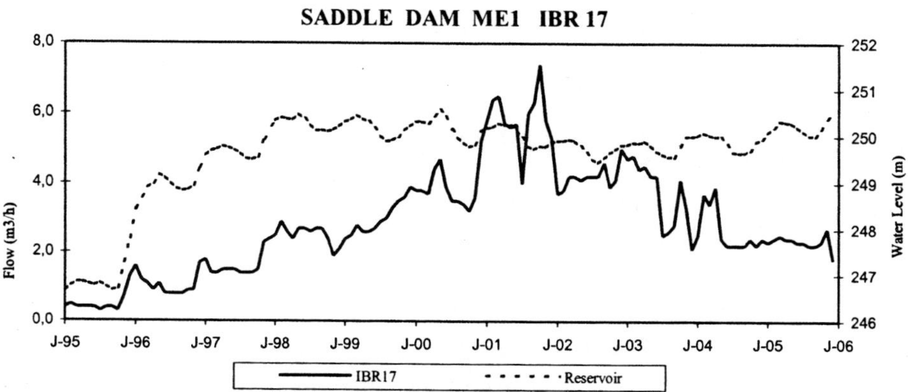

Figure 5. Seepage volumes with time for IBR17

Seepage volumes exhibited a steady increase during the initial deposition period. However, as tailings deposits developed across the storage area, consolidation reduced their vertical permeability, effectively sealing the basin. By 2001 total seepage volumes from the depository had begun to decline at each embankment, Figure 5, despite the difficulties experienced in creating a uniform deposit of tailings across the basin.

Seepage quality monitoring over this period revealed the increasing occurrence of acidic runoff from the embankment slopes. The seepage data, in particular, indicated a direct correlation between seepage, pH and rainfall intensity, and the occurrence of seasonal acidification of the downstream embankment faces, Figure 6. Investigation revealed that this acidification was as a result of previously unidentified rockfill zones incorporated into the embankment section and which included modest yet destructive quantities of pyrite. The precipitation-induced acidification (oxidation and flushing cycle) led to a marked seasonal reduction in the pH of the seepage and to the breakdown of rockfill particles, resulting in increasing fineness of the fill and reduced shear strength. The consequent lower permeability led to a small but noticeable rise in piezometric levels in the embankment shoulders. The reduced pH posed long-term environmental performance concerns, not only for seepage quality but also for closure and final surface rehabilitation.

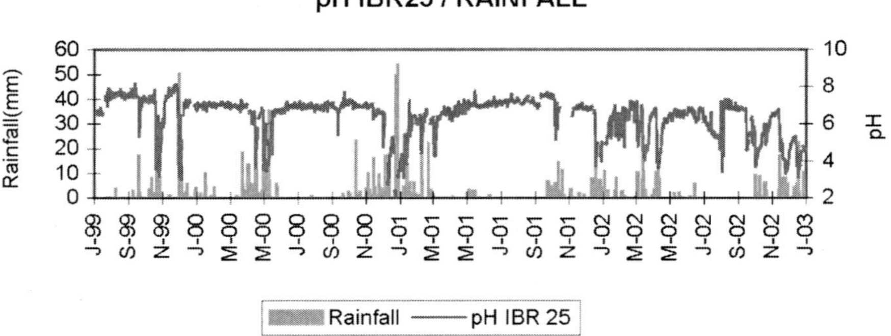

Figure 6. Correlation of seepage pH with precipitating events from IBR25

As a result, a series of boreholes were put down at selected locations throughout the embankment to identify the depth extent of any deterioration and to obtain samples for geotechnical and geochemical analysis. Further hydraulic piezometers were installed in these boreholes to enable any untoward rise in phreatic surface as a result of changes in the characteristics of the fill to be monitored. The investigation indicated that oxidation was apparently limited to the outer face of the embankment but that zones of potentially destructive quantities of pyritic fill had been included in the embankment section. Laboratory analysis indicated reduced shear strength of the oxidised fill and led to a stability review and further modification of the Phase 4 embankment cross-section. The revised section was designed to ensure long-term integrity and to prevent future deterioration of the existing rockfill. Percolation through the downstream face was to be minimised by utilising acid- and weather-resistant coarser, clean, competent rockfill from a nearby quarry. The seepage control system was also upgraded to maximise seepage capture and return and to minimise seasonal movement of piezometric levels within the embankment shoulders, thus inhibiting the oxidation/acidification cycles.

<u>Phase 4</u>
The most recent construction phase, completed in late 2005, involved raising the crest to the ultimate permitted level of 255mOD, giving a maximum embankment height of 42m and a peripheral length of 3327m. The reservoir at top level now contains some 20.4Mm^3 of storage and has a surface area of some 1.8km^2. The embankment raise entailed extending the HDPE membrane throughout the periphery of the embankment, raising the downstream face and modifying both toe drainage and seepage control systems.

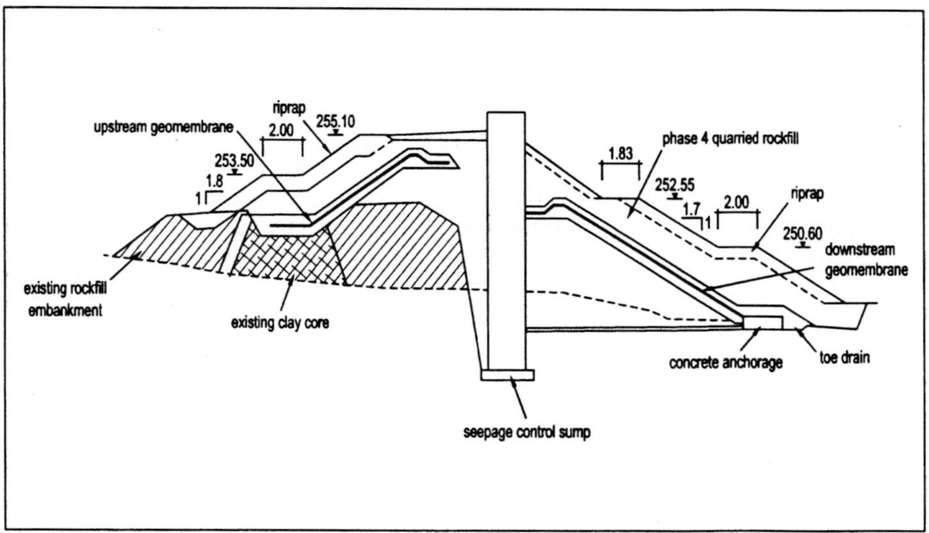

Figure 7. Cross-section through the Monte Branco embankment

During this phase of embankment raise, both the emergency overflow structure and siphon drawdown facility were relocated and the final diversion dam, Monte Branco, at the extremity of saddle dam ME2, completed. This small dam could be inundated on both upstream and downstream faces, and thus included seepage control via a deep sump located on the centre line of the embankment to prevent any potentially contaminated seepage from being discharged into the diversion system, Figure 7.

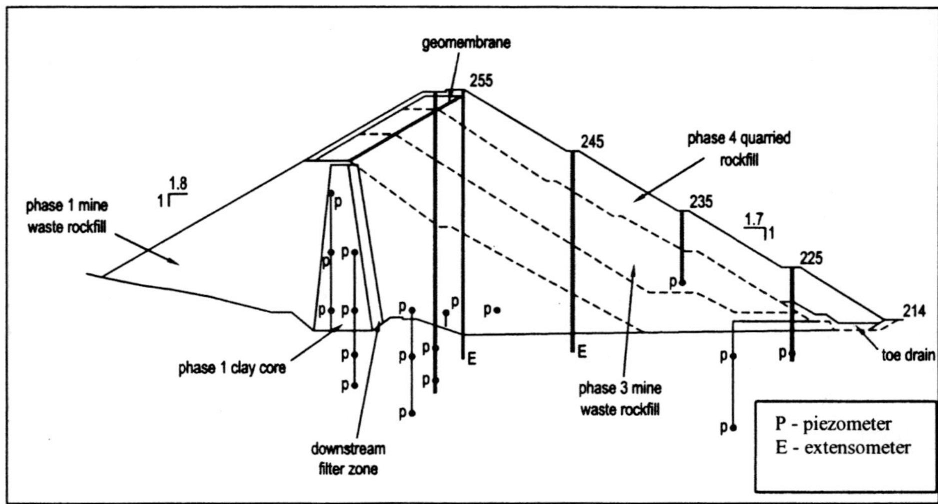

Figure 8. Phase 4 Embankment cross-section showing instrumentation

The final construction phase was undertaken under the management of Somincor by a local contractor, with site supervision by Cenorgeo, (Oliveira Toscano, Romeiro and Almeida, 2006). This phase included further instrumentation, Figure 8, comprising the upgrading of all seepage monitoring flow-meters, the installation of a further 15 No. piezometers, the final settlement beacons, 4 No. inclinometers and 2 No. seismographic stations.

Phases	Piezometers	Survey beacons	Neutral pressure cells	Seepage flow-meters	Inclino-meters	Seismo-graphs
1	19	18	6	4		
2	12	26		6		
3	4	36		12		
3/4	12					
4	15	41		19	4	2

Table 2. Summary of Barragem de Cerro do Lobo Instrumentation

SUMMARY

The Phase 4 construction works on the Barragem de Cerro do Lobo tailings disposal facility were completed late in 2005 and included the installation of the final instrumentation and monitoring equipment, Table 2. Throughout the last 18 years, the Barragem de Cerro do Lobo has been inspected and monitored, not only for operational reasons, but also to ensure that the facility performs in accordance with its design criteria. External expert and regulatory inspections have been undertaken for both engineering and environmental purposes, and have ensured that not only are instrumentation and monitoring data recorded but also that they are analysed and the results interpreted regularly. The monitoring data, together with a rigorous internal inspection system, have enabled the Company to adapt the construction and operation of the facility to meet the needs not only of operational efficiency, but also those of international standards of engineering and environmental compliance. The data collected have been used to great effect to improve the performance of the facility and to ensure that all untoward signs have been identified and corrected, either immediately or through subsequent construction phases. Throughout this period the facility has continued to receive tailings from the production plant and to provide a continuous supply of process water. A significant achievement over the operating period has been the use of water balance and tailings disposal modelling to fine-tune water usage. This has lead to a significant reduction (>20%) in freshwater demand from the Santa Clara Reservoir, a major achievement, particularly in the context of the adverse drought conditions recently experienced in southern Portugal. This tailings management facility, with all its modifications, has performed to its design criteria and can be considered to meet the standards demanded internationally for both water dams and mine waste storage facilities. The instrumentation installed will continue to be monitored both during mineral operations on the site and also post-closure until the facility is proven to be benign.

CONCLUSIONS

The Barragem de Cerro do Lobo has been designed to meet the tailings storage requirements emanating from copper production at the Neves Corvo Mine. The embankment and its appurtenant works have been constructed, inspected and monitored over an 18-year period to meet the demands of tailings storage and process water requirements, as well as of legislative and environmental standards. During this period, the successful operation of this facility has been aided by well-planned and rigorous instrumentation and monitoring routines to enable the performance of the facility to be optimised. Expert independent inspections and audits, combined with the internal monitoring system, have ensured the ongoing safety and integrity of the facility, its compliance with the best international standards and full environmental compliance.

As for all mineral operations, the existing orebody will eventually be depleted, but further mineral deposits may be identified. The future of this tailings management facility, should such new deposits prove economical to extract, may involve a further raise, more innovative thinking in design and construction and additional modifications to the embankment cross-section. The instrumentation and monitoring, together with the operator's experience over the last 18 years, will enable the engineering of any new extension or facility to be optimised.

ACKNOWLEDGEMENTS
The authors wish to thank Somincor S.A. for their permission to publish this paper.

REFERENCES

Cambridge, M. & Coulton, R. H. 1990. Geotechnical Aspects of the Construction of Tailings Dams – Two European Studies. The Embankment Dam. Proceedings of the 6th Conference of the British Dam Society

Cambridge, M. & Maranha de Neves, E. 1991. Textured Geomembrane for the Staged Raising of the Cerro do Lobo Dam. Water Power and Dam Construction, June 1991

Cambridge, M. 1993-1997. Neves Corvo Mine Laboratory test data for sub-aqueously deposited tailings, unpublished

Oliveira Toscano, M. & Cambridge, M. 2006. The Phased construction of the Barragem de Cerro do Lobo, 22nd International Congress on Large Dams, 2006

Oliveira Toscano, M. & Fonseca Almeida, L. 2004. Análise dos Resultados da Observação da Barragem do Cerro do Lobo. IX Congresso Nacional de Geotecnia, Aveiro, Portugal

Oliveira Toscano, M., Romeiro, M. & Almeida, N. 2006. Alteamento da Barragem de Rejeitados do Cerro do Lobo. X Congresso Nacional de Geotecnia, Lisboa, Portugal

Real, F. & Franco, A. 1990. Tailings Disposal at Neves Corvo Mine, Portugal. Acid Mine Water in Pyritic Environments, Lisboa 90